AF780633

Wave and Tidal Energy

Wave and Tidal Energy

Special Issue Editors

Carlos Guedes Soares

Matthew Lewis

MDPI • Basel • Beijing • Wuhan • Barcelona • Belgrade • Manchester • Tokyo • Cluj • Tianjin

Special Issue Editors
Carlos Guedes Soares
Centre for Marine Technology
and Ocean Engineering
(CENTEC), Instituto Superior
Técnico, Universidade de Lisboa
Portugal

Matthew Lewis
School of Ocean Sciences,
Bangor University
UK

Editorial Office
MDPI
St. Alban-Anlage 66
4052 Basel, Switzerland

This is a reprint of articles from the Special Issue published online in the open access journal *Energies* (ISSN 1996-1073) (available at: https://www.mdpi.com/journal/energies/special_issues/Wave_Tidal_Energy).

For citation purposes, cite each article independently as indicated on the article page online and as indicated below:

LastName, A.A.; LastName, B.B.; LastName, C.C. Article Title. *Journal Name* **Year**, *Article Number*, Page Range.

ISBN 978-3-03936-292-9 (Pbk)
ISBN 978-3-03936-293-6 (PDF)

© 2020 by the authors. Articles in this book are Open Access and distributed under the Creative Commons Attribution (CC BY) license, which allows users to download, copy and build upon published articles, as long as the author and publisher are properly credited, which ensures maximum dissemination and a wider impact of our publications.
The book as a whole is distributed by MDPI under the terms and conditions of the Creative Commons license CC BY-NC-ND.

Contents

About the Special Issue Editors

Carlos Guedes Soares is a Distinguished Professor of the Engineering Faculty (Instituto Superior Técnico) of the University of Lisbon. He is the Head of the Centre for Marine Technology and Ocean Engineering (CENTEC), which is a research centre classified by the Portuguese Foundation for Science and Technology as Excellent. He received his MSc and Ocean Engineer degrees from the Massachusetts Institute of Technology, USA. His Ph.D. degree was received from the Norwegian Institute of Technology of the University of Trondheim. His Doctor of Science degree was received from the Technical University of Lisbon, Portugal. He was conferred as Doctor honoris causa by the Technical University of Varna and University "Dunarea de Jos" of Galati. He is a Fellow of SNAME, RINA, IMarEST, ASME and Ordem dos Engenheiros; Member of ASCE, AGU, AMS and SRA; and Member of the Portuguese Academy of Engineering. He is actively involved in various research areas within Ocean Engineering, including wave energy.

Matt Lewis is an EPRSC Research Fellow at Bangor University. His marine renewable energy research aims to characterise the resource (e.g., suitable locations and electricity quality) and oceanographic conditions (e.g., to inform resilient design of devices), as well as likely environmental impacts. His research interests include understanding how systems interact, which he accomplishes by combining computer models of biological and physical processes to better understand the marine system. Matt is actively working in a number of research areas: future flood risk, estuarine processes modelling, aquaculture, tidal energy, wave–tide interaction modelling, BGM and larvae modelling.

Preface to "Wave and Tidal Energy"

The development of renewable energy around the world has been given increased importance due to concerns regarding energy supply and climate change. Wave and tidal energy are very important components of the renewable energies that can be extracted from the oceans. Different locations around the world have good natural conditions for one of these forms of energy, with some locations even being suitable for both. One of the important aspects related to the industrial development of the wave and tidal energy is the identification of the locations where the energy resource has levels that are high enough to justify commercial exploitation. This book contains three papers dealing with the evaluation of the wave energy resources around the world, covering areas like USA, Taiwan and Chile where important developments are being considered. Such energy resource predictions have been questioned in regard to their long-term consistency, in view of the climate changes that are being predicted. One of the papers addresses this topic specifically, presenting a methodology that is applied to the Black Sea but can also be applied to other areas. Wave energy farms not only absorb energy, but also change the wave fields in their vicinity; one of the papers addresses this issue by modelling the impact of offshore wave farms on coastal process and water users, with a special emphasis on beach morphology. Other impacts on the environment are also of concern and need to be addressed for the proper development of wave and tidal energy technologies. The potential environmental risks associated with wave energy converter deployments are addressed in a paper that describes the development of a spatial environmental assessment tool. Another paper deals with the efficiency of multi-type floating bodies for a novel heaving point absorber that can be applied in low-power unmanned ocean devices. This is one of the many devices that are under development, although in this case it covers a very specific type of application. The most common class of energy conversion devices are those that aim at producing large amounts of energy for supply to the electric grid, a problem that is treated in another paper that considers the energy produced by an oscillating water column device. Finally, the economic viability of this form of energy depends on economic considerations; this topic is covered in another paper that develops a method to assess the economic feasibility of wave energy farms, in the present case applied to coastal regions around Portugal. Two papers address the developments in tidal energy. One of them deals with a marine current energy converter for profiling floats, where the energy converter aims again at supplying very low levels of energy to an autonomous device, which is important to make it autonomous. The final paper addresses the issue of control of the devices, considering the effect of control strategy on tidal stream turbine performance in this particular case. Control is an important issue both for tidal energy and for wave energy devices, and it is essential for optimising their performance. Overall, this book covers a set of problems that cover different aspects of interest for wave and tidal energy conversion, and we hope it may be of interest to readers.

Carlos Guedes Soares, Matthew Lewis
Special Issue Editors

energies

Article

Wave Resource Characterization Using an Unstructured Grid Modeling Approach

Wei-Cheng Wu *, Zhaoqing Yang and Taiping Wang

Pacific Northwest National Laboratory, 1100 Dexter Ave North, Ste 500, Seattle, WA 98109, USA; zhaoqing.yang@pnnl.gov (Z.Y.); taiping.wang@pnnl.gov (T.W.)
* Correspondence: wei-cheng.wu@pnnl.gov; Tel.: +1-541-602-9879

Received: 28 December 2017; Accepted: 6 March 2018; Published: 9 March 2018

Abstract: This paper presents a modeling study conducted on the central Oregon coast for wave resource characterization, using the unstructured grid Simulating WAve Nearshore (SWAN) model coupled with a nested grid WAVEWATCH III® (WWIII) model. The flexibility of models with various spatial resolutions and the effects of open boundary conditions simulated by a nested grid WWIII model with different physics packages were evaluated. The model results demonstrate the advantage of the unstructured grid-modeling approach for flexible model resolution and good model skills in simulating the six wave resource parameters recommended by the International Electrotechnical Commission in comparison to the observed data in Year 2009 at National Data Buoy Center Buoy 46050. Notably, spectral analysis indicates that the ST4 physics package improves upon the ST2 physics package's ability to predict wave power density for large waves, which is important for wave resource assessment, load calculation of devices, and risk management. In addition, bivariate distributions show that the simulated sea state of maximum occurrence with the ST4 physics package matched the observed data better than with the ST2 physics package. This study demonstrated that the unstructured grid wave modeling approach, driven by regional nested grid WWIII outputs along with the ST4 physics package, can efficiently provide accurate wave hindcasts to support wave resource characterization. Our study also suggests that wind effects need to be considered if the dimension of the model domain is greater than approximately 100 km, or O (10^2 km).

Keywords: unstructured grid model; wave energy; resource characterization; WaveWatch III; SWAN

1. Introduction

Third-generation wave models have been significantly developed in recent years to capture nonlinear wave–wave interaction dynamics and nearshore shallow water hydrodynamics. The most popular third-generation phase-average spectral models, such as WAVEWATCH III® (WWIII) [1], the Wave Action Model (WAM) [2], Simulating WAve Nearshore (SWAN) [3], TOMAWAC [4], and MIKE-21 Spectral Wave (MIKE-21 SW) [5] models, have been widely validated in many coastal waters and open oceans worldwide. For example, the WWIII model has been maintained and used for operational ocean wave forecasts by the National Oceanic and Atmospheric Administration's (NOAA's) National Centers for Environmental Prediction (NCEP) [1,6,7]. The WAM is operated by the European Centre for Medium-Range Weather Forecasts (ECMWF), which provides wave hindcast data over the North Atlantic Ocean. WWIII results have been used to produce the first ocean wave energy resource assessment in the United States [8]. The WAM was also used to estimate wave energy resources in Europe. Both the WAM and WWIII are most commonly used to simulate wind-generated waves in deep waters and provide open boundary conditions for simulating wave dynamics in intermediate and shallow water areas. WWIII uses a time-splitting approach, calculating the solution of the energy balance equation differing from the WAM's numerical scheme [9]. Although WWIII includes curvilinear, structured, and unstructured grid options, the structured grid has been the most commonly used.

In contrast to structured grid models, unstructured grid models use flexible meshes that have high resolutions to represent the complicated bottom topography and irregular coastlines in nearshore areas (coarser resolutions are used for other areas). Hence, it is computationally more efficient to simulate wave climate with high spatial variability using one model grid without nesting [10]. For instance, TOMAWAC, a third-generation spectral wave model within the integrated TELEMAC modeling system, uses the finite element numerical method in an unstructured grid framework [11]. The TELEMAC system was originally developed by the Laboratoire National d'Hydraulique et Environnement in France and its primary users are based in European countries. UnSWAN is the unstructured grid version of the SWAN wave model, which has been widely used worldwide [10,12–14]. Please note that "unstructured grid SWAN" is often used in the literature, but that for the sake of brevity we shall use "UnSWAN" in this paper.

The main benefit of unstructured grids is that they can be applied at variable spatial resolutions while using the same flexible computational grid. Flexible meshes are useful in capturing the sharp gradients at varying water depths. MIKE 21 SW is the commercial 3G spectral wave sub-module of the MIKE 21 modeling system that solves action balance equations on an unstructured grid, using a finite volume method [15]. Similar to UnSWAN, it simulates the effects of various nonlinear physical effects. A recently developed unstructured grid version of WWIII remains under continuous development and validation [1,16–19]. Because the development of unstructured grid spectral wave models is relatively new compared to that of structured grid models, previous applications of unstructured grid wave models to characterize wave climate and resources are limited. Recently, Robertson et al. [20] applied UnSWAN to simulate wave resource characterization on the Pacific Northwest coast of Vancouver Island, British Columbia, which has very complex coastlines and a narrow continental shelf. Based on a review of the literature, we found that the UnSWAN is more popular and better validated than other unstructured grid wave models, including TOMAWAC, unstructured grid WWIII, and MIKE-21 SW. Therefore, UnSWAN was selected for this study because of its sophistication and popularity.

The International Electrotechnical Commission (IEC) released a Technical Specification (TS) for wave energy resource assessment and characterization that includes a set of standards and methods for consistent and accurate assessment of wave energy resources [21]. Following IEC TS recommendations, a number of wave energy resource assessment studies have been conducted. Lenee-Bluhm et al. assessed and characterized the wave energy resource of the U.S. Pacific Northwest using six characteristic quantities, compared with the archived spectral records from 10 wave measurement buoys from the National Data Buoy Center (NDBC) and the Coastal Data Information Program (CDIP) [22]. Akpinar et al. presented a potential wave energy assessment in the Black Sea and showed spatial distribution maps based on monthly, seasonal, and annual averages for the establishment and design of a wave energy converter (WEC) system [23]. García-Medina et al. conducted a seven-year hindcast, using nested grid WWIII and structured grid SWAN models to assess the temporal and spatial variability as well as the trend of wave resources in Oregon and southwestern Washington [24,25]. Yang et al. conducted a wave resource assessment at a test bed off the central Oregon coast, using structured grid WWIII and SWAN with a four-level nested grid approach. The physics packages were also compared to better understand the effects of the ST4 physics model for predicting wave characteristics in the frequency and directional 2D domain [26]. Structured grid WWIII and SWAN were used to evaluate wave energy resources on the United Kingdom's southwest coast and on France's west coast by Soares et al. and Goncalves et al. They found that the model performance of simulating significant wave height is better than that of mean period [27,28]. Silva et al. used WWIII and SWAN to assess wave energy resources with high resolution. They identified the differences in wave energy resources between the Iberian north and west coast with the consistent parameterization of the SWAN model setup and wind forcing [29]. Bento et al. assessed potential wave energy resources at Galway Bay using WWIII and SWAN. They found that the model performance was not good in the bay, because it is difficult for the model to generate local wind waves in a small fetch within the enclosed bay area [30]. Morim et al. assessed the wave energy resource along the southeastern coast of Australia, using the structured grid WWIII and the curvilinear SWAN. They indicated that the wind field observations from the coastal ocean to overland wind would be considered for the future research [31]. Akpinar et al. recommended applying unstructured grid systems and validating wind fields against observed wind data to further improve the wave model [32].

This paper describes how the model skills of WWIII and UnSWAN for predicting the wave energy resource parameters recommended by the IEC TS were calculated and compared with those calculated from the WWIII results. Because wave models and the quality of wave hindcasts are highly dependent on the quality of the wind field [32–35], the sensitivity of wind effects on wave predictions was also investigated. A detailed validation of physics packages using the modeled wave spectra and bivariate distributions is described.

2. Methods

2.1. Wave Models

Both WWIII and UnSWAN were used in this study. UnSWAN includes source terms for linear and exponential wind input growth and the formulation for wind input parameterization. Dissipation terms due to whitecapping, depth-induced wave breaking, and quadruplet and triad wave interactions were considered in the simulation. WWIII consists of source terms with different physics package options that consider sea ice and various wind–wave interaction and dissipation effects [6,36,37]. Specifically, the ST2 physics package is based on previously developed wind input and nonlinear interaction source terms and a new dissipation source term consisting of high- and low-frequency constituents [36]. The ST4 physics package consists of new parameterizations for swell, wave breaking, and short-wave dissipations of winds-generated waves, which are consistent under a wide range of conditions and at scales from the global ocean to coastal regions [37]. Both ST2 and ST4 physics packages were evaluated in this study.

2.2. Wave Model Grids

The model domain chosen off the central Oregon coast is shown in Figure 1. Real-time wave and meteorological data were collected from NDBC Buoy 46050 inside the model domain, using representative water depth and high-quality, long-term wave measurements.

A nested grid modeling approach was employed to drive the UnSWAN model in this study. Three levels of structured grid WWIII models provide the open boundaries for the unstructured grid model. The model domain coverage, spatial resolution, and grid size (number of grid points) for Global Grid L1 and two Intermediate Grids L2 and L3, are summarized in Table 1. The model output from the Intermediate Grid L3 provides wave open boundary conditions for the unstructured grid model domain.

Table 1. Grid information for WAVEWATCH III® (WWIII).

Grid Name	Coverage	Resolution (Long., Lat.)	Grid Size
Global Grid L1	77.5° S–77.5° N	$0.5° \times 0.5°$ ($30' \times 30'$)	223,920
Intermediate Grid L2	35°–50° N; 128°–120° W	$0.1° \times 0.1°$ ($6' \times 6'$)	12,231
Intermediate Grid L3	43.6°–45.9° N; 125.6°–123.8° W	$1' \times 1'$	15,151

The model bathymetry for all grid levels was interpolated using three NOAA bathymetry data sets: (1) 1 arc-minute ETOPO1 Global Relief Model, (2) 3 arc-second Coastal Relief Model, and (3) 1/3 arc-second tsunami high-resolution bathymetry data. The 1 arcminute ETOPO1 Global Relief Model was used for the outer shelf region and the deep ocean basins. The 3 arcsecond (~90 m) Coastal Relief Model for the inner shelf region was used for the model bathymetry and for the L2 to UnSWAN model domain. The resolution of the Coastal Relief Model data set was sufficient for the inner shelf region, because the local model grid resolution is approximately 300 m. The model bathymetry was further interpolated from NOAA's high-resolution (1/3 arcsecond) tsunami bathymetry data. The dimension of the unstructured grid for the model domain is 12 arcsec by 10 arcsec. The unstructured grid of the model domain includes 44,974 nodes and 89,100 elements.

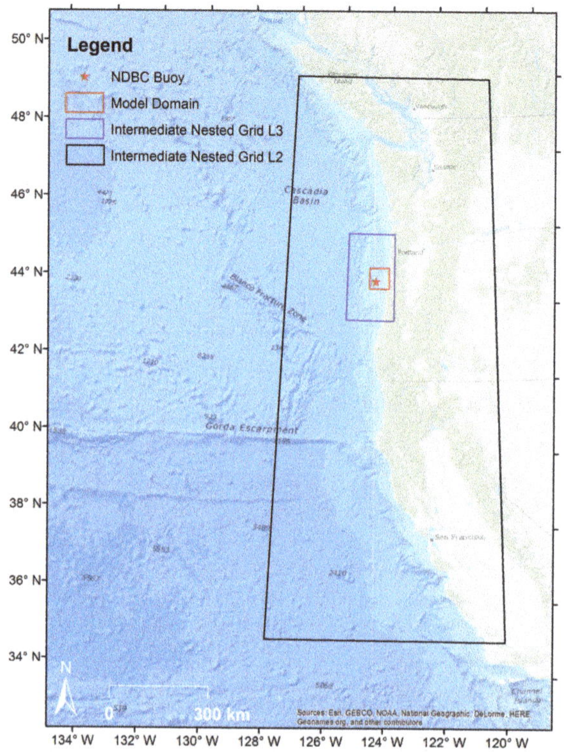

Figure 1. The UnSWAN model domain (red box) off the central Oregon coast, along with the location of the National Oceanic and Atmospheric Administration National Data Buoy Center (NDBC) Buoy 46050. The nested WWIII model domains include a global grid (L1, not indicated on the map) and two intermediate grids (L2 and L3).

An extended UnSWAN model grid, UnSWAN (L), was generated to support the conduct of additional sensitivity tests of unstructured grid flexibility and wind effect. UnSWAN (L) covers a much larger region with flexible meshes—from approximately 125.6° W to 124° W in the longitudinal direction and from 43.7° N to 45.6° N in the latitudinal direction—to demonstrate the flexibility and efficiency of unstructured grid models. The UnSWAN (L) model grid and the bathymetric features are shown in Figure 2. For the internal region that overlays the UnSWAN model domain (water depth less than 500 m), the grid resolution is very fine and has an average element area of 82,066 m^2, which is equivalent to a side length of 435 m for an equilateral triangle. Outside the UnSWAN domain, the grid resolution gradually decreases to about 8000 m toward the open boundary, where the maximum water depth is about 3000 m (Figure 2).

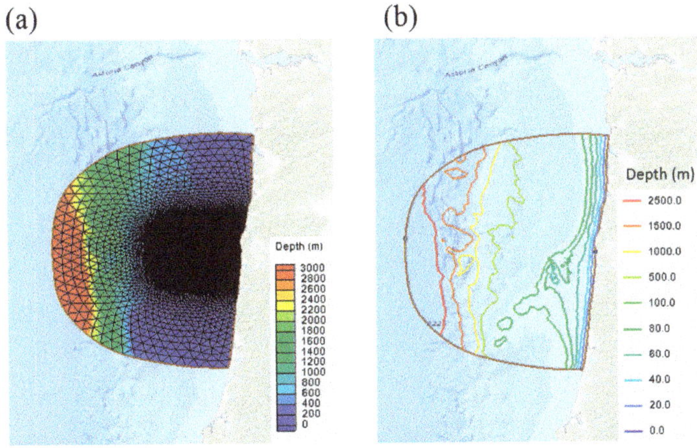

Figure 2. (**a**) The UnSWAN (L) model grid and (**b**) the bathymetric contours.

2.3. Wave Model Forcing

Sea surface wind forcing is an important factor for accurately simulating wave growth, propagation, and dissipation. In this study, hourly wind vectors from the Climate Forecast System Reanalysis (CFSR) product were interpolated onto the UnSWAN model grid. The CFSR data use temporal intervals of 1 h and a resolution of 0.5 degree, which roughly meets the requirements of the IEC standards for wind forcing resolution (50 km) for feasibility class.

3. Results and Discussion

3.1. Model Simulations

UnSWAN was applied to all model simulations in this study by using the non-stationary mode with spherical coordinates. The model configuration uses 29 spectral frequency bins ranging from 0.035 to 0.505 Hz with a logarithmic increment factor of 1.1, and 24 directional bins with a resolution of 15 degrees. This spectral resolution meets the minimum requirements specified by the IEC TS [21]: a minimum of 25 frequency components, 24 to 48 directional components, and a frequency range covering at least 0.04 to 0.5 Hz. The same spectral and directional resolutions were used in the WWIII model configuration. Default parameter settings were applied to all model simulations presented herein.

The calendar year 2009 was selected as the model validation period because of the availability and completeness of the met-ocean data at NDBC Buoy 46050; directional spectral data were available from 5 March 2008 to 2016 at NDBC Buoy 46050. A full-year simulation was first conducted to evaluate the seasonal variations in wave resource parameters. The significant wave height in 2009 at NDBC Buoy 46050 showed strong seasonal variations, a few storms occurring in the winter, and relatively calm conditions with small wave heights in the summer.

The six wave resource parameters recommended by the IEC TS [21] were calculated from simulated directional wave spectra to characterize the wave energy resource. These six simulated wave resource parameters are as follows: omnidirectional wave power, J_{omni}; significant wave height, H_{m0}; energy period, T_e; spectral width, ϵ_0; direction of maximum directionally resolved wave power, θ; and the directionality coefficient, d_θ. The formulations of these six wave resource parameters are defined below.

The omnidirectional wave power, J_{omni}, sums the contributions to energy flux from each of the components of the wave spectrum that qualifies the total sea state,

$$J = \rho g \sum_{i,j} c_{g,i} S_{ij} \Delta f_i \Delta\theta_j \quad (1)$$

where

ρ = the density of sea water,
g = the gravitational constant,
c_g = the group velocity,
S = the frequency–direction wave spectrum,
Δf_i = the frequency bin width at each discrete frequency, and
$\Delta\theta_j$ = the incident direction bin width at each discrete direction.

Assuming that waves are Rayleigh-distributed, the significant wave height may be estimated from spectral data based on the zeroth frequency spectral moment as

$$H_s \sim H_{m0} = 4.004\sqrt{m_0} \quad (2)$$

where the nth spectral moments of the variance spectrum are calculated as

$$m_n = \sum_i f_i^n S_i \Delta f_i \quad (3)$$

The energy period, T_e, is widely used in the wave energy community and it is given by

$$T_e = \frac{m_{-1}}{m_0} \quad (4)$$

The spectral width, ϵ_0, given by

$$\epsilon_0 = \sqrt{\frac{m_0 m_{-2}}{(m_{-1})^2} - 1}, \quad (5)$$

describes the spreading of wave energy over the frequency spectrum. Note that directions play an important role in WEC designs and deployment. To evaluate the directionality of the wave energy resource, the directionally resolved wave power is the sum of the wave power at each spectral direction θ:

$$J_\theta = \rho g \sum_{i,j} c_{g,i} S_{ij} \Delta f_i \Delta\theta_j \cos(\theta - \theta_j)\delta$$
$$\begin{cases} \delta = 1, & \cos(\theta - \theta_j) \geq 0 \\ \delta = 0, & \cos(\theta - \theta_j) < 0 \end{cases} \quad (6)$$

where J_θ is the directionally resolved wave power in spectral direction θ. The maximum directional resolved wave power, $J_{\theta_{max}}$, and associated direction, θ_J, can possibly be qualified for detailed WEC performance investigations. In addition, the directionality coefficient, d_θ, represents the direction of the wave power preference, and it is defined as:

$$d_\theta = \frac{J_{\theta_{max}}}{J}. \quad (7)$$

3.2. Model Performance Metrics

The six IEC TS wave resource parameters were calculated from both UnSWAN model results and measured data at NDBC Buoy 46050. Figure 3 shows the comparisons of three representative wave resource parameters—omnidirectional wave power, J_{omni}; significant wave height, H_{m0}; and energy period, T_e—between UnSWAN model results forced by WWIII-ST2 simulation and the measurements in 2009. Overall, model predictions for these three wave resource parameters match the observed data well and closely capture the seasonal variations in the measured data. Wave power density and significant wave height are much smaller from June to September than those from November to April, corresponding to the seasonal wind variations in the region.

Model performance was also examined using XY scatter plots for all six wave resource parameters in Figure 4. The XY scatter plots show good correlations between simulated and observed omnidirectional wave power, significant wave height, and energy period, similar to the time-series comparisons in Figure 3. Furthermore, simulated omnidirectional wave power and significant wave height exhibit more scattering for large waves (Figure 4a,b), indicating that the model's function of predicting large waves under extreme events is less accurate than that under the normal sea-state conditions. In addition, the simulated wave energy periods tend to be slightly larger compared to observed data (Figure 4c). Furthermore, the simulated spectral width falls within the range of 0.2 to 0.7; the small value corresponds to a narrow spectral spread in winter, whereas the large value corresponds to a broad spread in summer [22] (Figure 4d). However, correlations for the direction of the maximum directionally resolved wave power and directionality coefficient are not very strong, as shown in Figure 4e,f.

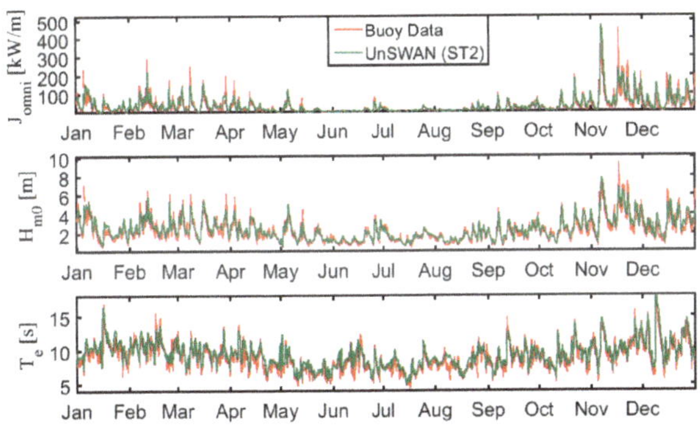

Figure 3. Hourly time-series comparison of three representative International Electrotechnical Commission (IEC) wave resource parameters between UnSWAN (ST2) predictions and observed data over the one-year period of 2009 at NDBC Buoy 46050. The open-boundary condition of UnSWAN is forced with the model output of WWIII, using the ST2 physics package.

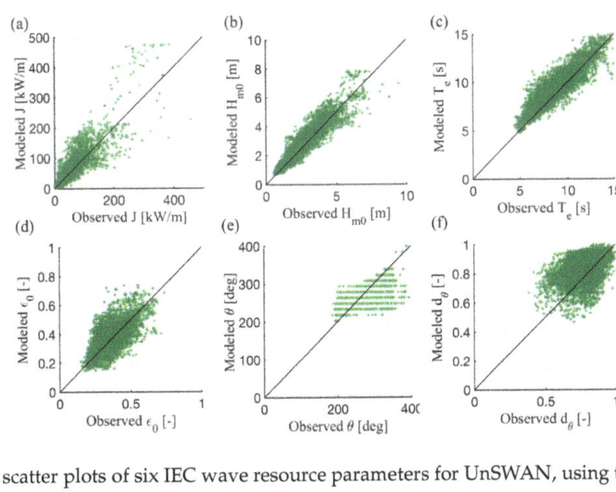

Figure 4. XY scatter plots of six IEC wave resource parameters for UnSWAN, using the ST2 physics package versus observed data. The solid black line indicates the 1:1 line.

For each of six wave resource parameters recommended by the IEC TS, the following model performance metrics were computed to show the model skills, which are commonly used in other modeling studies [24,26]. All of these metrics represent an average estimate of the difference between predicted and measured values over a defined period of simulation. The root-mean-square-error (RMSE) is defined as

$$\mathrm{RMSE} = \sqrt{\frac{\sum_{i=1}^{N}(X_i - Y_i)^2}{N}} \tag{8}$$

where N is the number of observed data, X_i is model predictions, and Y_i is the observed data.

RMSE represents the sample standard deviation of the differences between modeled data and measured data.

The percentage error (PE) is defined as

$$\mathrm{PE}(\%) = \frac{100}{N}\sum_{i=1}^{N}\left(\frac{X_i - Y_i}{Y_i}\right) \tag{9}$$

The scatter index (SI) is the RMSE normalized by the average of all measured data $\overline{Y}$

$$\mathrm{SI} = \frac{\mathrm{RMSE}}{\overline{Y}}. \tag{10}$$

Model bias and percentage bias represent the average difference between the predicted and measured data, which are defined as

$$\mathrm{Bias} = \frac{1}{N}\sum_{i=1}^{N}(X_i - Y_i). \tag{11}$$

and

$$\mathrm{Bias}(\%) = \frac{\sum_{i=1}^{N} X_i - \sum_{i=1}^{N} Y_i}{\sum_{i=1}^{N} Y_i}\cdot 100 \tag{12}$$

The linear correlation coefficient (R) is defined as

$$R = \frac{\sum_{i=1}^{N}(X_i - \overline{X})(Y_i - \overline{Y})}{\sqrt{\left(\sum_{i=1}^{N}(X_i - \overline{X})^2\right)\left(\sum_{i=1}^{N}(Y_i - \overline{Y})^2\right)}} \tag{13}$$

The model performance metrics for WWIII simulation (independent runs with ST2 and ST4 packages) and UnSWAN simulation (runs with WWIII boundary conditions when ST2 and ST4 physics were applied respectively) for the six IEC TS parameters are listed in Table 2. The error statistics for all four model runs are very similar to those in the previous studies conducted in the same region. This indicates that all model runs perform very well and that the results are in good agreement with the observed data at NDBC Buoy 46050. The RMSEs for J_{omni}, H_s, and T_e are approximately 20 and 21 (kW/m), 0.43 and 0.46 m, 0.99 and 0.94 s, respectively, for WWIII and UnSWAN with the ST2 package. In comparison, the RMSEs for J_{omni}, H_s, and T_e are about 15 and 15 (kW/m), 0.36 and 0.35 m, 1.21 and 1.09 s, respectively, for WWIII and UnSWAN with the ST4 package. This suggests that WWIII and UnSWAN with the ST4 physics package perform better in simulating J_{omni} and H_s, but slightly worse in simulating the wave energy period T_e (Table 2). Overall, UnSWAN results have better linear correlation coefficients than WWIII for all six IEC wave resource parameters. Also, for both models, spectral width (ϵ_0), direction of maximum directionally resolved wave power (θ), and the directionality coefficient (d_θ), show generally low correlation coefficients. The low correlation coefficient of ϵ_0 is caused by the higher-order moments of the variance spectrum. For θ and d_θ, the low correlation coefficients are due to the uncertainties in both modeled and measured directional metrics [20,22,24].

Table 2. Performance metrics for WWIII and UnSWAN using the ST2 and ST4 physics packages.

Parameter	Model	RMSE	PE (%)	SI	Bias	Bias (%)	R
J (kW/m)	WWIII-ST2	20	32	0.66	6.1	19.7	0.91
	WWIII-ST4	15	25	0.48	2.0	6.5	0.93
	UnSWAN-ST2	21	38	0.68	6.7	21.5	0.90
	UnSWAN-ST4	15	28	0.47	2.9	9.4	0.93
H_s (m)	WWIII-ST2	0.43	9	0.19	0.17	7.3	0.94
	WWIII-ST4	0.36	4	0.16	0.01	0.4	0.95
	UnSWAN-ST2	0.46	12	0.20	0.19	8.5	0.93
	UnSWAN-ST4	0.35	6	0.16	0.05	2.3	0.95
T_e (s)	WWIII-ST2	0.99	7	0.11	0.50	5.6	0.90
	WWIII-ST4	1.21	11	0.14	0.86	9.7	0.90
	UnSWAN-ST2	0.94	6	0.11	0.50	5.6	0.91
	UnSWAN-ST4	1.09	9	0.12	0.77	8.6	0.92
ϵ_0	WWIII-ST2	0.07	3	0.20	0.01	1.6	0.67
	WWIII-ST4	0.07	4	0.21	0.01	2.5	0.65
	UnSWAN-ST2	0.07	2	0.20	0.00	1.1	0.71
	UnSWAN-ST4	0.07	6	0.20	0.02	4.7	0.70
θ	WWIII-ST2	22.87	−2	0.08	−6.86	−2.4	0.74
	WWIII-ST4	23.33	−2	0.08	−7.62	−2.7	0.73
	UnSWAN-ST2	22.04	−2	0.08	−6.44	−2.3	0.76
	UnSWAN-ST4	22.17	−2	0.08	−7.26	−2.5	0.76
d_θ (-)	WWIII-ST2	0.10	7	0.13	0.05	6.2	0.48
	WWIII-ST4	0.10	7	0.13	0.05	5.8	0.44
	UnSWAN-ST2	0.10	7	0.13	0.05	5.6	0.51
	UnSWAN-ST4	0.10	6	0.12	0.04	5.4	0.49

3.3. Evaluation of Physics Packages

To compare the model skills of the ST2 (UnSWAN-ST2) and ST4 (UnSWAN-ST4) physics packages, wave spectra are presented here in terms of radian frequency (ϖ) in Hz and variance density (S) in m^2s/rad. Figure 5a compares the predicted and measured wave spectrum at NDBC Buoy 46050 at 5 a.m. on 17 November 2009, the time at which large waves occurred in November. The buoy data show a single swell peak at 0.0875 Hz, and the UnSWAN-ST4 result compares better with the data than the UnSWAN-ST2 result. Figure 5b shows the monthly averaged wave spectrum in November 2009, which confirms that the ST4 package performs better than the ST2 package, as relative to matching the buoy data. The ST4 physics package can improve the model's ability to predict the significant wave height and timing of large waves because of the better ST4 representation of peak frequency, as shown in Figure 5b. This is consistent with the performance metrics for simulated omnidirectional power in Table 2; the UnSWAN-ST4 result has a smaller over-predicted bias (9.4%) than the UnSWAN-ST2 result (21.5%).

In addition, the modeled and measured wave spectra at NDBC Buoy 46050 at 1 a.m. on 15 July 2009 and the July monthly average are shown in Figure 6a,b, respectively. Model results from UnSWAN-ST2 and UnSWAN-ST4 show a trend similar to the measured data, but slight under-prediction for the spectrum peak (Figure 6b). Figure 6b shows that the swell components (the first peak) are over-predicted while the wind sea components (the second peak) are under-predicted by both ST2 and ST4 physics packages. The ST4 physics package also appears to perform better in simulating swell growth and dissipation.

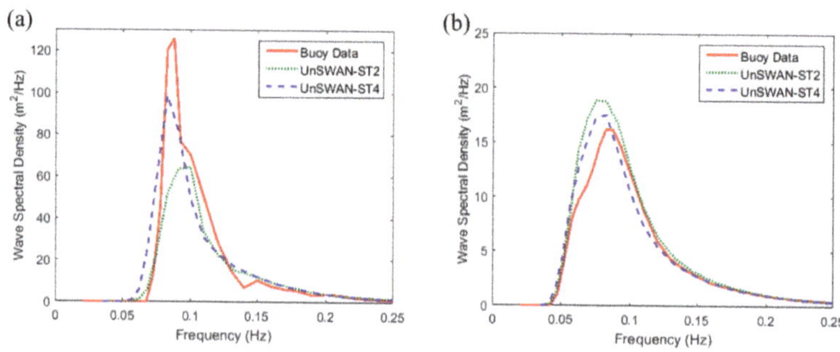

Figure 5. Measured and modeled wave spectra at NDBC Buoy 46050 (**a**) at 5 a.m. on 17 November 2009; and (**b**) the November monthly average, 2009.

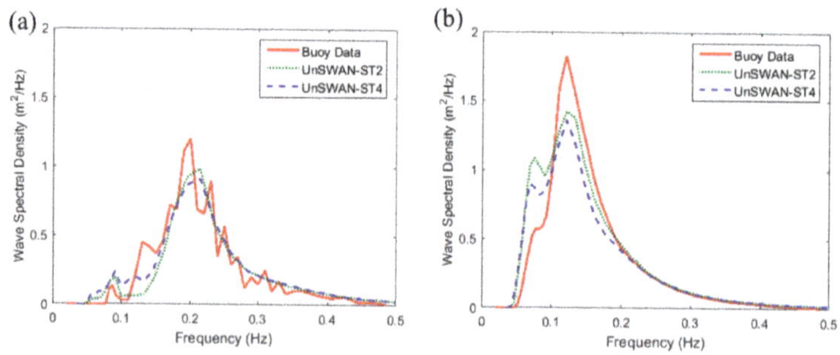

Figure 6. Measured and modeled wave spectra at NDBC Buoy 46050 (**a**) at 1 a.m. on 15 July 2009; and (**b**) the July monthly average, 2009.

3.4. Bivariate Distributions

Bivariate distributions are often used to convey the occurrence frequency of the wave climate characteristic defined by significant wave height and energy period [22], and to show the details of the mean annual frequency and energy transport characteristics of wave height and energy period combination [20]. Figure 7 shows the number of hours and the proportion of annual incident energy transport characteristics expected from sea states featuring significant wave height and energy period for buoy observations in the year 2009. Significant wave height is divided into 1 m bins over a range of 0–10 m, and the wave energy period is discretized by 1 s bins over the range of 5–16 s. The frequency in number of hours per year is illustrated by each discrete combination of significant wave height and frequency bins, while the percent of total energy is indicated by the contour color ramp. The maximum-occurring wave state is at H_{m0} = (2–3) m and T_e = (11–12) s with occurrence number 61 (Figure 7). The sea states with the maximum significant wave height do not correspond to the percent of total energy, because they only occur at limited times [31]. In comparison, the maximum occurring wave state predicted by UnSWAN-ST2 is 87, which occurred at H_{m0} = (3–4) m and T_e = (11–12) s (Figure 8a), while the maximum occurrence number predicted by UnSWAN-ST4 is 62, at H_{m0} = (2–3) m and T_e = (11–12) s (Figure 8b). Therefore, UnSWAN-ST4 has an overall better performance than UnSWAN-ST2 in predicting the bivariate distribution of sea state. However, UnSWAN-ST4 predicted the maximum occurrence sea state at the same significant wave height as the measurements, while UnSWAN-ST2 predicted it for the same energy period. This explains why UnSWAN-ST4 has a better modeling ability to predict significant wave height [34], while UnSWAN-ST2 performs slightly better in simulating energy period.

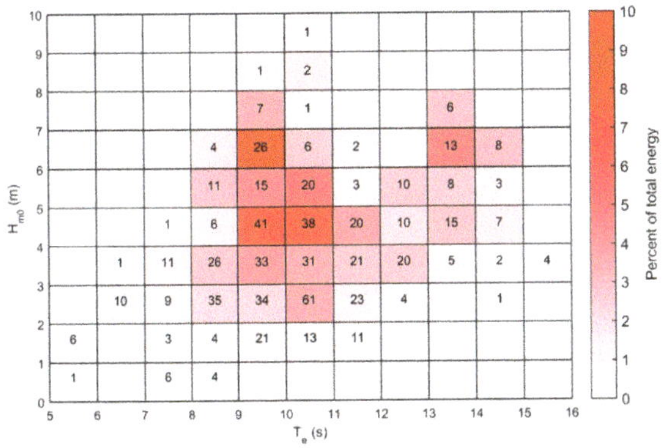

Figure 7. Bivariate distribution of occurrence and energy showing the probability and energy distribution defined by significant wave height and energy period. The results are calculated from the field data at NDBC 46050 in 2009.

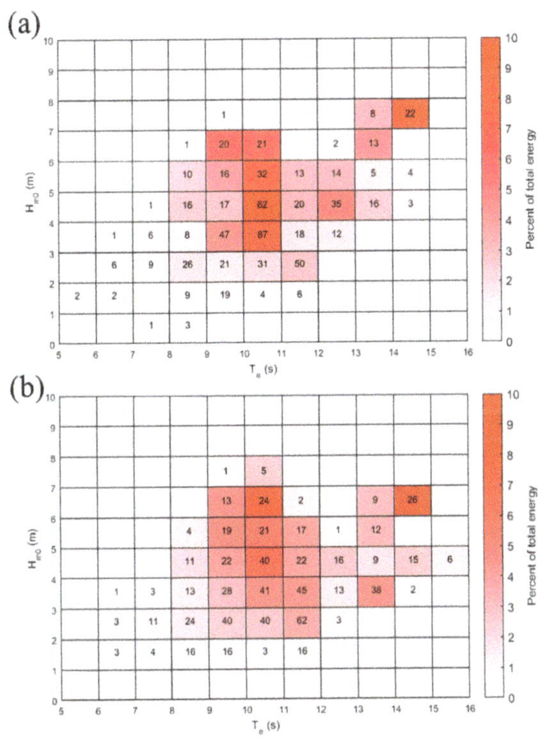

Figure 8. (**a**) UnSWAN-ST2 and (**b**) UnSWAN-ST4 bivariate distribution of occurrence and energy showing the probability and energy distribution defined by significant wave height and energy period.

The maximum occurring sea state does not correspond to the sea state with the highest percentage of total energy, because of the contribution from waves with lower significant wave height and energy periods, as shown in Figures 7 and 8. These findings are consistent with Lenee-Bluhm et al. [22]. Red shading is used to differentiate the percent of total energy; dark red color means the greatest contribution to energy. In Figure 7, the maximum percentage of total energy derived from data occurs at H_{m0} = (6–7) m and T_e = (9–10) s. On the other hand, the maximum percentage of total energy simulated by UnSWAN-ST2 occurs at H_{m0} = (4–5) m and T_e = (10–11) s, and at H_{m0} = (7–8) m and T_e = (14–15) s (Figure 8a), while the maximum simulated by UnSWAN-ST4 occurs at H_{m0} = (6–7) m and T_e = (11–5) s, and at H_{m0} = (6–7) and T_e = (14–15) (Figure 8b). This is indicative of the ST4 physics package being more skilled at simulating significant wave heights and bivariate distributions of maximum energy.

3.5. Wind Effects

Surface wind forcing plays an important role in the accuracy of the wave model hindcast, particularly under extreme wave conditions. To demonstrate the wind effects, a model simulation with UnSWAN in a much larger domain was also conducted. Model performance metrics for simulating the six wave resource parameters using UnSWAN with both the small and larger domains were compared. The UnSWAN (L) has performance metrics (Table 3) that are similar to those of UnSWAN in the model domain (UnSWAN–ST4, Table 2). Note that for the six IEC parameters, UnSWAN–ST4 results without wind forcing simulation were similar to UnSWAN–ST4 results with wind simulation, as listed in Table 2. This suggests that for the small domain, the wind effect has no significant impact on the accuracy of the numerical wave hindcast. This finding is consistent with [35].

Table 3. Performance matrix for UnSWAN (L), considering wind effects using the ST4 physics package.

Parameter	UnSWAN (L)–ST4	RMSE	PE (%)	SI	Bias	Bias (%)	*R*
J (kW/m)	wind	20	39	0.63	6.0	19.2	0.91
	no wind	22	30	0.71	3.7	11.7	0.86
H_s (m)	wind	0.44	12	0.19	0.18	7.8	0.94
	no wind	0.54	5	0.24	0.03	1.3	0.88
T_e (s)	wind	0.98	7	0.11	0.56	6.2	0.91
	no wind	1.50	14	0.17	1.05	11.8	0.84
ϵ_0	wind	0.06	2	0.19	0.00	0.7	0.72
	no wind	0.09	−10	0.25	−0.04	−12.2	0.57
θ	wind	21.83	−2	0.08	−7.28	−2.6	0.77
	no wind	26.75	−2	0.09	−8.20	−2.9	0.62
d_θ (-)	wind	0.10	6	0.12	0.04	5.1	0.51
	no wind	0.11	9	0.14	0.06	7.7	0.42

However, UnSWAN (L)–no wind has a significantly different result than UnSWAN (L) in Table 3. As calculated by Akpinar et al. [32], with a fetch of over 100 km and at a 10 m/s wind speed, a 1.63 m wave height can be fully developed. This is consistent with our model results. Therefore, for a model domain at a scale of 100 km or greater, it is important to consider the wind effect. Figure 9 shows the comparison of monthly averaged significant wave height distributions for no wind and with wind UnSWAN (L) simulations in November 2009. As one can see in Figure 9a, if the wave model does not consider wind effects, the spatial distribution of the significant wave height corresponds to the bathymetry contour shown in Figure 2b, indicating that wave model results are sensitive to bathymetry [34]. The wind forcing significantly changes the wave height spatial distribution (Figure 9b). Similarly, if UnSWAN (L) does not consider the wind effects in summer, when wind speed is generally weaker, the UnSWAN (L)–no wind run also underestimates the significant wave height, as shown in Figure 10 and Table 3. Akpinar et al. [32] showed that low wind speed would underestimate significate wave heights, which also indicates that wind effects play an important role in wave modeling.

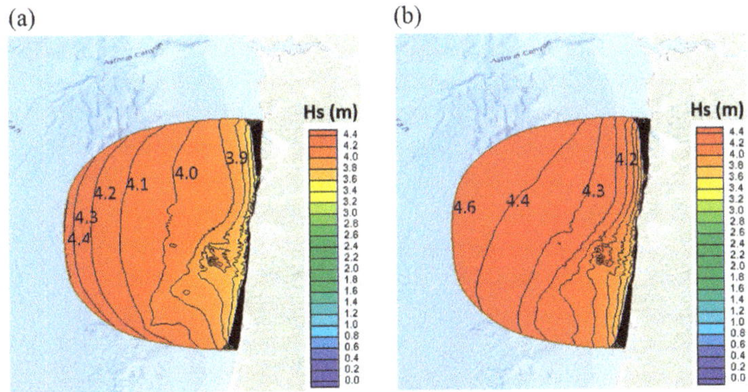

Figure 9. Modeled 2D significant wave height distributions with UnSWAN (L): (**a**) no wind—November average, 2009; (**b**) with wind—November average, 2009.

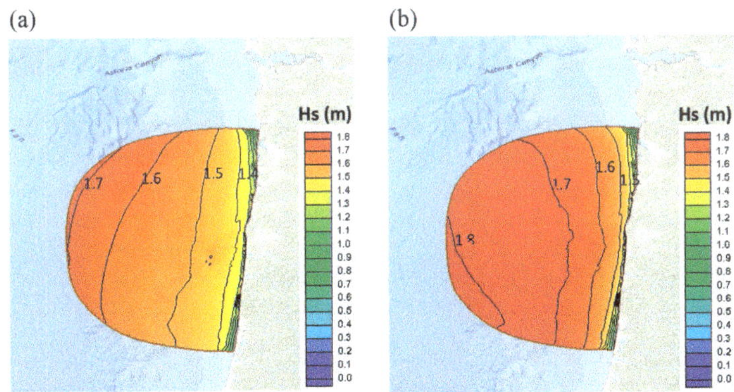

Figure 10. Modeled 2D significant wave height distribution with UnSWAN (L) (**a**) no wind—July average, 2009; (**b**) with wind—July average, 2009.

4. Conclusions

This paper presents the results of using the unstructured grid SWAN model to simulate the six IEC wave resource parameters at a model domain off the central Oregon coast. The results demonstrate the advantages of unstructured grid wave models in terms of their computational efficiency and grid flexibility for wave resource characterization. The model performance was evaluated using standard performance metrics based on a comparison of the simulations of six wave resource parameters, derived by model hindcasts to those derived from measured data. UnSWAN results show a good agreement with buoy measurements for omnidirectional wave power, significant wave height, and wave energy period. In addition, with better representations of wave growth and dissipation in the WWIII ST4 physics package to provide more accurate open boundary conditions, the UnSWAN model performs generally better than the WWIII model.

Notably, spectral analysis shows that use of the ST4 physics package improves the model's ability to predict wave spectral density for large waves. Also, a comparison of simulated and measured bivariate distributions of the maximum occurrence of sea state indicated that the ST4 physics package performed better than the ST2 physics package [34], although ST4 and ST2 showed an increased ability to predict significant wave height and wave energy period, respectively.

The sensitivity analysis of wind effect also showed that UnSWAN (L)–no wind has significantly worse results than UnSWAN (L) with wind, indicating that wave modeling with a domain at a scale of O (100 km) or greater needs to consider wind effects. Additional research is needed to investigate wind effects in wave modeling, especially under extreme wave conditions. In summary, this study demonstrated that unstructured grid wave modeling with the ST4 physics package provides advantages for assessing the temporal and spatial variability of wave climates, using high grid resolution near complex geometries, especially within a large model domain.

Acknowledgments: This study was funded by the U.S. Department of Energy, the Office of Energy Efficiency and Renewable Energy, the Water Power Technology Office, under contract DE-AC05-76RL01830 to the Pacific Northwest National Laboratory.

Author Contributions: Wei-Cheng Wu, Zhaoqing Yang and Taiping Wang conceived the study. The final manuscript has been read and approved by all the authors.

Conflicts of Interest: The authors declare no conflict of interest.

References

1. Tolman, H.L.; WAVEWATCH III Development Group. *User Manual and System Documentation of Wavewatch III® Version 4.18*; National Oceanic and Atmospheric Administration, National Weather Service, National Centers for Environmental Prediction: College Park, MD, USA, 2014; 311p.
2. WAMDI Group. The WAM model—A third generation ocean wave prediction model. *J. Phys. Oceanogr.* **1988**, *18*, 1775–1810.
3. SWAN. *SWAN: User Manual, Cycle III Version 41.01AB*; Delft University of Technology: Delft, The Netherlands, 2015.
4. Benoit, M.; Marcos, F.; Becq, F. TOMAWAC: A prediction model for offshore and nearshore storm waves. In *Environmental and Coastal Hydraulics: Protecting the Aquatic Habitat; Proceedings of the Theme B, Volumes 1 & 2*; American Society of Civil Engineers: New York, NY, USA, 1997; Volume 27, pp. 1316–1321, ISBN 0-7844-0272-8.
5. *DHI Mike 21 Spectral Waves FM e Short Description*; DHI: Horsholm, Denmark, 2012; Volume 16.
6. Tolman, H.L. Treatment of unresolved islands and ice in wind wave models. *Ocean Model.* **2003**, *5*, 219–231. [CrossRef]
7. Tolman, H.L. A new global wave forecast system at NCEP. *Ocean Wave Meas. Anal.* **1997**, *2*, 777–786.
8. EPRI. *Mapping and Assessment of the United States Ocean Wave Energy Resource*; Electric Power Research Institute: Palo Alto, CA, USA, 2011.
9. Yang, Z.; Copping, A. (Eds.) *Marine Renewable Energy—Resource Characterization and Physical Effects*; Springer International Publishing: Cham, Switzerland, 2017; pp. 37–70, ISBN 978-3-319-53534-0.
10. Zijlema, M. Computation of wind-wave spectra in coastal waters with SWAN on unstructured grids. *Coast. Eng.* **2010**, *57*, 267–277. [CrossRef]
11. Gagnaire-Renou, E.; Benoit, M.; Forget, P. Ocean wave spectrum properties as derived from quasi-exact computations of nonlinear wave-wave interactions. *J. Geophys. Res. Ocean.* **2010**, *115*. [CrossRef]
12. Cobell, Z.; Zhao, H.; Roberts, H.J.; Clark, F.R.; Zou, S. Surge and Wave Modeling for Louisiana 2012 Coastal Master Plan. *J. Coast. Res.* **2013**, *67*, 88–108. [CrossRef]
13. Robertson, B.; Bailey, H.; Clancy, D.; Ortiz, J.; Buckham, B. Influence of wave resource assessment methodology on wave energy production estimates. *Renew. Energy* **2016**, *86*, 1145–1160. [CrossRef]
14. Yuk, J.H.; Kim, K.O.; Jung, K.T.; Choi, B.H. Swell Prediction for the Korean Coast. *J. Coast. Res.* **2015**, *317*, 705–712. [CrossRef]
15. Sørensen, O.R.; Kofoed-Hansen, H.; Rugbjerg, M.; Sørensen, L.S. A third-generation spectral wave model using an unstructured finite volume technique. In Proceedings of the 29th International Conference on Coastal Engineering, Lisbon, Portugal, 19–24 September 2004; pp. 894–906. [CrossRef]
16. Dodet, G.; Bertin, X.; Bruneau, N.; Fortunato, A.B.; Nahon, A.; Roland, A. Wave-current interactions in a wave-dominated tidal inlet. *J. Geophys. Res. Ocean.* **2013**, *118*, 1587–1605. [CrossRef]
17. Roland, A.; Cucco, A.; Ferrarin, C.; Hsu, T.W.; Liau, J.M.; Ou, S.H.; Umgiesser, G.; Zanke, U. On the development and verification of a 2-D coupled wave-current model on unstructured meshes. *J. Mar. Syst.* **2009**, *78*, S244–S254. [CrossRef]

18. Roland, A.; Zhang, Y.J.; Wang, H.V.; Meng, Y.; Teng, Y.C.; Maderich, V.; Brovchenko, I.; Dutour-Sikiric, M.; Zanke, U. A fully coupled 3D wave-current interaction model on unstructured grids. *J. Geophys. Res. Ocean.* **2012**, *117*. [CrossRef]
19. Gallagher, S.; Tiron, R.; Whelan, E.; Gleeson, E.; Dias, F.; McGrath, R. The nearshore wind and wave energy potential of Ireland: A high resolution assessment of availability and accessibility. *Renew. Energy* **2016**, *88*, 494–516. [CrossRef]
20. Robertson, B.R.D.; Hiles, C.E.; Buckham, B.J. Characterizing the near shore wave energy resource on the west coast of Vancouver Island, Canada. *Renew. Energy* **2014**, *71*, 665–678. [CrossRef]
21. IEC (International Electrotechnical Commission). *Marine Energy—Wave, Tidal and Other Water Current Converters—Part 101: Wave Energy Resource Assessment and Characterization*; IEC TS 62600-101; Edition 1.0.2015-06; International Electrotechnical Commission: Geneva, Switzerland, 2015.
22. Lenee-bluhm, P.; Paasch, R.; Özkan-haller, H.T. Characterizing the wave energy resource of the US Pacific Northwest. *Renew. Energy* **2011**, *36*, 2106–2119. [CrossRef]
23. Akpinar, A.; Kömürcü, M.I. Assessment of wave energy resource of the Black Sea based on 15-year numerical hindcast data. *Appl. Energy* **2013**, *101*, 502–512. [CrossRef]
24. García-Medina, G.; Özkan-Haller, H.T.; Ruggiero, P. Wave resource assessment in Oregon and southwest Washington, USA. *Renew. Energy* **2014**, *64*, 203–214. [CrossRef]
25. García-Medina, G.; Özkan-Haller, H.T.; Ruggiero, P.; Oskamp, J. An Inner-Shelf Wave Forecasting System for the U.S. Pacific Northwest. *Weather Forecast.* **2013**, *28*, 681–703. [CrossRef]
26. Yang, Z.; Neary, V.S.; Wang, T.; Gunawan, B.; Dallman, A.R.; Wu, W. A wave model test bed study for wave energy resource characterization. *Renew. Energy* **2017**, 1–13. [CrossRef]
27. Soares, C.G.; Bento, A.R.; Gonçalves, M.; Silva, D.; Martinho, P. Numerical evaluation of the wave energy resource along the Atlantic European coast. *Comput. Geosci.* **2014**, *71*, 37–49. [CrossRef]
28. Gonçalves, M.; Martinho, P.; Guedes Soares, C. Wave energy conditions in the western French coast. *Renew. Energy* **2014**, *62*, 155–163. [CrossRef]
29. Silva, D.; Bento, A.R.; Martinho, P.; Guedes Soares, C. High resolution local wave energy modelling in the Iberian Peninsula. *Energy* **2015**, *91*, 1099–1112. [CrossRef]
30. Bento, A.R.; Martinho, P.; Guedes Soares, C. Numerical modelling of the wave energy in Galway Bay. *Renew. Energy* **2015**, *78*, 457–466. [CrossRef]
31. Morim, J.; Cartwright, N.; Etemad-Shahidi, A.; Strauss, D.; Hemer, M. Wave energy resource assessment along the Southeast coast of Australia on the basis of a 31-year hindcast. *Appl. Energy* **2016**, *184*, 276–297. [CrossRef]
32. Akpinar, A.; van Vledder, G.P.; Kömürcü, M.I.; Özger, M. Evaluation of the numerical wave model (SWAN) for wave simulation in the Black Sea. *Cont. Shelf Res.* **2012**, *50–51*, 80–99. [CrossRef]
33. Moeini, M.H.; Etemad-Shahidi, A.; Chegini, V.; Rahmani, I. Wave data assimilation using a hybrid approach in the Persian Gulf. *Ocean Dyn.* **2012**, *62*, 785–797. [CrossRef]
34. Amrutha, M.M.; Kumar, V.S.; Sandhya, K.G.; Nair, T.M.B.; Rathod, J.L. Wave hindcast studies using SWAN nested in WAVEWATCH III—Comparison with measured nearshore buoy data off Karwar, eastern Arabian Sea. *Ocean Eng.* **2016**, *119*, 114–124. [CrossRef]
35. Strauss, D.; Mirferendesk, H.; Tomlinson, R. Comparison of two wave models for Gold Coast, Australia. *J. Coast. Res.* **2007**, *50*, 312–316.
36. Tolman, H.L.; Chalikov, D. Source Terms in a Third-Generation Wind Wave Model. *J. Phys. Oceanogr.* **1996**, *26*, 2497–2518. [CrossRef]
37. Ardhuin, F.; Rogers, E.; Babanin, A.V.; Filipot, J.-F.; Magne, R.; Roland, A.; van der Westhuysen, A.; Queffeulou, P.; Lefevre, J.-M.; Aouf, L.; et al. Semiempirical Dissipation Source Functions for Ocean Waves. Part I: Definition, Calibration, and Validation. *J. Phys. Oceanogr.* **2010**, *40*, 1917–1941. [CrossRef]

© 2018 by the authors. Licensee MDPI, Basel, Switzerland. This article is an open access article distributed under the terms and conditions of the Creative Commons Attribution (CC BY) license (http://creativecommons.org/licenses/by/4.0/).

energies

MDPI

Article

Identifying the Optimal Offshore Areas for Wave Energy Converter Deployments in Taiwanese Waters Based on 12-Year Model Hindcasts

Hung-Ju Shih, Chih-Hsin Chang, Wei-Bo Chen * and Lee-Yaw Lin

National Science and Technology Center for Disaster Reduction, New Taipei City 23143, Taiwan; shihruby713@ncdr.nat.gov.tw (H.-J.S.); chang.c.h@ncdr.nat.gov.tw (C.-H.C.); yaw@ncdr.nat.gov.tw (L.-Y.L.)
* Correspondence: wbchen@livemail.tw or wbchen@ncdr.nat.gov.tw; Tel.: +886-2-8195-8612

Received: 8 January 2018; Accepted: 23 February 2018; Published: 27 February 2018

Abstract: A 12-year sea-state hindcast for Taiwanese waters, covering the period from 2005 to 2016, was conducted using a fully coupled tide-surge-wave model. The hindcasts of significant wave height and peak period were employed to estimate the wave power resources in the waters surrounding Taiwan. Numerical simulations based on unstructured grids were converted to structured grids with a resolution of 25 × 25 km. The spatial distribution maps of offshore annual mean wave power were created for each year and for the 12-year period. Waters with higher wave power density were observed off the northern, northeastern, southeastern (south of Green Island and southeast of Lanyu) and southern coasts of Taiwan. Five energetic sea areas with spatial average annual total wave energy density of 60–90 MWh/m were selected for further analysis. The 25 × 25 km square grids were then downscaled to resolutions of 5 × 5 km, and five 5 × 5 km optimal areas were identified for wave energy converter deployments. The spatial average annual total wave energy yields at the five optimal areas (S1)–(S5) were estimated to be 64.3, 84.1, 84.5, 111.0 and 99.3 MWh/m, respectively. The prevailing wave directions for these five areas lie between east and northeast.

Keywords: tide-surge-wave model; Taiwanese waters; sea-state hindcast; wave power; wave energy

1. Introduction

Oceans cover over 70% of the Earth's surface and represent a wealth of renewable energy resources [1]. Many potential candidates for ocean energy sources exist, including wave energy, tide energy, tidal current energy, thermal gradients and salinity gradients [2,3]. Of these, wave power is currently the most researched area; it represents the largest ocean energy resource and is the most widely distributed marine renewable energy around the world. Many methods of wave energy conversion have been developed to extract wave power from oceans. The overtopping/terminator type of wave energy converter (WEC) consists of two wave reflectors that collect overtopping water and convert the pressure head into power [4]. The raft-type WEC uses relative rotation around a hinge to drive an electrical generation system to convert wave energy to electricity [5]. Another approach involves a point-oscillating absorber type WEC, which utilizes relative translational motion in which the oscillating motion of the floater is converted into electricity by a power take off (PTO) system for the oscillating-body WEC. This type of WEC is widely used for offshore deployments [6]. The two-floater WEC also uses a PTO system to convert ocean wave-induced motion into electricity [7]. The attenuator type of WEC captures energy from the relative motions of two arms as waves pass.

The use of numerical models to assess ocean wave power has been widely adopted on both regional [8–11] and global scales [12–15] and has been implemented for several islands, including the Canary Islands, Madeira, and the Azores in the Atlantic Ocean [16–19], Hawaii and Taiwan in the Pacific Ocean [20–23], Fuerte Island in the Caribbean Sea [24], and Sardinia and Menorca in the Mediterranean Sea [25–27].

Hashemi and Neill [28] used a wave-tide coupled model called the Simulating WAves Nearshore-Regional Ocean Modeling System (SWAN-ROMS) to investigate wave energy resources in the seas along the northwest European shelf. Their results suggested that tidal impact is significant and that the contributions from the effects of tidal currents on wave power resources are greater than the contribution from variations in tide levels. The maximum current speeds of the Kuroshio in the eastern offshore sea of Taiwan range between 0.6 and 1.2 m/s [29]. Therefore, using a tide-surge-wave coupled model is essential for accurately assessing the distribution of wave power in Taiwanese waters.

There are substantial ocean energy resources around the Island of Taiwan. However, the statistical data issued from the Ministry of Economic Affairs, Taiwan, revealed that the annual total electricity generating capacity of Taiwan was 264.1 billion kWh in 2016, of which fuel-fired power accounted for 82% (216.56 billion kWh), nuclear power accounted for 12% (31.69 billion kWh), hydroelectric power accounted for 1.2% (3.17 billion kWh), and renewable power accounted for 4.8% (12.68 billion kWh). Regarding renewable energy, only wind power and solar power are currently utilized in Taiwan. However, exploitation of energy resources in the ocean is urgently needed in Taiwan because they are renewable and do not contribute to atmospheric pollution.

Assessments of optimal locations or hotspots for deploying wave energy converters in previous studies have usually been conducted by reporting the location of selected "points". For instance, Morim et al. [30] evaluated the wave energy resources and optimal locations along the southeast coast of Australia, and Su et al. [23] assessed the distribution of wave power and hotspots for the surrounding waters of Taiwan. Although this method is quite straightforward, the possible energetic "sea area" where wave energy converters can be deployed is not definite. This is an important issue; wave energy converters should be placed in an "area" rather than at a "point". In the present study, numerical simulations based on unstructured grids were converted to structured (square) grids to identify energetic "sea areas" using the gridding method available through ArcGIS software. The annual mean wave energy yields of the final optimal areas for WEC deployments were acquired by spatial averaging over 5 km square regions. The approach proposed in the present study is both innovative and helpful for assessing the most appropriate offshore sea areas in which to deploy wave energy converters in Taiwanese waters.

The primary objectives of the present study were as follows: (1) implement a tide-surge-wave fully coupled high-resolution model for Taiwanese waters; (2) validate the fully coupled model with available observations; (3) create spatial distribution maps for annual and 12-year-average wave power; (4) using the gridding method, identify the most energetic and optimal areas in the offshore seas of Taiwan for deploying wave energy converters; and (5) assess the annual total wave energy density and the dominant wave direction in each optimal area.

2. Methods and Data

2.1. Ocean Circulation Model

The sea surface elevations and currents of the waters surrounding Taiwan were simulated with the community ocean model Semi-implicit Cross-scale Hydroscience Integrated System Model (SCHISM). SCHISM is an upgraded and evolved version of the original Semi-implicit Eulerian-Lagrangian Finite-Element/volume (SELFE [31]) model that includes many enhancements and improvements such as a new extension to simulate large-scale eddying regimes and a seamless cross-scale capability ranging from creeks to oceans [32]. SCHISM and SELFE have been widely used to simulate the propagation of tsunami waves [33], to predict water quality and ecosystem dynamics [34–36], analyze oil spill diffusion and transport [37], generate flooding maps [38,39], evaluate the extent of inundation induced by extreme river flows, typhoons or hurricanes [40,41] and to estimate tidal-current power [42]. In Taiwanese waters, the surface circulation feature is highly similar to that on the bottom during the spring [43], and weak winter stratification is caused by vertical mixing [44]; therefore, a two-dimensional SCHISM model (SCHISM-2D) vertically integrated with the barotropic mode was employed in this study. The governing equations in the Cartesian coordinate system and two-dimensional form are as follows:

$$\frac{\partial \eta}{\partial t} + \frac{\partial uH}{\partial x} + \frac{\partial vH}{\partial y} = 0, \tag{1}$$

$$\frac{Du}{Dt} = fv - \frac{\partial}{\partial x}\left\{g(\eta - \alpha\hat{\psi}) + \frac{P_A}{\rho_0}\right\} + \frac{\tau_{sx} + \tau_{rx} - \tau_{bx}}{\rho_0 H}, \tag{2}$$

$$\frac{Dv}{Dt} = -fu - \frac{\partial}{\partial y}\left\{g(\eta - \alpha\hat{\psi}) + \frac{P_A}{\rho_0}\right\} + \frac{\tau_{sy} + \tau_{ry} - \tau_{by}}{\rho_0 H}, \tag{3}$$

where $\eta(x, y, t)$ is the free-surface elevation, $H = \eta + h$ is the total water depth, h is the bathymetric depth, $u(x, y, t)$ and $v(x, y, t)$ are the horizontal velocity components in the x, y direction, respectively, f is the Coriolis factor, g is the acceleration due to gravity, $\hat{\psi}$ is the Earth's tidal potential, α is the effective earth elasticity factor, ρ_0 is the reference density of water, and $P_A(x, y, t)$ is the atmospheric pressure at the free surface. In Equations (2) and (3), τ_{sx} and τ_{sy} are the wind stress components in the x, y direction, respectively, and can be expressed as follows:

$$\tau_{sx} = \rho_a C_s \sqrt{{W_x}^2 + {W_y}^2} W_x \tag{4a}$$

$$\tau_{sy} = \rho_a C_s \sqrt{{W_x}^2 + {W_y}^2} W_y \tag{4b}$$

where C_s is the wind drag coefficient; ρ_a is the air density; and W_x, W_y are the wind-speed components 10 m above the sea surface in the x, y directions, respectively. C_s is often regarded as an increasing function of wind velocity; however, Powell et al. [45] suggested that C_s should be limited at high wind speeds. The formula to calculate C_s in SCHISM-2D is:

$$C_s = 1.0^{-3}\begin{cases} (0.61 + 0.063 \times 6.0), W < 6.0 \\ (0.61 + 0.063W),\ 6.0 \le W \le 50.0 \\ (0.61 + 0.063 \times 50.0), W > 50.0 \end{cases} \tag{5}$$

where W is the resultant wind speed at 10 m above the sea surface.

Here, τ_{bx} and τ_{by} are the bottom shear stress components in the x, y directions and are computed by the following formulas:

$$\tau_{bx} = \rho_0 C_d \sqrt{u^2 + v^2} u \tag{6a}$$

$$\tau_{by} = \rho_0 C_d \sqrt{u^2 + v^2} v \tag{6b}$$

where C_d is the bottom drag coefficient, which can be parameterized as follows:

$$C_d = gn^2 / H^{1/3} \tag{7}$$

where n is the Manning coefficient, which was set to 0.025 in the model due to the lack of knowledge of the sea-bottom material type. However, the bottom drag coefficient C_d varies with H based on Equation (7).

Here, τ_{rx} and τ_{ry} are the radiation stress components in the x, y directions, respectively, and can be computed following Longuet-Higgins and Stewart [46,47], as shown below:

$$\tau_{rx} = -\frac{\partial S_{xx}}{\partial x} - \frac{\partial S_{xy}}{\partial y} \tag{8a}$$

$$\tau_{ry} = -\frac{\partial S_{xy}}{\partial x} - \frac{\partial S_{yy}}{\partial y} \tag{8b}$$

where S_{xx}, S_{xy} and S_{yy} are the wave radiation stress components and are represented according to Battjes [48]:

$$S_{xx} = \int_0^{2\pi}\int_0^{\infty} N\sigma \frac{C_g}{C_p}\sin(\theta)\mathrm{d}\theta\mathrm{d}\sigma, \tag{9a}$$

$$S_{xy} = S_{yx} = \int_0^{2\pi}\int_0^{\infty} N\sigma\left[\frac{C_g}{C_p}\left(\cos^2(\theta)+1\right) - \frac{1}{2}\right]\mathrm{d}\theta\mathrm{d}\sigma \tag{9b}$$

$$S_{yy} = \int_0^{2\pi}\int_0^{\infty} N\sigma\left[\frac{C_g}{C_p}\left(\sin^2(\theta)+1\right) - \frac{1}{2}\right]\mathrm{d}\theta\mathrm{d}\sigma \tag{9c}$$

where N is the spectral wave action density, and the independent variables are the wave relative frequency, σ, and the wave direction, θ. Here, C_g and C_p are the wave group velocity and the wave phase velocity, respectively.

2.2. Spectral Wind Wave Model

A third-generation spectral wind wave model, Wind Wave Model III (WWM-III, 3.0, College of William & Mary, Williamsburg, VA, USA), was adopted to predict the significant wave heights (SWH) and wave peak period (Tp) in the coastal waters of Taiwan. The wave-action equation that governs WWM-III is given as:

$$\frac{\partial N}{\partial t} + \frac{\partial (C_{gx}+u)N}{\partial x} + \frac{\partial (C_{gy}+v)N}{\partial y} + \frac{\partial (C_\sigma N)}{\partial \sigma} + \frac{\partial (C_\theta N)}{\partial \theta} = \frac{S_{tot}}{\sigma} \tag{10}$$

where C_{gx} and C_{gy} are the wave group velocity components in the x, y direction, C_σ and C_θ are the propagation velocities in σ, θ space, respectively, and S_{tot} is the sum of the source terms for wave variance, including non-linear interactions, wind growth and dissipation by white-capping, bottom friction and wave breaking. The bottom friction and peak enhancement factors are set to 0.067 and 3.3, respectively, based on the formulation of the Joint North Sea WAve Project (JONSWAP) [49]. WWM-III computes waves breaking in shallow-water areas using the method proposed by [50]. A constant wave-breaking coefficient of 0.78 was adopted. The maximum wave direction is 360°, which is discretized into 36 bins in WWM-III. The low- and high-frequency limits of the discrete wave frequency are 0.03 and 1.0 Hz, respectively and are also divided into 36 bins.

SCHISM-2D (College of William & Mary, Williamsburg, VA, USA) computes the depth-averaged current and water surface elevation and then passes the results to the wave model, WWM-III, which sends the radiation stress to SCHISM-2D after solving the wave action equation. More details about the model coupling procedures of SCHISM-WWM-III can be found in [51].

2.3. Meteorological and Boundary Conditions

The wind forcing data used to drive SCHISM-WWM-III were the wind field at 10 m above sea level and the sea level pressure produced by ERA-Interim, which were acquired from public datasets available from the European Center for Medium-Range Weather Forecasts (ECMWF). ERA-Interim is the latest global atmospheric reanalysis, which is normally updated monthly with a two-month delay [52]. The 10-m wind speed in the x, y components were extracted from the ERA-Interim reanalysis results with a temporal resolution of 6 h (four analysis fields per day, at 00:00, 06:00, 12:00 and

18:00 UTC) and a spatial resolution of 0.125° × 0.125° and converted to the unstructured grids of SCHISM-WWM-III using the Inverse Distance Weighting (IDW) method. The IDW formula is:

$$p_e(x,y) = \frac{\sum_{i=1}^{n} \frac{1}{d_i} p_m(x_i, y_i)}{\sum_{i=1}^{n} \frac{1}{d_i}} \tag{11}$$

where $p_e(x, y)$ is the value to be estimated (the wind speed in SCHISM-WWM-III), $p_m(x, y)$ is the known value (the wind speed in ERA-Interim), and d_i represents the distances from the n data points to the point estimated n.

The wave boundary conditions of the regional SCHISM-WWM-III, including the significant wave height, the mean wave direction, the mean directional spreading, the peak frequency, and the zero down-crossing frequency, were obtained from a global WaveWatch III (WW III) model developed by the French Research Institute for Exploitation of the Sea (IFREMER) with a spatial resolution of 0.5° × 0.5° and a temporal resolution of 3 h. To obtain the tide boundary conditions, the harmonic constants of eight tidal constituents (M2, S2, N2, K2, K1, O1, P1, and Q1) were extracted from a regional inverse tidal model, China Seas & Indonesia 2016 [53], with a resolution of 1/30° and were set at each node on the open boundary to drive the model.

2.4. Estimation of Wave Power

The extractable wave power per unit of the wave crest length can be calculated through the spectral output of SCHISM-WWM-III and is given in kW per meter as shown below:

$$P = \rho g \int_0^{2\pi} \int_0^{\infty} c_g(\sigma, d) S(\sigma, \theta) \mathrm{d}\sigma \mathrm{d}\theta \tag{12}$$

where $S(\sigma, \theta)$ is the directional wave variance density spectrum, ρ is the density of seawater, g is the acceleration due to gravity, and d is the water depth. The wave group velocity, $c_g(\sigma, \theta)$, can be expressed as follows:

$$c_g(\sigma, d) = \frac{g}{4\pi\sigma}\left[1 + \frac{2\kappa d}{\sin h(2\kappa d)}\right] \tan[h(\kappa d)] \tag{13}$$

where $\kappa = 2\pi/L$ is the number of waves, and L is the wave length. In deep water conditions ($d > 0.5L$) and $c_g(\sigma, \theta) = g/4\pi\sigma$, Then:

$$P = \frac{\rho g^2}{4\pi} \int_0^{2\pi} \int_0^{\infty} \sigma^{-1} S(\sigma, \theta) \mathrm{d}\sigma \mathrm{d}\theta \tag{14}$$

The spectral moments of order, n, are defined as follows:

$$m_n = \frac{\rho g^2}{4\pi} \int_0^{2\pi} \int_0^{\infty} \sigma^{n} S(\sigma, \theta) \mathrm{d}\sigma \mathrm{d}\theta \tag{15}$$

The energy period T_e and the significant wave height H_s in terms of spectral moment are as follows:

$$T_e = \frac{m_{-1}}{m_0} = \frac{\int_0^{2\pi} \int_0^{\infty} \sigma^{-1} S(\sigma, \theta) \mathrm{d}\sigma \mathrm{d}\theta}{\int_0^{2\pi} \int_0^{\infty} S(\sigma, \theta) \mathrm{d}\sigma \mathrm{d}\theta} \tag{16}$$

$$H_s = 4(m_0)^{1/2} = 4\left(\int_0^{2\pi}\int_0^{\infty} S(\sigma,\theta)\mathrm{d}\sigma\mathrm{d}\theta\right)^{1/2} \quad (17)$$

Therefore, Equation (11) can be simplified to:

$$P = \frac{\rho g^2}{64\pi} H_s^2 T_e \quad (18)$$

If ρ is assumed to be 1025 kg/m^3, Equation (18) becomes:

$$P(\text{kilowatt/meter, kw/m}) = 0.49 H_s^2 T_e \quad (19)$$

It has been suggested that T_e is rarely specified and must be estimated from other variables, while the peak period T_p is often specified by measurements of sea states [23,49,50]. T_e is equal to T_p when multiplied by a coefficient α:

$$T_e = \alpha T_p \quad (20)$$

in which α = 0.9 was adopted from the standard JONSWAP spectrum with a peak enhancement factor of γ = 3.3 [23,27,54,55]. Thus, the theoretical wave power at any point can be computed using the following expression:

$$P(\text{kilowatt/meter, kw/m}) = 0.44 H_s^2 T_p. \quad (21)$$

The present study used Equation (15) to evaluate the wave power resources in Taiwanese waters, even when the deep-water hypothesis was not strictly satisfied [56].

Thus, the total annual wave energy per unit length can be generated for a given time interval and is calculated as follows:

$$E_t(\text{kilowatt hour/meter, kwh/m}) = \sum_i P_i \times \Delta t_i, \quad (22)$$

where Δt_i is the temporal sampling interval; Δt_i = 1 h in the present study.

2.5. Model Configuration

The computational domain covers Taiwan in its entirety as well as its main offshore islands. The region covers the area from a longitude of 114° E to 130° E and a latitude of 19° N to 29° N, as shown in Figure 1a. Two gridded bathymetry datasets were employed in this study. One dataset consists of global data obtained from the General Bathymetric Chart of the Oceans (GEBCO) and has a resolution of 30 arc-seconds. It was generated by combining quality-controlled ship depth soundings with interpolations between the sounding points guided by satellite-derived gravitational data. The other dataset is local data provided by the Department of Land Administration, Ministry of the Interior in Taiwan, with a resolution of 200 m and is distributed over a longitude of only 100° E to 128° E and a latitude of 4° N to 29° N. The bathymetry data for the coupled model were produced by merging the GEBCO and local bathymetry datasets. The method for combining two bathymetry sets is to merge the values in some order of priority using Surface-water Modeling System (SMS) software. A priority must be adopted because data overlaps exist in the source datasets. In the present study, when overlap occurs, the GEBCO data points in a set are considered as lower priority than the local data points; thus, they are not included in the final merged set. When no overlap occurs, all data points are merged. This approach ensured the preservation of the high-resolution local data. Our modeling domain is composed of 278,630 triangular cells and 142,041 non-overlapping, unstructured grids. Coarse meshes with 30 km resolution were arranged on the open ocean boundaries, while fine meshes with 300 m resolution were used along the coastline of Taiwan and the small offshore islands (Figure 1b). After the unstructured grids were created, the final merged, gridded bathymetry dataset was interpolated to each node to represent the bottom elevations in the coupled model (Figure 1a) using the linear interpolation method in SMS software.

In the ocean circulation model, a time step of 120 s was used for the present unstructured-grid system with numerical stability. SCHISM and WWM-III exchange the computational results every five hydrodynamic time steps (i.e., the time step of WWM-III is 600 s).

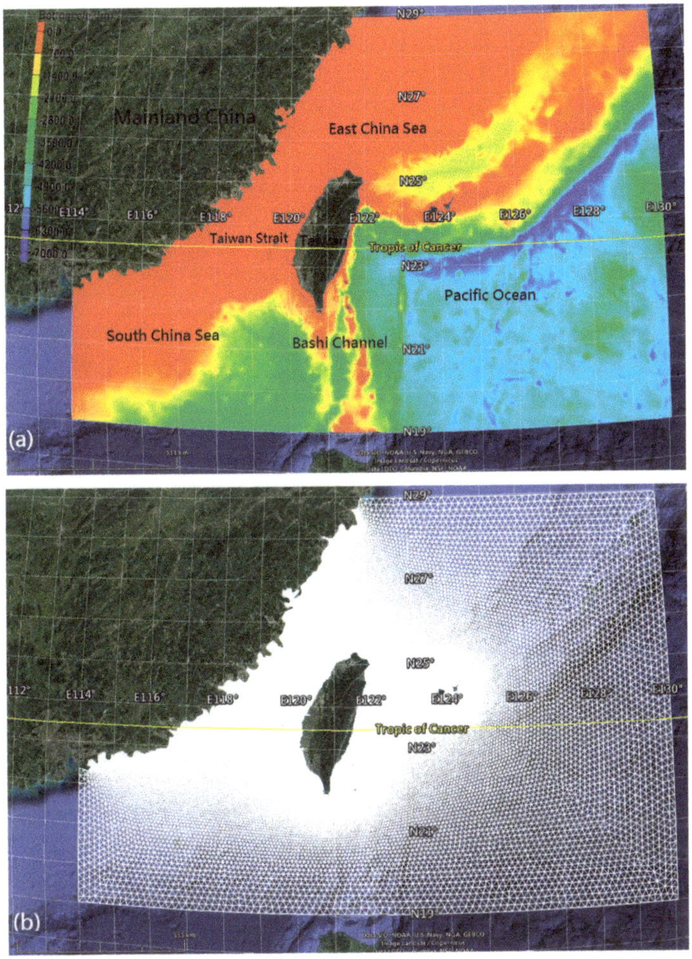

Figure 1. (**a**) Bathymetry and (**b**) unstructured grid of the computational domain.

3. Model Validation

The observations adopted to validate SCHISM-WWM-III were measured significant wave heights (SWH) and peak periods (*Tp*) provided by the Harbor and Marine Technology Center (HMTC) in Taiwan at four buoys. The distribution map of these four wave buoys around Taiwanese waters is shown in Figure 2. The Keelung, Suao, and Hualien buoys are situated along the northeastern and eastern coast of Taiwan, while the Penghu buoy lies on the western coast of Taiwan (the Taiwan Strait). Figures 3 and 4 present the comparisons of SWH (Figure 3) and *Tp* (Figure 4) between the model hindcasts and the measurements at the four buoys from 31 January 2016 to 30 December

2016. This period includes normal wind-generated waves (*Tp*), monsoon-induced waves (*Tp*) and the abnormal waves (*Tp*) caused by strong winds from Typhoon Nepartax (a severe typhoon, from 6 July 2016 to 9 July 2016), Typhoon Meranti (a severe typhoon, from 12 September 2016 to 15 September 2016) and Typhoon Megi (a moderate typhoon, from 25 September 2016 to 28 September 2016). The maximum SWH and *Tp* reached approximately 5.5 m and 14 s, 9.5 m and 16 s, 7.5 m and 15 s and 3.5 m and 13 s at the Keelung, Suao, Hualien, and Penghu buoys, respectively, due to the approach of Typhoon Megi. Peak SWH occurred at Keelung, Suao, and Hualien buoys on 28 September 2016, with a lag of almost one day at the Penghu buoy (as shown in Figures 3 and 4). This result occurred because Typhoon Megi crossed Taiwan from east to west.

The statistical errors of the differences between the model hindcasts and the observations at the four wave buoys are also estimated. The correlation coefficients of the SWH and *Tp* are 0.86 and 0.74, 0.83 and 0.84, 0.87 and 0.83 and 0.80 and 0.73 for the Keelung, Suao, Hualien, and Penghu buoys, respectively. Even though the hindcasted SWH and *Tp* agree well with the measurements during both ordinary and extraordinary meteorological conditions, slight discrepancies still exist. It must be emphasized that the numerical wave models fail to capture *Tp* well, especially during normal meteorological conditions, for instance, April, May and June. The possible reasons include the nature of wave models. Some of the setup assumptions and numerical solutions within the models affect their accuracy [57]. In addition, the lower spatial and temporal resolutions of the wind field data from the ERA-Interim dataset affected model hindcasting, which is more sensitive to *Tp*. The adoption of higher spatial and temporal resolution wind data could improve the performance [58]. Yet a third reason for poor *Tp* hindcasts may be due to observational error. However, based on the model-data comparison, the hindcasts of SWH and *Tp* by SCHISM-WWM-III are relatively reliable and can be further applied to assess the distribution of wave power and wave energy in the waters surrounding Taiwan.

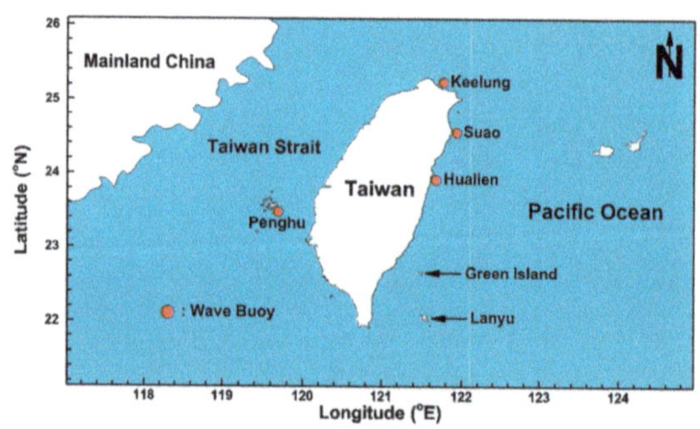

Figure 2. Distribution of wave buoys around Taiwanese waters. The cyan area represents ocean and the white areas represent land.

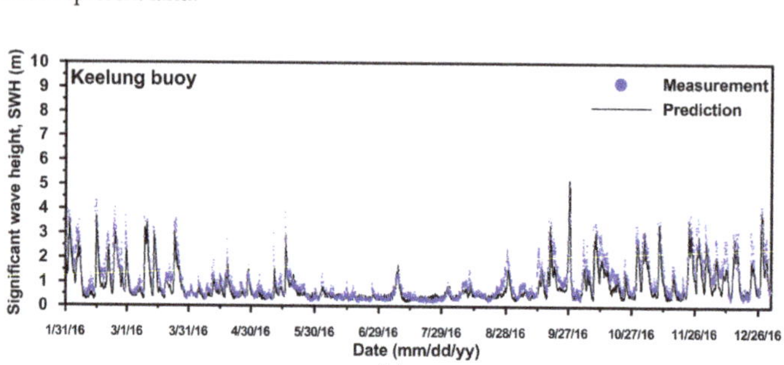

(**a**)

Figure 3. *Cont.*

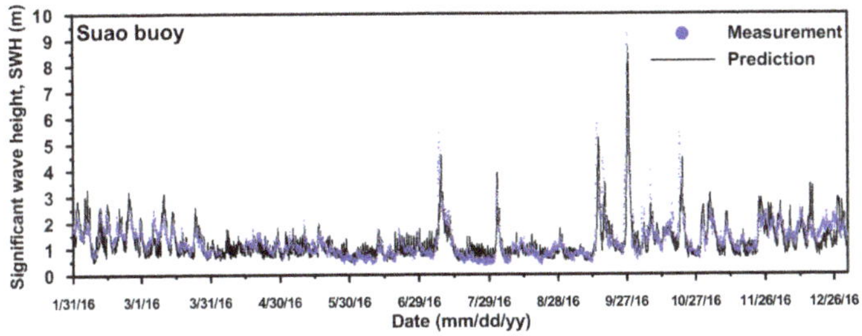

(b)

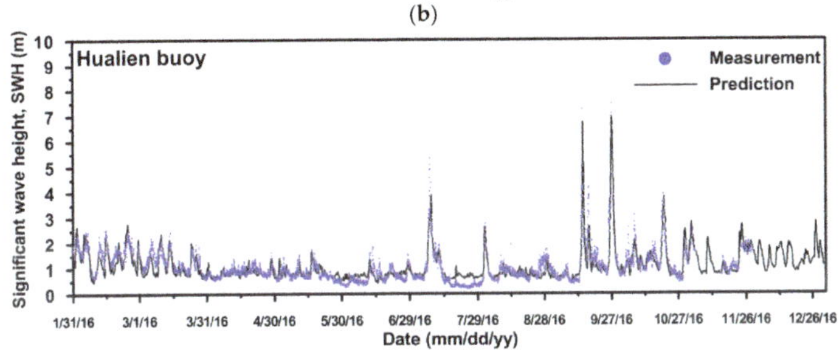

(c)

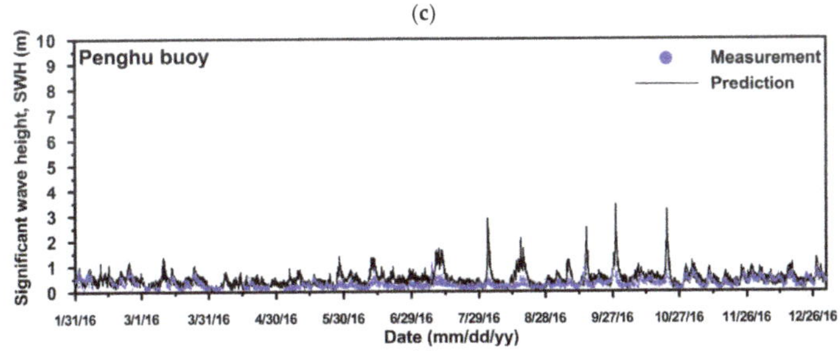

(d)

Figure 3. Mode-data comparison of significant wave height (SWH) at four wave buoys. (**a**) Keelung; (**b**) Suao; (**c**) Hualien and (**d**) Penghu.

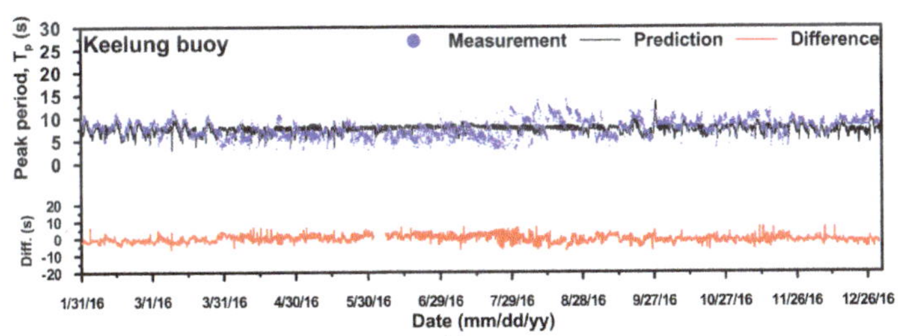

(a)

Figure 4. *Cont.*

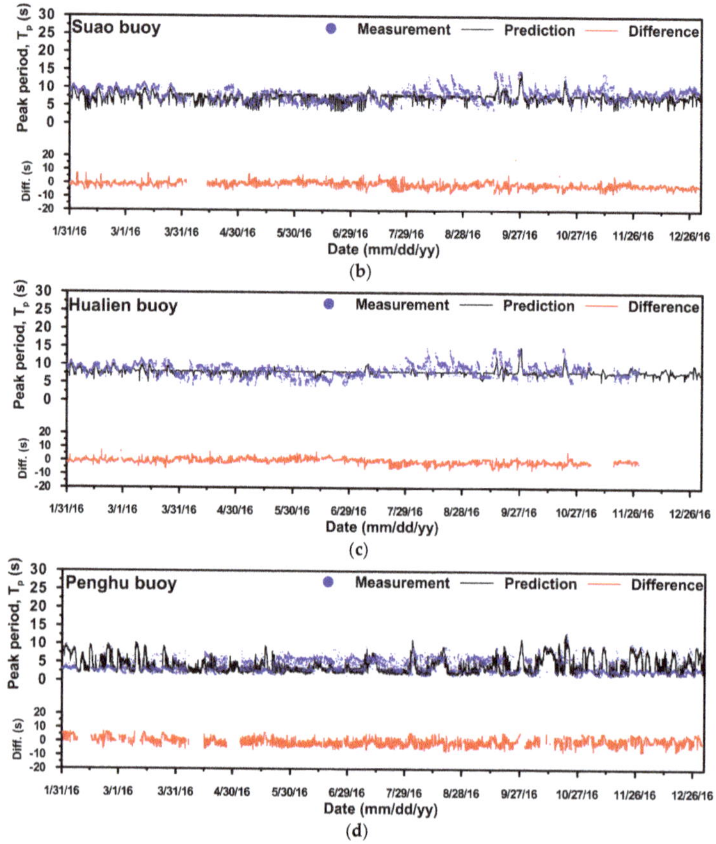

Figure 4. Mode-data comparison of peak period (*Tp*) at four wave buoys. (**a**) Keelung; (**b**) Suao; (**c**) Hualien and (**d**) Penghu.

4. Results and Discussion

Sea-state hindcasts covering the period from 2005 to 2016 were conducted to estimate the wave power and wave energy output. Additionally, numerical simulations based on unstructured grids were converted to structured grids with a resolution of 25 × 25 km to locate the energetic sea areas.

4.1. Spatial Distribution of Annual Mean Significant Wave Height and Wave Power

Figure 5 illustrates the spatial distribution of offshore annual mean SWH in Taiwanese waters for each year from 2005 to 2016 and shows highly similar patterns of mean SWH distribution. The mean SWH gradually increased from the nearshore region (shallower water) to the offshore areas (deeper water). This predictable phenomenon is due to the dissipation of SWH by depth effects [59]. The offshore sea areas with the higher SWH lie off the northeastern, eastern and southern coasts of Taiwan. The SWH values range from 1.2 m to 1.6 m along the northeastern coast and 1.4 m–1.8 m along the eastern and southern coasts. Figure 5 also shows that the mean SWH values are relatively low in the Taiwan Strait, ranging from only 0.6 m to 1.2 m. The spatial distribution of offshore SWH was averaged over 12 years (2005–2016) and is presented in Figure 6. The offshore sea areas southeast of Taiwan exhibit the highest mean SWH values, ranging from 1.2 m to 1.8 m. Although the prevalence of the strong northeast monsoon plays an important role in the higher SWH values, the deeper waters in this area also contribute to the SWH increases.

The SWH and *Tp* were output hourly from SCHISM-WWM-III to compute the wave power for each grid using Equation (21). Figure 7 demonstrates the spatial distribution of offshore annual mean wave power (in kW/m) in the waters surrounding Taiwan from 2005 to 2016. The more energetic sea areas are along the northeastern, eastern and southern coasts of Taiwan, and these locations are consistent with the distributions of the higher SWH. The mean wave power values were 6–12 kW/m for the northeastern coast and 8–16 kW/m for the eastern and the southern coast. The offshore wave power distribution for the 12-year annual mean is displayed in Figure 8. The most energetic areas are the eastern and the southeastern waters of Lanyu, with mean wave power values of 12–14 kW/m.

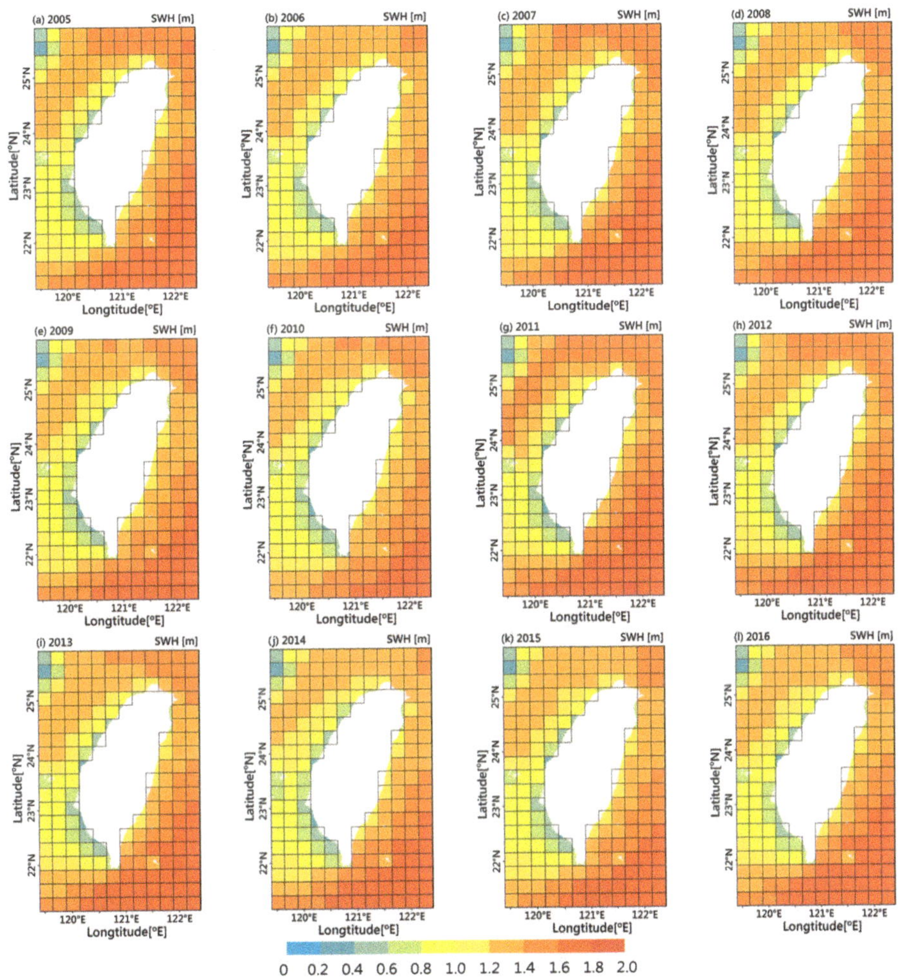

Figure 5. Spatial distribution map of offshore annual average significant wave height (SWH) for each year from 2005 to 2016 in Taiwanese waters.

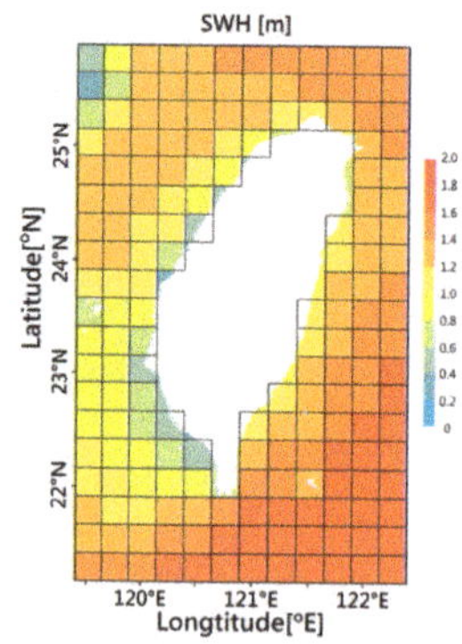

Figure 6. Spatial distribution map of offshore 12-year (2005 to 2016) annual average significant wave height (SWH) in Taiwanese waters.

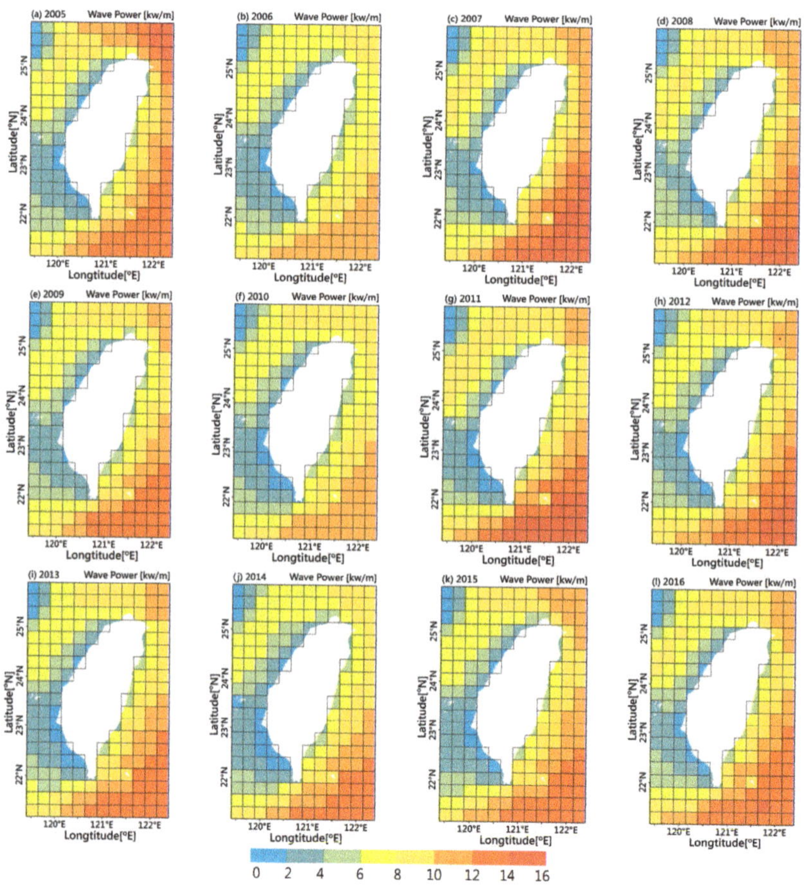

Figure 7. Spatial distribution map of offshore annual average wave power for each year from 2005 to 2016 in Taiwanese waters.

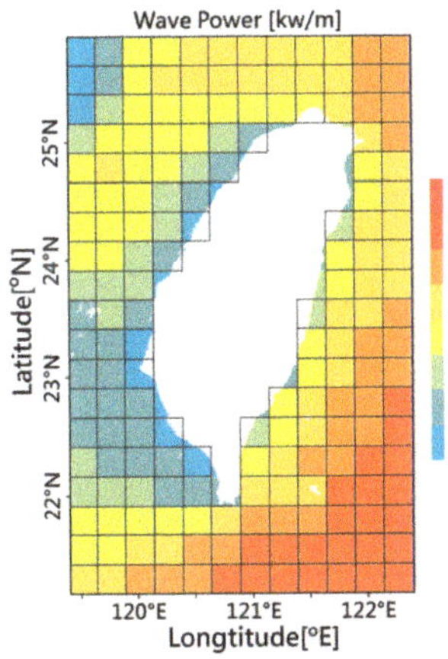

Figure 8. Spatial distribution map of offshore 12-year (2005–2016) annual average wave power in Taiwanese waters.

4.2. Identifying the Optimal Offshore Areas for Wave Energy Converter Deployment

As shown in Figure 8, the offshore areas with the highest wave power density were observed off the northern, northeastern, southeastern (south of Green Island and southeast of Lanyu) and southern coasts of Taiwan. The annual total wave energy yields based on the 12-year average for each 25 × 25 km square grid were calculated by applying Equation (22) and are shown in Figure 9 (left panel). Even though higher wave energies exist in several sea areas, especially in the southeast (105–120 kW/m), the distance between coast and wave energy converter (WEC) should also be considered because of the cabling costs [19]. The five 25 × 25 km energetic sea areas marked with deep blue lines, with spatial average annual total wave energy density of 60–90 kW/m were selected for further analysis (as shown in the left panel of Figures 9–13). The selected 25 × 25 km grid for the northern coast of Taiwan was downscaled to a resolution of 5 × 5 km. A total of 25 square grids are exhibited in the right panel of Figure 9. The grids are within 5 km of the coasts, have higher wave energies and were determined to be optimal areas for WEC deployments (see S1 in the right panel of Figure 9). The same approach was adopted to select an optimal area for the northeastern coast of Taiwan (S2 in the right panel of Figure 10), for the southern coast of Green Island (S3 in the right panel of Figure 11), for the southeastern coast of Lanyu (S4 in the right panel of Figure 12) and for the southern coast of Taiwan (S5 in the right panel of Figure 13). The spatial average (average over a square of 5 × 5 km) annual total wave energy densities are estimated to be 64.3, 84.1, 84.5, 111.0 and 99.3 MWh/m at S1, S2, S3, S4 and S5, respectively. The central longitudes, latitudes, and average water depths for these five optimal areas are listed in Table 1. In other words, the sea areas within 2.5 km of the central coordinate (as listed in Table 1) are considered to be appropriate sites for WEC deployments.

Tsai et al. [21] utilized the SWAN (Simulating WAves Nearshore) driven by NECP (National Centers for Environmental Prediction) global reanalysis wind field to hindcast waves and periods around Taiwanese waters. They revealed that higher significant wave heights were found near Lanyu. Chiu et al. [22] analyzed the temporal and spatial characteristics of wave power density in the coastal areas of Taiwan wave parameter hindcasts (significant wave height and period) based on the TaiCOMS model and observational data. Their results also indicated that the maximum annual mean wave power was found in Lanyu. Su et al. [23] employed the WWM-III model to assess the distribution of wave power in the waters surrounding Taiwan and suggested that the most energetic sea area is southeast of Lanyu. Thus, our results are consistent with those of previous studies.

Wave direction is another important parameter that should be considered in WEC placements [60]. This is particularly true for attenuator-type and terminator-type Weeks: attenuators should be parallel to the predominant wave direction while the principal axis of terminator devices should be perpendicular to the predominant wave direction [61]. Point absorbers are buoy-type WECs that harvest incoming wave energy from all directions, however, arrays of this type of WEC can be sensitive to wave direction when they are deployed in offshore regions [6]. Figure 14 shows wave rose plots for the 12-year average at the centers of S1, S2, S3, S4 and S5. The dominant wave directions for these five optimal areas lie between east and northeast.

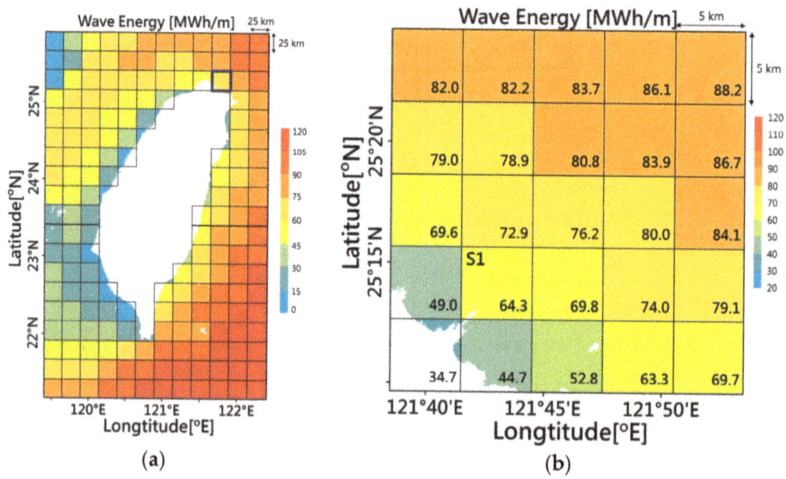

Figure 9. Spatial distribution map of offshore 12-year annual total wave energy output with (**a**) a resolution of 25 × 25 km and (**b**) a downscaled 5 × 5 km for optimal area 1 (S1).

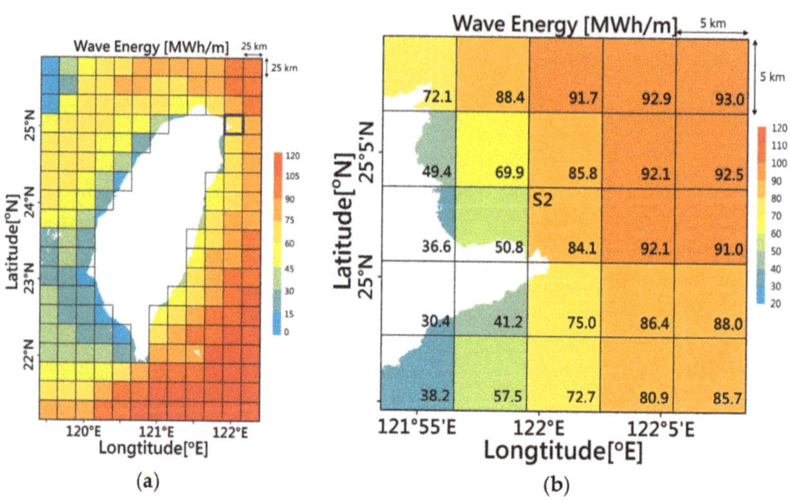

Figure 10. Spatial distribution map of offshore 12-year annual total wave energy output with (**a**) a resolution of 25 × 25 km and (**b**) a downscaled 5 × 5 km for optimal area 2 (S2).

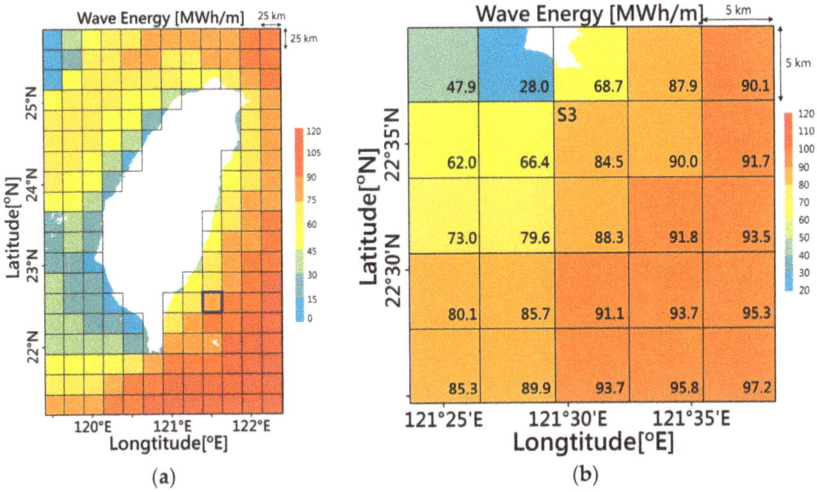

Figure 11. Spatial distribution map of offshore 12-year annual total wave energy output with (**a**) a resolution of 25 × 25 km and (**b**) a downscaled 5 × 5 km for optimal area 3 (S3).

Table 1. Central latitudes, longitudes, and average water depths of five optimal areas for wave energy converter (WEC) deployment.

Area	Longitude (°)	Latitude (°)	Depth * (m)
S1	121.71	25.23	100.01
S2	122.01	25.03	123.39
S3	121.51	22.58	1008.28
S4	121.61	21.91	122.51
S5	120.86	21.83	208.86

* Depth is below the sea level.

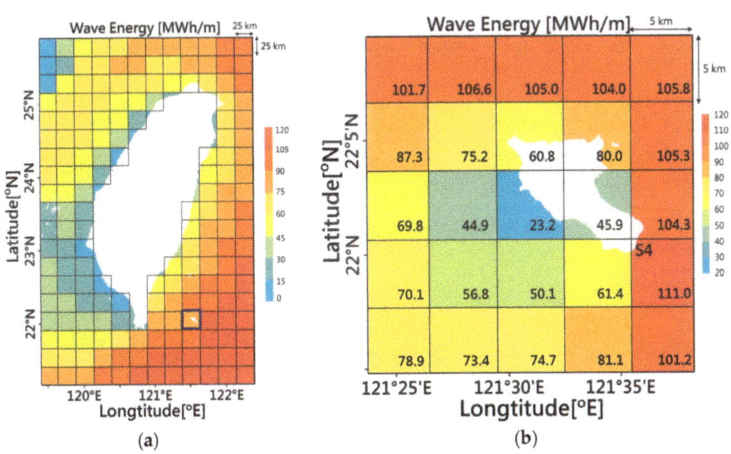

Figure 12. Spatial distribution map of offshore 12-year annual total wave energy output with (**a**) a resolution of 25 × 25 km and (**b**) a downscaled 5 × 5 km for optimal area 4 (S4).

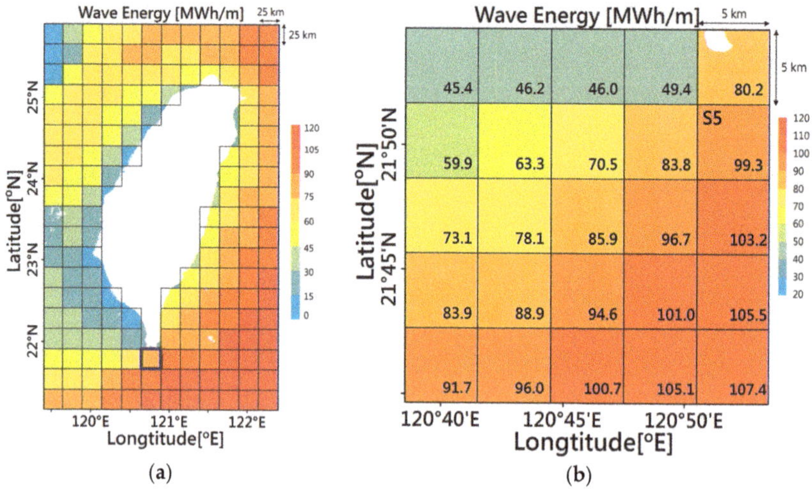

Figure 13. Spatial distribution map of offshore 12-year annual total wave energy output with (**a**) a resolution of 25 × 25 km and (**b**) a downscaled 5 × 5 km for optimal area 5 (S5).

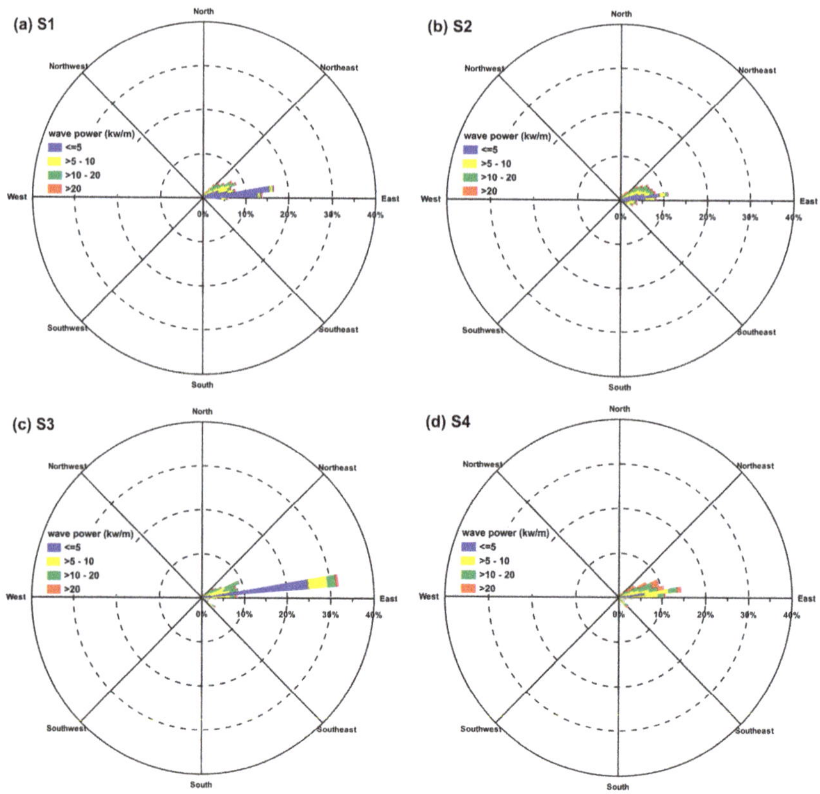

Figure 14. *Cont.*

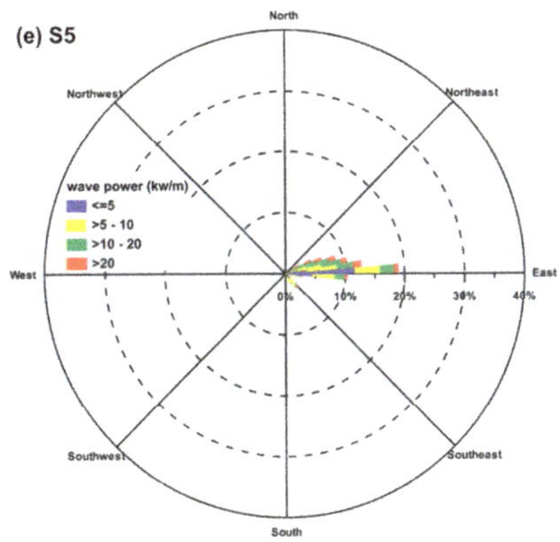

Figure 14. Directional distributions of average wave power for five optimal areas (**a**) S1; (**b**) S2; (**c**) S3; (**d**) S4; (**e**) S5 for 2005–2016.

4.3. Discussion

In many coastal areas, nearshore regions have been considered to be the ideal locations for deploying wave energy converters due to their lower costs. However, by comparing wave lengths to wave energy conversion, the nearshore areas can be regarded as shallow-water areas. The effect of water depth on wave propagation is significant and cannot be neglected for wave energy assessment in shallow-water regions [62]. However, the effects of water depths are typically ignored because studies usually employ the formulas for deep-water waves. This simplification may underestimate the wave energy in finite water depths. Sheng and Li [63] conducted wave power assessment at a water depth of 50 m using the deep-water formula and the proposed method in their study and compared the results to the actual values. They found that the deep-water formula underestimated the annual mean wave power by up to 10.18% at the studied finite water depth (50 m). However, when the water depth modification factor (C_h) was considered, the underestimation of the annual mean wave power fell to 1.47%. Their results suggested that a modification of the deep-water formula is necessary to accurately assess wave power, especially in cases that involve finite water depth effects. Additionally, tide levels and wind intensity may influence the wave characteristics in shallow-water locations [64]. These effects are considered in our tide-surge-wave fully coupled model.

Additional uncertainties in wave power density assessments might be introduced by using a deep-water assumption (Equations (14)–(22)). However, this assumption causes obvious errors only in shallow water areas (i.e., those with water depths <0.5 × wavelength) [65]. The more energetic sea areas such as S1–S5 are located in the northern, eastern and southern offshore waters of Taiwan and can be regarded as deep-water areas. Therefore, the uncertainty of wave power density using Equation (21) can be ignored.

Guillou [66] employed SWAN (Simulating Waves Nearshore model) and TOAWAC (TELEMAC-based Operational Model Addressing Wave Action Computation) to evaluate wave power for the Iroise Sea. The results revealed that SWAN calculates a 15% lower mean wave power than does TOAWAC in offshore waters. Therefore, inter-model comparisons of wave power computation will be performed for Taiwanese waters in the near future.

The wind field data with a temporal resolution of 6 h and a spatial resolution of 0.125° by 0.125° from the ERA-interim dataset seems too low to accurately force a tide-surge-wave coupled model, although it is convenient for providing meteorological boundary conditions and producing an acceptable sea-state hindcast. Consequently, a three-layer nested WRF (Weather Research and Forecast, [67]) model with a temporal resolution of 1 h and a spatial resolution of 45 × 45 km for domain 1, 15 × 15 km for domain 2 and 5 × 5 km for domain 3 will be considered in the future as an alternative source of meteorological information to produce more accurate wave power estimations.

5. Conclusions

A tide-surge-wave fully coupled model based on an unstructured grid system, SCHISM-WWM-III, was implemented to simulate the sea states in the waters surrounding Taiwan. The wind field 10 m above sea level from the ERA-Interim reanalysis data and harmonic constants of eight tidal constituents extracted from a regional inverse tidal model were used to drive SCHISM-WWM-III. The fully coupled model was verified against available significant wave height and peak period measurements. Twelve-year model hindcasts for 2005–2016 were conducted to evaluate the spatial distributions of wave power and wave energy and to determine the most energetic areas in the offshore seas of Taiwan. Numerical simulations with unstructured grids were converted to 25 × 25 km structured grids to identify the most energetic sea areas. An offshore wave power and wave energy distribution map was created for the 12-year annual mean. Five 25 × 25 km energetic sea areas with spatial average annual total wave energy densities of 60–90 kW/m were selected and subsequently downscaled to resolutions of 5 × 5 km. The 5 × 5 km grids are within 5 km of the coasts and have higher wave energies and were determined to be optimal areas for WEC deployments. Five 5 × 5 km square areas off the northern, northeastern and southern coasts of Taiwan (S1, S2 and S5), the southern coast of Green Island (S3) and the southeastern coast of Lanyu (S4) were selected as the optimal areas for WEC deployments. The spatial average annual total wave energy densities are estimated to be 64.3, 84.1, 84.5, 111.0 and 99.3 MWh/m for the areas S1, S2, S3, S4 and S5, respectively. The dominant wave directions for these five optimal areas lie between east and northeast.

Acknowledgments: This research was supported by the Ministry of Science and Technology (MOST), Taiwan, grant Nos. MOST 106-2625-M-865-001. The authors would like to thank the Harbor and Marine Technology Center, Taiwan, for providing the observational data and Joseph Zhang at the Virginia Institute of Marine Science, College of William & Mary, and Aron Roland at the Institute for Hydraulic and Water Resources Engineering, Technische Universität Darmstadt, for kindly sharing their experiences concerning the use of the numerical model.

Author Contributions: Hung-Ju Shih, Chih-Hsin Chang, Lee-Yaw Lin and Wei-Bo Chen conceived the study, and Wei-Bo Chen performed the model simulations. The final manuscript has been read and approved by all the authors.

Conflicts of Interest: The authors declare no conflict of interest.

References

1. U.S. National Renewable Energy Laboratory. *Ocean Energy Technology Overview*; U.S. Department of Energy Federal Energy Management Program; U.S. National Renewable Energy Laboratory: Lakewood, CO, USA, 2009.
2. Defne, Z.; Hass, K.A.; Fritz, H.M. Numerical modeling of tidal currents and the effects of power extraction on estuarine hydrodynamics along the Georgia coast, USA. *Renew. Energy* **2011**, *36*, 3461–3471. [CrossRef]
3. Lewis, M.J.; Angeloudis, A.; Robins, P.E.; Evans, P.S.; Neill, S.P. Influence of storm surge on tidal range energy. *Energy* **2017**, *122*, 25–36. [CrossRef]
4. Kofoed, J.P.; Frigaard, P.; Friis-Madsen, E.; Sørensen, H.C. Prototype testing of the wave energy converter wave dragon. *Renew. Energy* **2006**, *31*, 181–189. [CrossRef]
5. Zheng, S.M.; Zhang, Y.H.; Zhang, Y.L.; Sheng, W.A. Numerical study on the dynamics of a two-raft wave energy conversion device. *J. Fluids Struct.* **2015**, *58*, 271–290. [CrossRef]
6. Zhang, X.; Yang, J. Power capture performance of an oscillating-body WEC with nonlinear snap through PTO systems in irregular waves. *Appl. Ocean Res.* **2015**, *52*, 261–273. [CrossRef]
7. Zhang, X.; Lu, D.; Guo, F.; Gao, Y.; Sun, Y. The maximum wave energy conversion by two interconnected floaters: Effects of structural flexibility. *Appl. Ocean Res.* **2018**, *71*, 34–47. [CrossRef]
8. Zhou, G.; Huang, J.; Zhang, G. Elevation of the wave energy conditions along the coastal waters of Beibu Gulf, China. *Energy* **2015**, *85*, 449–457. [CrossRef]

9. Bento, R.A.; Martinho, P.; Soares, C.G. Numerical modelling of the wave energy in Galway Bay. *Renew. Energy* **2015**, *78*, 457–466. [CrossRef]
10. Mirzaei, A.; Tangang, F.; Juneng, L. Wave energy potential assessment in the central and southern regions of the South China Sea. *Renew. Energy* **2015**, *80*, 454–470. [CrossRef]
11. Akpmar, A.; Bingolbali, B.; Van Vledder, G.P. Long-term analysis of wave power potential in the Black Sea, Based on 31-year SWAN simulations. *Ocean Eng.* **2017**, *13*, 482–497. [CrossRef]
12. Mork, G.; Barstow, S.; Kabuth, A.; Pontes, M.T. Assessing the global wave energy potential. In Proceedings of the ASME 2010 29th International Conference on Ocean, Offshore and Arctic Engineering, Shanghai, China, 6–11 June 2010; pp. 447–454. [CrossRef]
13. Arinaga, R.A.; Cheung, K.F. Atlas of global wave energy from 10 years of reanalysis and hindcast data. *Renew. Energy* **2012**, *39*, 49–64. [CrossRef]
14. Gunn, K.; Stock-Williams, C. Quantifying the global wave power resource. *Renew. Energy* **2012**, *44*, 296–304. [CrossRef]
15. Reguero, B.G.; Losada, I.J.; Mendez, F.J. A global wave power resource and its seasonal, interannual and long-term variability. *Appl. Energy* **2015**, *148*, 366–380. [CrossRef]
16. Iglesias, G.; Carballo, R. Wave power for La Isla Bonita. *Energy* **2010**, *35*, 513–521. [CrossRef]
17. Iglesias, G.; Carballo, R. Wave resource in El Hiero—An island towards energy selfe-suficiency. *Renew. Energy* **2011**, *36*, 689–698. [CrossRef]
18. Sierra, J.P.; Gonzale-Marco, D.; Sospedra, J.; Gironella, X.; Moss, A.; Sanchez-Arcilla, A. Wave energy resource assessment in Lanzarote (Spain). *Renew. Energy* **2013**, *55*, 480–489. [CrossRef]
19. Goncalves, M.; Martinho, P.; Soares, C.G. Assessment of wave energy in the Canary Islands. *Renew. Energy* **2013**, *68*, 774–784. [CrossRef]
20. Stopa, J.E.; Cheung, K.F.; Chen, Y.-L. Assessment of wave energy resources in Hawaii. *Renew. Energy* **2011**, *36*, 554–567. [CrossRef]
21. Tsai, C.P.; Hwang, C.H.; Chuen, C.; Cheng, H.Y. Study on the wave climate variation to the renewable wave energy assessment. *Renew. Energy* **2012**, *38*, 50–60.
22. Chiu, F.-C.; Huang, W.-Y.; Tiao, W.-C. The spatial and temporal characteristics of the wave energy resources around Taiwan. *Renew. Energy* **2013**, *52*, 218–221. [CrossRef]
23. Su, W.-R.; Chen, H.; Chen, W.-B.; Chang, C.-H.; Lin, L.-Y.; Jang, J.-H.; Yu, Y.-C. Numerical investigation of wave energy resources and hotspots in the surrounding waters of Taiwan. *Renew. Energy* **2018**, *118*, 814–824. [CrossRef]
24. Ortega, S.; Osorio, A.F.; Agudelo, P. Estimation of the wave power resource in the Caribbean Sea in areas with scarce instrumentation. Case study: Isla Fuerte, Colombia. *Renew. Energy* **2013**, *57*, 240–248. [CrossRef]
25. Vicinanza, D.; Contestabile, P.; Ferrante, V. Wave energy potential in the northwest of Sardinia (Italy). *Renew. Energy* **2013**, *50*, 506–521. [CrossRef]
26. Sierra, J.P.; Moss, A.; Gonzale-Marco, D. Wave energy resource assessment in Menorca (Spain). *Renew. Energy* **2014**, *71*, 51–60. [CrossRef]
27. Sierra, J.P.; Casas-Prat, M.; Campins, E. Impact of climate change on wave energy resource: The case of Menorca (Spain). *Renew. Energy* **2017**, *101*, 275–285. [CrossRef]
28. Hashemi, M.R.; Neill, S.P. The role of tides in shelf-scale simulations of the wave energy resource. *Renew. Energy* **2014**, *69*, 300–310. [CrossRef]
29. Hsin, Y.C.; Wu, C.R.; Shaw, P.T. Spatial and temporal variations of the Kuroshio east of Taiwan, 1982–2005: A numerical study. *J. Geophys. Res.* **2008**, *113*, C04002. [CrossRef]
30. Morim, J.; Cartwright, N.; Etemad-Shahidi, A.; Strauss, D.; Hemer, M. Wave energy resource assessment along the Southeast coast of Australia on the basis of a 31-year hindcast. *Appl. Energy* **2016**, *184*, 276–297. [CrossRef]
31. Zhang, Y.J.; Baptista, A.M. SELFE: A semi-implicit Eulerian-Lagrangian finite-element model for cross-scale ocean circulation. *Ocean Model.* **2008**, *21*, 71–96. [CrossRef]
32. Zhang, Y.; Ye, F.; Stanev, E.V.; Grashorn, S. Seamless cross-scale modelling with SCHISM. *Ocean Model.* **2016**, *102*, 64–81. [CrossRef]
33. Zhang, Y.J.; Witter, R.C.; Priest, G.R. Tsunami-tide interaction in 1964 Prince William Sound tsunami. *Ocean Model.* **2011**, *40*, 246–259. [CrossRef]

34. Rodrigues, M.; Oliveira, A.; Queiroga, H.; Fortunato, A.B.; Zhang, Y.J. Three-dimensional modeling of the lower trophic levels in the Ria de Aveiro (Portugal). *Ecol. Model.* **2009**, *220*, 1274–1290. [CrossRef]
35. Rodrigues, M.; Oliveira, A.; Guerreior, M.; Fortunato, A.B.; Menaia, J.; David, L.M.; Cravo, A. Modeling fecal contamination in the Aljezur coastal stream (Portugal). *Ocean Dyn.* **2011**, *61*, 841–856. [CrossRef]
36. Chen, W.B.; Liu, W.C. Investigating the fate and transport of fecal coliform contamination in a tidal estuarine system using a three-dimensional model. *Mar. Pollut. Bull.* **2017**, *116*, 365–384. [CrossRef] [PubMed]
37. Azevedo, A.; Oliveira, A.; Fortunato, A.B.; Bertin, X. Application of an Eulerian-Lagrangian oil spill modeling system to the prestige accident: Trajectory analysis. *J. Coast. Res.* **2009**, *1*, 777–781.
38. Fortunato, A.B.; Rodrigues, M.; Dias, J.M.; Lopes, C.; Oliveira, A. Generating inundation maps for a coastal lagoon: A case study in the Ria de Aveiro (Portugal). *Ocean Eng.* **2013**, *64*, 60–71. [CrossRef]
39. Chen, W.B.; Liu, W.C. Assessment of storm surge inundation and potential hazard maps for the southern coast of Taiwan. *Nat. Hazards* **2016**, *82*, 591–616. [CrossRef]
40. Chen, W.B.; Liu, W.C. Modeling flood inundation Induced by river flow and storm surges over a river basin. *Water* **2014**, *6*, 3182–3199. [CrossRef]
41. Wang, H.V.; Loftis, J.D.; Liu, Z.; Forrest, D.; Zhang, J. The storm surge and sub-grid inundation modeling in New York City during hurricane Sandy. *J. Mar. Sci. Eng.* **2014**, *2*, 226–246. [CrossRef]
42. Chen, W.B.; Chen, H.; Lin, L.-Y.; Yu, Y.-C. Tidal Current Power Resource and Influence of Sea-Level Rise in the Coastal Waters of Kinmen Island, Taiwan. *Energies* **2017**, *10*, 652. [CrossRef]
43. Wu, C.R.; Chao, S.Y.; Hsu, C. Transient, seasonal and interannual variability of the Taiwan Strait current. *J. Oceanogr.* **2007**, *63*, 821–833. [CrossRef]
44. Oey, L.Y.; Hsin, Y.C.; Wu, C.R. Why does the Kuroshio northeast of Taiwan shift shelfward in winter? *Ocean Dyn.* **2010**, *60*, 413–426. [CrossRef]
45. Powell, M.D.; Vickery, P.J.; Reinhold, T.A. Reduced drag coefficient for high wind speeds in tropical cyclones. *Nature* **2003**, *422*, 279–283. [CrossRef] [PubMed]
46. Longuet-Higgins, M.S.; Stewart, R.W. Radiation stress and transport in gravity wave, with application to 'surf beat'. *J. Fluid Mech.* **1962**, *13*, 485–504. [CrossRef]
47. Longuet-Higgins, M.S.; Stewart, R.W. Radiation stress in water waves: A physical discussion with applications. *Deep Sea Res. Oceanogr. Abstr.* **1964**, *11*, 529–562. [CrossRef]
48. Battjes, J.A. Computation of Set-up, Longshore Currents, Run-up and Overtopping Due to Wind-Generated Waves. Communications on Hydraulics No. 74-2. Ph.D. Thesis, Department of Civil Engineering, Delft University of Technology, Delft, The Netherlands, 1974.
49. Hasselmann, K.; Barnett, T.P.; Bouws, E.; Carlson, H.; Cartwright, D.E.; Enke, K. *Measurements of Wind-Wave Growth and Swell Decay during the Joint North Sea Wave Project (JONSWAP)*; Deutsche Hydrographische Institut (DHI): Hamburg, Germany, 1973; Volume 12.
50. Battjes, J.A.; Janssen, J.P.F.M. Energy loss and set-up due to breaking of random waves. In Proceedings of the 16th International Conference on Coastal Engineering (ASCE), Hamburg, Germany, 27 August–3 September 1978; pp. 569–587.
51. Roland, A.; Zhang, Y.J.; Wang, H.V.; Meng, Y.; Teng, Y.-C.; Maderich, V.; Brovchenko, I.; Dutour-Sikiric, M.; Zanke, U. A fully coupled 3D wave-current interaction model on unstructured grids. *J. Geophys. Res.* **2012**, *117*, C00J33. [CrossRef]
52. Dee, D.P.; Uppala, S.M.; Simmons, A.J.; Berrisford, P.; Poli, P.; Kobayashi, S.; Andrae, U.; Balmaseda, M.A.; Balsamo, G.; Bauer, P.; et al. The ERA-Interim reanalysis: Configuration and performance of the data assimilation system. *Q. J. R. Meteorol. Soc.* **2011**, *137*, 53–597. [CrossRef]
53. Zu, T.; Gana, J.; Erofeevac, S.Y. Numerical study of the tide and tidal dynamics in the South China Sea. *Deep Sea Res. Part I Oceanogr. Res. Pap.* **2008**, *55*, 137–154. [CrossRef]
54. Cornett, A.M. A global wave energy resource assessment. In Proceedings of the International Offshore and Polar Engineering Conference, Vancouver, BC, Canada, 6–11 July 2008; pp. 318–326.
55. Hadadpour, S.; Etemad-Shahidi, A.; Jabbari, E.; Kamranzad, B. Wave energy and hot spots in Anzali port. *Energy* **2014**, *74*, 529–536. [CrossRef]
56. Barbariol, F.; Benetazzo, A.; Carniel, S.; Sclavo, M. Improving the assessment of wave energy resources by means of coupled wave-ocean numerical modeling. *Renew. Energy* **2013**, *60*, 462–471. [CrossRef]
57. Cavaleri, L. Wave modeling—Missing the peaks. *J. Phys. Oceanogr.* **2009**, *39*, 2757–2778. [CrossRef]

58. Lavidas, G.; Venugopal, V.; Friedrich, D. Wave energy extraction in Scotland through an improved nearshore wave atlas. *Int. J. Mar. Energy* **2017**, *17*, 64–83. [CrossRef]
59. Wang, Z.; Dong, S.; Li, X.; Soares, C.G. Assessment of wave energy in the Bohai Sea, China. *Renew. Energy* **2016**, *90*, 145–156. [CrossRef]
60. Atan, R.; Goggins, J.; Nash, S. A Detailed Assessment of the Wave Energy Resource at the Atlantic Marine Energy Test Site. *Energies* **2016**, *9*, 967. [CrossRef]
61. Drew, B.; Plummer, A.R.; Sahinkaya, M.N. A review of wave energy converter technology. *Proc. Inst. Mech. Eng. A J. Power Energy* **2009**, *223*, 887–902. [CrossRef]
62. Sheng, W.; Li, H.; Murphy, J. An improved method for energy and resource assessment of waves in finite water depths. *Energies* **2017**, *10*, 1188. [CrossRef]
63. Sheng, W.; Li, H. A method for energy and resource assessment of wave in finite water depths. *Energies* **2017**, *10*, 460. [CrossRef]
64. Silva, D.; Rusu, E.; Soares, C.G. High-resolution wave energy assessment in shallow water accounting for tides. *Energies* **2016**, *9*, 761. [CrossRef]
65. Liang, B.; Shao, Z.; Wu, G.; Shao, M.; Sun, J. New equation of wave energy assessment accounting for the water depth. *Appl. Energy* **2017**, *188*, 130–139. [CrossRef]
66. Guillou, N. Evaluation of wave energy potential in the Sea of Iroise with two spectral models. *Ocean Eng.* **2015**, *106*, 141–151. [CrossRef]
67. Michalakes, J.; Chen, S.; Dudhia, J.; Hart, L.; Klemp, J.; Middlecoff, J.; Skamarock, W. Development of a next generation regional weather research and forecast model. In *Developments in Teracomputing, Proceedings of the Ninth ECMWF Workshop on the Use of High Performance Computing in Meteorology, Reading, UK, 13–17 November 2000*; Zwieflhofer, W., Kreitz, N., Eds.; World Scientific: Singapore, 2001; pp. 269–276.

© 2018 by the authors. Licensee MDPI, Basel, Switzerland. This article is an open access article distributed under the terms and conditions of the Creative Commons Attribution (CC BY) license (http://creativecommons.org/licenses/by/4.0/).

energies

Article

Historical Evolution of the Wave Resource and Energy Production off the Chilean Coast over the 20th Century

Alain Ulazia [1,*,†], Markel Penalba [2,†], Arkaitz Rabanal [3,†], Gabriel Ibarra-Berastegi [4,5,†], John Ringwood [2,†] and Jon Sáenz [5,6,†]

1 Department of NE and Fluid Mechanics, University of the Basque Country (UPV/EHU), Otaola 29, 20600 Eibar, Spain
2 Centre for Ocean Energy Research, Maynooth University, Maynooth, Co. Kildare, Ireland; mpenalba@eeng.nuim.ie (M.P.); john.ringwood@eeng.nuim.ie (J.R.)
3 University of the Basque Country (UPV/EHU), Otaola 29, 20600 Eibar, Spain; arabanal004@ikasle.ehu.eus
4 Department of NE and Fluid Mechanics, University of the Basque Country (UPV/EHU), Alda. Urkijo, 48013 Bilbao, Spain; gabriel.ibarra@ehu.eus
5 Joint Research Unit (UPV/EHU-IEO) Plentziako Itsas Estazioa (PIE), University of Basque Country (UPV/EHU), Areatza Hiribidea 47, 48620 Plentzia, Spain
6 Department of Applied Physics II, University of the Basque Country (UPV/EHU). B. Sarriena s/n, 48940 Leioa, Spain; jon.saenz@ehu.eus
* Correspondence: alain.ulazia@ehu.eus; Tel.: +034-943-033-051
† These authors contributed equally to this work.

Received: 22 July 2018; Accepted: 27 August 2018; Published: 30 August 2018

Abstract: The wave energy resource in the Chilean coast shows particularly profitable characteristics for wave energy production, with relatively high mean wave power and low inter-annual resource variability. This combination is as interesting as unusual, since high energetic locations are usually also highly variable, such as the west coast of Ireland. Long-term wave resource variations are also an important aspect when designing wave energy converters (WECs), which are often neglected in resource assessment. The present paper studies the long-term resource variability of the Chilean coast, dividing the 20th century into five do-decades and analysing the variations between the different do-decades. To that end, the ERA20C reanalysis of the European Centre for Medium-Range Weather Forecasts is calibrated versus the ERA-Interim reanalysis and validated against buoy measurements collected in different points of the Chilean coast. Historical resource variations off the Chilean coast are compared to resource variations off the west coast in Ireland, showing a significantly more consistent wave resource. In addition, the impact of historical wave resource variations on a realistic WEC, similar to the Corpower device, is studied, comparing the results to those obtained off the west coast of Ireland. The annual power production off the Chilean coast is demonstrated to be remarkably more regular over the 20th century, with variations of just 1% between the different do-decades.

Keywords: wave energy trends; reanalysis wave data; Chilean coast; renewable energy; wave energy converters; annual mean power production

1. Introduction

Ocean waves store a tremendous potential that is still untapped. The main reason why wave energy has not been yet exploited is the difficulty to economically harvest energy from ocean waves. Different technological solutions have been suggested in recent decades [1], but none of these technologies has achieved yet economical viability to compete in the energy market with other more established energy sources. The Mutriku wave energy plant, based on the oscillating water column principle, is one of the examples of wave energy converters (WECs), which is the first wave energy plant connected to the electricity grid. Despite its aim to be an experimental plant, the plant has achieved a capacity factor (CF) of 11% in the first years of operation [2], showing WECs can be an additional alternative to reduce greenhouse emissions.

In the way towards designing technically and economically viable WECs, there are several aspects to be improved, such as the survivability of the devices, the cost reduction in the construction stage, minimisation of maintenance operations, energy maximisation via control strategies and design of efficient power take-off (PTO) systems. A precise assessment of the wave energy resource is also essential to accurately design WECs and maximise the power extraction from ocean waves. Wave energy converters, based on different working principles and deployed in different locations, are analysed in [3], where the CF of each WEC varies significantly depending on the wave climate in each location.

In particular, the variability of the resource, both inter- and intra-annual variability, is a critical aspect that can affect the design of the WEC. Intra-annual variations of the resource have been analysed in different spatial scales [4–6]. However, historical long-term variations of the wave climate have been often neglected. The authors have studied these long-term variations in the Bay of Biscay and off the west coast of Ireland in [7] and [8], respectively, analysing wave trends over the 20th century and the influence of these trends in the performance of specific WECs. In both locations, positive trends have been found, with wave height (H_s), peak period (T_p) and wave energy flux (WEF) increasing significantly over the 20th century.

The evaluation of the wave energy potential in a specific location usually involves a combination of observations (buoys and satellites), reanalyses and, sometimes, numerical modelling with models such as SWAN or Mike 21. Using these techniques, several studies have recently focused on different parts of the American continent for wave energy potential assessment [9–11]. The Chilean coast is one of the interesting areas in South America for the implementation of WEC farms, with an estimated resource of about 165 GW along its 5000 km [12]. This background has encouraged researchers to study the potential of specific converters at specific locations [13].

Chile and the west coast of Ireland are considered paradigmatic examples, with respect to their wave energy resource. Both present high mean wave power, but the wave resource off the west coast of Ireland is highly variable, while the resource off the Chilean coast is more consistent. This contrast is shown in [14], based on measurements collected by several buoys around the world, via the coefficient of variation (COV), which is given as follows,

$$COV = \frac{\sigma(WEF)}{\overline{WEF}}, \tag{1}$$

where $\sigma(WEF)$ is the standard deviation (SD) of the WEF and $\overline{WEF}$ is the average of the WEF.

Ringwood and Brandle [14] showed that the maximum COV for Chile is 0.9, which is observed in the south of the Chilean coast, where the WEF is about 121 kW/m. In contrast, the COV for Ireland is 1.8 (twice as high as the maximum COV for Chile), where the mean wave power is lower than for the southern Chilean coast: 95 kW/m. Therefore, Chile shows a high energetic and consistent wave resource, which is important in the process to harness and produce energy from ocean waves.

The most recent validation of wave energy assessment in Chile covers the 1989–2013 period [15], where a third generation wave model is used, validated against buoy measurements, to generate three-hourly sea-state parameters (H_s-T_p pairs). These data are then propagated from offshore to nearshore locations by means of the SWAM model, to characterise the wave climate along the entire coast. Validation results show good agreement between modelled and measured data, with errors of less than 10%. Wave power and seasonal variability in Chile increase with the latitude, which fluctuates between 20 and 35 kW/m in the areas close to shore, and where the most energetic sea-states happen in winter. These results and others [16] show lower wave resource potential, about 5–10 kW/m lower, compared to previous studies, which suggests that previous assessments may have overestimated the wave resource [17].

However, these studies have been developed within a period of 30–40 years, and, thus, historical trends of wave energy are not considered. In the present study, the objective was to analyse the long-term variability of the wave energy resource in Chile, studying the wave energy resource in Chile over the 20th century and comparing the results from the Chilean coast to other locations. Following the recommendations by the World Meteorological Organization [18,19], reliable estimation of climate variables requires at least 30 years of data. These data may be obtained by means of different techniques: buoy measurements [20], observations from ships [21,22], satellite altimeter [23,24] or models and reanalysis datasets [5,25–31]. The latter method was used in the present study, using the ERA20 reanalysis from the European Centre for Medium-Range Weather Forecasts (ECMWF), calibrated with the ERA-Interim (ERAI) reanalysis.

Among the different techniques to estimate climate variables, the satellite altimeter has considerably improved in the recent decades, generating very interesting data for long-term wind or wave trends. For instance, Young et al. [24] presented a study of wind speed and wave height trends in a global scale, which is the only study, to the best of authors' knowledge, that covers the same area of study analysed in the present paper (the Chilean shoreline). However, the time period analysed in [24] is shorter, 23 years between 1985 and 2008, and only the trends of wave heights are analysed, which may be insufficient to draw strong conclusions on the historical wave energy trends, since the influence of the wave period is demonstrated to be essential [7]. Young et al. [24] showed a significant statistical trend for wave heights, analysing mean monthly values in the east of Africa, in the east of North America and the west of South America (the Chilean coast and south-eastern Pacific Ocean). Furthermore, the increase of wind speed is very important in Chile, due to the effect of a hot spot (a positive trend of 15%/decade) in the middle of the Pacific Ocean at equatorial latitudes. Thus, according to [24], wave height trends are large in Chile, showing a positive trend between 0% and 0.25% per year, meaning that a typical wave of 2 m height would increase about 5 cm per decade.

A similar analysis is carried out in [30,31], where the global wave energy resource is assessed via the ERA40 reanalysis, the previous version of the ERAI reanalysis used in this paper. Wave trends in [30,31] are calculated by means of a linear regression. As for the study carried out in [24], positive wave trends are found in [30,31] for the Chilean coast, where these trends are also shown to increase with the latitude. More precisely, wave trends of about 1 kW/m/decade are found in the north of the Chilean coast, while trends of 2 kW/m/decade are observed in the south, which implies about 3.3%/decade for a mean wave power of 60kW/m. Additionally, other studies, such as [32], also show long-term wave energy resource variations in several areas.

Therefore, a precise characterisation of the resource requires the analysis of long-term variations, including a time-evolving description of the resource, to accurately understand the resource in which the WEC is deployed. However, wave energy resource assessment studies commonly rely on recent past data (typically, between 10 and 30 years of past data), and analyse the resource as a static element, even though it has been widely demonstrated that the ocean is a highly dynamic environment [7,14,30,31]. The objective of this paper is to fill this gap for the Chilean coast, analysing the long-term trends of wave energy over the 20th century. In addition, the impact of these variations is evaluated on a realistic point absorber (PA) type WEC, similar to the real Corpower device.

The remainder of the paper is as follows: Section 2 presents the datasets and the methodology to calibrate and study the wave trend, Section 3 describes the realistic WEC and the hydrodynamic model employed to evaluate its power absorption, Section 4 shows the results related to the resource variations and the power absorptions, Section 5 discusses the results and Section 6 presents the conclusions of the study.

2. Wave Data and Methodology

2.1. Wave Data

Two different sources of wave data are used in this paper. On the one hand, two reanalyses of the ECMWF are employed and, on the other hand, wave data collected via buoy measurements is used. Further information about both ECMWF reanalyses and buoy measurements is given in Sections 2.1.1 and 2.1.2.

2.1.1. ERA20C and ERA-Interim Reanalyses

The two reanalyses of the ECMWF used in this study are the ERA20C and ERA-Interim reanalyses (ERA20 and ERAI henceforth). The ERA20 reanalysis is ECMWF's first atmospheric reanalysis, which covers the whole 20th century. In the ERA20 reanalysis, observations of surface pressure and surface marine winds are assimilated [33] by means of a 4D-Var analysis. More observations are available, the more reliable are the data generated via the ERA20 reanalysis [34]. Consequently, the data provided by the ERA20 reanalysis are more accurate in the northern hemisphere, although they have also been used for the study of historical wave trends and coastal evolution in different locations of the southern hemisphere [35–37]. The spatial resolution of the ERA20 reanalysis is approximately 125 km and wave parameters can be obtained three-hourly [34].

The ERAI reanalysis is also a global reanalysis, but only covers the time period since 1979. The wave model implemented in the ERAI reanalysis is the Wave Modelling Project (WAM) approach [38], which reduces the error in the wave period assimilation, compared to previous ECMWF reanalyses, such as the ERA40 [39]. The ERAI reanalysis also assimilates data via a 4D-Var method, but using a 75 km spatial resolution and providing wave parameters every six hours [40].

Therefore, the ERA20 reanalysis was calibrated against the ERAI reanalysis, providing more precise wave data for the whole 20th century. Due to the discrepancy of the temporal resolution between ERA20 and ERAI, only six-hourly data were used in the calibration and the calibration was carried out in the intersection period between both reanalyses, which spans 32 years from 1979 to 2010.

2.1.2. Buoy Measurements

Buoy data used in the present study were provided by the SHOA (Spanish acronym of the Chilean Navy Hydrographic and Oceanographic Service). Data from two specific locations were used, i.e., Iquique and Valparaiso, depicted in Figure 1 together with the complete study area, from the Magellan Strait to the Peruvian border in the southern Pacific Ocean.

Table 1 shows the characteristics of the two buoys: the exact position (longitude, latitude), the distance to the nearest gridpoint in the reanalyses and the time period in which data were collected.

Table 1. Main characteristics of the buoys.

Buoy	Longtitude	Latitude	Distance (km)	Period
Iquique	−70.25	− 20.25	38	2004–2008
Valparaiso	−71.65	−32.97	33	2000–2003

Figure 1. The complete area of study and the exact location of the two buoys used in this study.

2.2. Methodology

2.2.1. Computation of the Wave Energy Flux

Although the WEF is directly obtained from the parameters available in the reanalysis, it can be given as a function of H_s and the energy period (T_e) as follows [41],

$$WEF = 0.49 H_s^2 T_e. \tag{2}$$

However, the wave period data provided by SHOA include only T_p for our validation. Therefore, to calculate the WEF at the buoy, Equation (2) needs to be adapted by including a correction coefficient between T_e and T_p, similar to the $\alpha = T_e/T_p$ coefficient described in [42]. According to this study, for a standard JONSWAP spectrum, α reaches 0.90. We took into account the correction factor α for T_p given by the reanalysis data in the validation procedure at the buoy's nearest gridpoint. The power matrix of the WECs is given also for T_p and this period was therefore used for the analysis of the production of our device.

On the other hand, Wave Period Ratio (WPR) described in [7,8,41] which relates the energy period to the average zero-crossing period was obtained for the trend computation of wave period. The WPR changes depending on the area of study, which, in the case of Iquique and Valparaiso, lies between 0.8 and 1. Furthermore, mean wave period and average zero-crossing period are very similar, within an error of 10% for monthly means values used for the trends. In any case, this kind of scale factors does not affect the results of the trends, since they constitute the relative slopes of the absolute values. Thus, T_m is the chosen period for the representation of maps, because it is one of the most frequently used parameters for the description of the wave period.

2.2.2. Directional Quantile-Matching Calibration

The calibration or bias correction is usually referred to as the classified quantile-matching method, although it has also been named in the literature as probability mapping [43], quantile-quantile mapping [44,45], statistical downscaling [46] or histogram equalization [47]. This calibration method is commonly used in the literature to calibrate temperature, precipitation, wind speed or other parameters [48–52].

The classified quantile-matching method presented in this paper allows for a more sophisticated classified calibration, where the relevance of each variable in the calibration can be considered. Hence, instead of using a single transfer function for the whole time-series, as in the previous studies of the authors [7,8], different transfer functions are used for each time-interval. In the case of wind speed, the variable used for the classification can be directionality, using four main wind directions (northeast, southeast, northwest and southwest). That way, a different transfer function is obtained for each direction interval, for a total of four transfer functions.

Another option is selecting irregular directional intervals to match the specific characteristics, i.e. the predominant wind directions, of the location under analysis. This type of classification is known as wind rose bias correction [53], which is an interesting approach for wave resources, due to their directional characterisation. The closer to the shore, the more defined this directionality is, which in the case of the Chilean coast, is significantly dominated by the western waves. More specifically, the predominant wave direction in Iquique and Valparaiso, illustrated by the wave roses depicted in Figure 2, is the southwest direction. Therefore, the strong directionality of the wave resource in Iquique and Valparaiso justifies the use of the classified quantile-matching technique for the calibration, using wave direction as the variable for the classification.

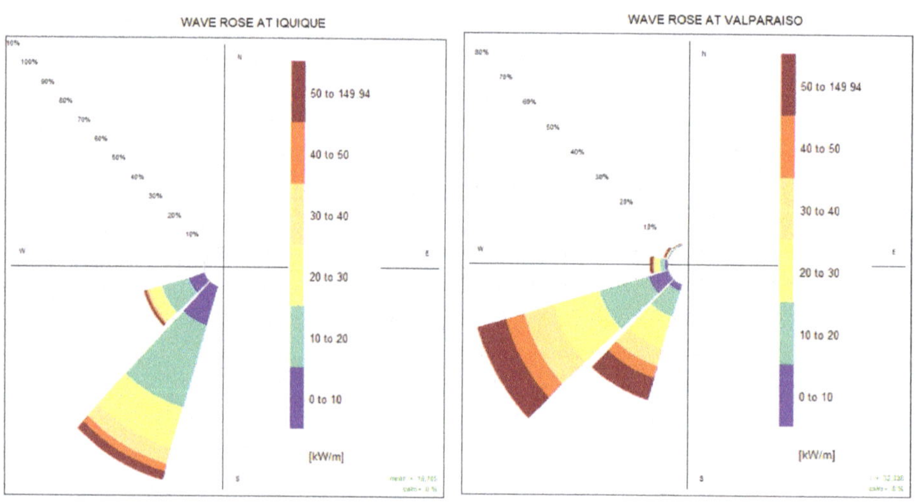

(a) (b)

Figure 2. Wave roses of the two buoys: (**a**) Iquique; and (**b**) Valparaiso.

To consider the directionality of the wave resources, seven transfer functions for seven intervals of wave direction, following the wave roses in Figure 2, have been created. Hence, the northeast, southeast and northwest quadrants are represented by a single interval of 90°, while the southeast quadrant is divided into four different intervals of 22.5°. Although the comparative of different types of distributions is out of the scope of this paper, the seven-interval distribution has been compared via proof and error to other distribution schemes, such as four quadrant or eight octant regular distributions, where the seven-interval distribution scheme provides the best results. In any case, the visual intuition provided by the wind roses is enough to justify the decision.

In previous studies by the authors in the Bay of Biscay and west coast of Ireland [7,8], no classification technique is used in the calibration, and, consequently, this directionally-classified quantile-matching technique can be considered as a novel contribution of this paper in the context of wave energy. To the best of authors' knowledge, this is the first time that a wave rose bias correction is used, analogous to the aforementioned "wind rose bias correction".

This method is a statistical method that matches the values with the same quantile between the ERA20 and ERAI reanalyses. Hence, by calibrating the ERAI reanalysis against the ERA20 reanalysis, using the directionally-classified quantile-matching technique, directionally-calibrated wave data (dcERA20) can be obtained. The process is divided into the following steps:

1. Classify sea events according to the previously selected direction intervals.
2. Compute the WEF of each event for the ERAI and ERA20 reanalyses in their intersection period (1979–2010).
3. Calculate the cumulative probability functions for both reanalyses.
4. Obtain a transfer function between the couple of WEF values with the same quantile, for each direction interval in the intersection period.
5. Apply these transfer functions to all the historical values of ERA20 (1900–2010) to obtain the calibrated dcERA20 time-series.
6. Verify the calibrated values against buoy measurements collected at the closest point.

2.2.3. Evaluation Metrics

Six statistical metrics were used for the validation of the calibration procedure against the two buoys, that is to say, to compare the non-calibrated data from the ERA20 reanalysis with the directionally-calibrated dcERA20 wave data and the ERAI reanalysis, used as the basis of the calibration:

1. Pearson's correlation of the WEF, represented by the exterior arc of a Taylor Diagram [54].
2. The root mean square error (RMSE) for the WEF, represented by the interior arc of a Taylor Diagram centred on the Observation point.
3. The SD of the WEF series represented by the interior arc of a Taylor Diagram that passes from the observation point on the X axis. This allowed for a visual comparison of the variability given by the SD in the observations and the wave models.
4. The variability of the data in relative terms was also analysed by the previously mentioned COV.
5. The bias of the WEF with respect to the buoy measurements, which can be more relevant than the RMSE or other absolute errors, since it facilitates to identify under- and over-estimation.
6. The mean absolute percentage error (MAPE) of the WEF, which represents the absolute error to be reduced by the calibration procedure.

2.2.4. Wave Resource Maps

The historical evolution of the resource is also represented by means of maps, showing wave trends of the entire area of study. The trends were computed using the non-parametric Theil-Sen [55,56] method, which fits a line using the median of the slopes. The significance of the trend at each grid point was evaluated at a 95% confidence level using bootstrap resampling with 1000 samples.

Thus, different variables are illustrated using maps:

1. The average H_s, T_m and WEF values for the entire area of study, based on the ERAI reanalysis, which provides a picture of the wave resource in the recent decades. In addition, the map with the average WEF is useful to identify the highest energetic locations (see Section 4.2.1).

2. The COV over the whole study area, also based on the ERAI reanalysis. Together with the average WEF map, the COV map can help to identify interesting locations for the implementation of WEC farms (see Section 4.2.1).
3. Decadal trends of the average H_s, T_m and WEF values over the 20th century, using the dcERA20 reanalysis, to show resource variations (Section 4.2.2).
4. Decadal trends of the seasonal WEF for the four seasons. The seasonal analysis provides more insight into the contribution of each season to the annual wave energy trend (Section 4.2.3).

3. Hydrodynamic Modelling

Several different WECs have been suggested in the literature to extract energy from ocean waves. Based on their working principles, all WEC can be categorised into four main groups [1]: overtopping devices, oscillating water column devices, oscillating wave surge converters and PAs. Although none of the prototypes suggested so far has yet achieved commercial viability, a large part of the most developed prototypes are PAs, such as the CETO [57], Seabased [58] and the Corpower [59] devices. The PA used in this study is inspired by the Corpower device, referred in the following as the cPA, illustrated in Figure 3a, adapted from [60]. Further details about the characteristics of the WEC implemented in this study are provided in [8].

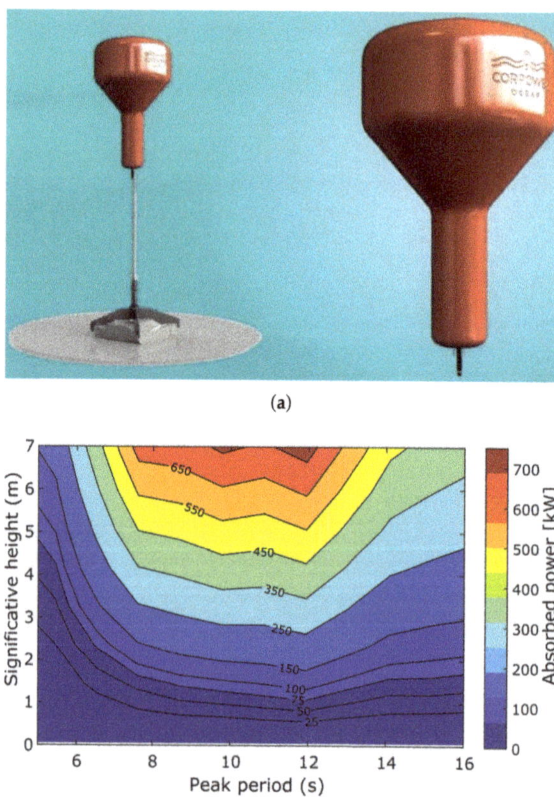

(**a**)

(**b**)

Figure 3. The illustration (**a**); and the power-matrix (**b**) of the Corpower device.

Power production capabilities of the cPA for a wide range of sea-states, represented using irregular-wave time-series based on the JOSNWAP spectrum [61], were estimated via numerical simulation, computing the power-matrix depicted in Figure 3b. The behaviour of the cPA in this study was evaluated by analysing its motion in two degrees of freedom, as in [59], following the linear Cummins' equation [62] as follows,

$$(M + \mu_\infty)\ddot{x}(t) = F_{ex} - K_H x - \int_0^t K_{rad}(t-\tau)\dot{x}(\tau)dt + F_{visc} + F_{PTO} + F_{MOO} + F_{EndStop}, \quad (3)$$

where M and μ_∞ are the mass and infinite added-mass matrices; x, $\dot{x}$ and $\ddot{x}$ are the displacement, velocity and acceleration of the device, respectively; F_{ex} is the excitation force; K_H is the hydrostatic stiffness; K_{rad} is the radiation impulse response; F_{PTO} is the PTO force; F_{MOO} is the mooring force; and $F_{EndStop}$ is the force that reproduces the end-stop effect of the PTO mechanism. The only nonlinearity included in the model is the viscous force (F_{visc}), which is modelled using a quadratic damping based on a Morison-like equation [63] as follows,

$$F_{visc} = \frac{1}{2}\rho C_D A_D(\dot{x} - V_0)|\dot{x} - V_0| \quad (4)$$

where ρ is the water density, C_D is drag coefficient, A_D is the characteristic area of the WEC and V_0 is the velocity of the undisturbed water flow.

The objective of the present study was to study the impact of wave energy resource variations, assessing the power produced by the cPA over the 20th century off the Chilean coast. Nonlinear effects, such as nonlinear Froude–Krylov forces, are shown to be essential for accurately estimating power production capabilities of PAs [64] and, more specifically, for the Corpower device [65]. However, since the authors are only interested in the relative values of the comparison, the mathematical model based on the linear Cummins's equation is considered adequate.

Similarly, the need for including the most relevant PTO dynamics, losses and constraints to accurately estimate power production capabilities of a WEC is demonstrated in [66]. Nevertheless, the present paper focuses on the impact of wave energy resource variations, for which analysing power absorption is found sufficient. However, constraints of the PTO system significantly affect the power absorption of a WEC. Therefore, three main constraints usually included by any PTO system, i.e., displacement, velocity and force constraints, are also considered, similar to [67].

4. Results

4.1. Evaluation Versus Buoys

Figure 4a,b illustrates the Taylor Diagrams for the Iquique and Valparaiso locations, respectively, where Pearson's correlation, the RMSE and the SD between the ERA20, ERAI and dcERA20 reanalyses are shown. The correlation of the dcERA20 reanalysis is shown to be very similar to that of the original ERA20, meaning that the correction did not improve the correlation. The correlation of the ERAI reanalysis, which is the upper limit for the calibration, is shown to be about 0.9 in both locations, while the ERA20 and dcERA20 show correlations of about 0.7–0.8. However, the dcERA20 reanalysis shows reasonable improvement in RMSE and SD. The RMSE for the ERA20 reanalysis is above 10 kW/m for Iquique, while reducing below that threshold after calibration. The RMSE reduction is even more significant for Valparaiso, reducing the WEF from 20 kW/m to 15 kW/m. In the case of the SD ratio, which shows the ratio between SDs of the wave models and buoy measurements, the calibration corrects the SD from 4/14 to 11/14 in Iquique, and from 6/22 to 18/22 in Valparaiso, which mean an improvement of over 50% in both cases. Hence, despite the low impact of the calibration on the correlation, the calibration is shown to significantly improve the wave data, approaching the more reliable ERAI reanalysis.

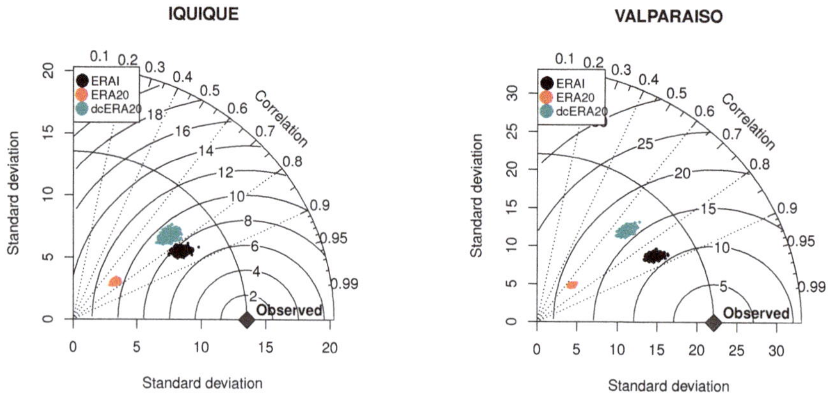

(**a**) (**b**)

Figure 4. Taylor Diagrams for Iquique (**a**) and Valparaiso (**b**) buoys.

Analysing other metrics described in Section 2.2.3, the effect of the calibration is even more evident, as shown in Tables 2 and 3. In Iquique, the ERA20 reanalysis underestimates the WEF, compared to the observations, which is corrected in the dcERA20 reanalysis, as presented in Table 2. This underestimation results in a negative bias of the ERA20 and a MAPE of only 37%. In contrast, the MAPE of the dcERA20 is halved due to the calibration. Important underestimation of the COV is also shown in Table 2 for the ERA20 reanalysis, which is remarkably corrected by the calibration.

Table 2. Mean WEF, bias and MAPE metrics for Iquique.

IQUIQUE	Mean WEF (kW/m)	COV	Bias (kW/M)	MAPE (%)
ERAI	17.4	0.64	1.7	10.8
ERA20	9.9	0.41	−5.9	37.3
dcERA20	18.6	0.62	2.9	18.5
Buoy	19.7	0.69	-	-

In the case of Valparaiso, the improvement of wave data due to the directional calibration method is even more important, as shown in Table 3, reaching mean WEF, COV, bias and MAPE values very close to the ERAI reanalysis, which sets the upper limit of the calibration. The mean WEF is improved substantially, reducing the MAPE from 56% to 4% and correcting the strong underestimation of the ERA20 reanalysis. The bias is also significantly reduced, from −17.6 to −1.2, and the COV given by the the dcERA20 reanalysis is identical to the COV obtained from the ERAI reanalysis. It should be noted that the COVs obtained from the buoy measurements in Iquique and Valparaiso are similar to the COV metrics presented in [14] for the same area, although data from different buoys were used: 0.6 in northern Chile [14], which is slightly lower than the 0.69 observed in this paper for Iquique; and 0.8 in central Chile [14], slightly higher than the 0.7 observed in this study for Valparaiso.

Table 3. Mean *WEF*, bias and MAPE metrics for Valparaiso.

VALPARAISO	Mean *WEF* (kW/m)	*COV*	Bias (kW/M)	MAPE (%)
ERAI	31.0	0.60	−0.3	1.17
ERA20	13.8	0.42	−17.6	56.0
dcERA20	30.1	0.60	−1.2	4.1
Buoy	32.3	0.70	-	-

4.2. Representation of Maps in the Study Area

4.2.1. Mean Values

Figure 5a–c shows, respectively, the mean values of H_s, T_m and *WEF* obtained from the ERAI reanalysis between 1979 and 2010. Mean H_s reaches very significant values in the south (up to 3.5 m) and progressively diminishes towards the north, with a minimum mean H_s of 1.7 m shown in the north of the country. The case of the T_m is exactly the opposite, where the mean T_m increases towards the north, with areas of long wave periods (up to 10 s) in the north and relatively short periods (about 7 s) in the south. Since the *WEF* is proportional to the square of H_s, as shown in Equation (2), the *WEF* follows the same spatial distribution as the H_s, meaning that wave power is highest in the south of the country. The results of the nearshore *WEF*, as shown in Figure 5c, are consistent with the recent study by Lucero et al. [15].

The map that illustrates the *COV* along the Chilean coast is shown in Figure 5d, where the spatial distribution of the *COV* is similar to that shown in [14], with *COV* values increasing towards the south. Additionally, one can observe a relation between the most energetic area in the south and the highest *COV*, and the decrease of both parameters towards the North. Consequently, the two locations under analysis in this paper (Iquique and Valparaiso) show reasonably low *COV* values (about 0.6). Note that, in the map illustrated in Figure 5d, and the maps shown in the following sections, purple cells with an **x** symbol mean that there is no significant variation at a 95% confidence level in that location.

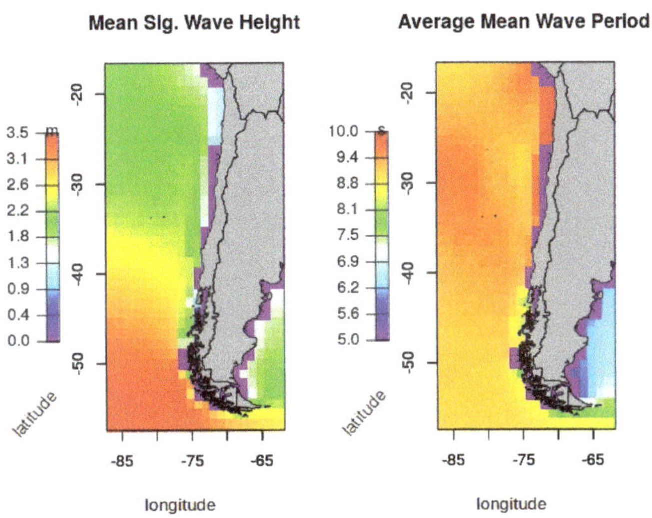

(**a**) Mean H_s (**b**) Mean T_m

Figure 5. *Cont.*

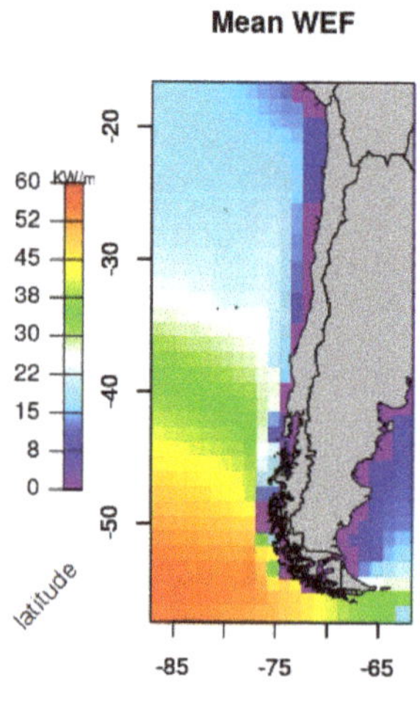

(c) Mean WEF

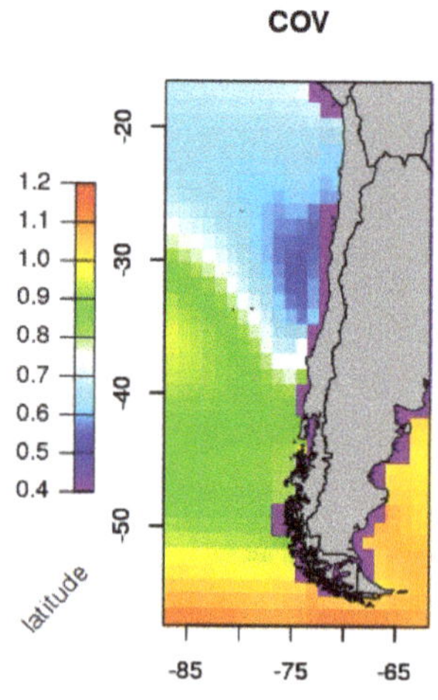

(**d**) Mean COV

Figure 5. Mean values of H_s (**a**), T_m (**b**), WEF (**c**) and COV (**d**) in the study area.

According to the spatial distribution of the mean WEF and the COV, Valparaiso shows interesting characteristics for the deployment of a WEC and, therefore, wave data from the dcERA20 reanalysis for the closest gridpoint to Valparaiso were used to evaluate historical wave resource variations and their impact on the power production of the cPA presented in Section 3.

4.2.2. Decadal Wave Trends

Decadal trends of the wave resource in Valparaiso are shown in Figure 6, where the evolution of the WEF, H_s and T_p are given in kW/m/decade, cm/decade and centiseconds per decade (cs/decade), respectively. The H_s is shown to increase slightly in the central and northern latitudes of Chile (1 cs/decade), while more significant increases (up to 5 cs/decade) are observed in the southern latitudes, as illustrated in Figure 6a. A similar pattern is also observed for WEF variations, as shown in Figure 6c, where the WEF increases up to 2 kW/m/decade. In contrast, variations of the mean T_p seem to be negligible in the north of the Chilean coast, while the T_p increases significantly (up to 4 cs/decade) in the central and southern latitudes. These results are consistent with the results shown in [24] for the H_s and in [30] with respect to the WEF. Trends of the COV are also studied over the 20th century, but are not shown in Figure 6, because the results do not show any significant variation at a 95% confidence level.

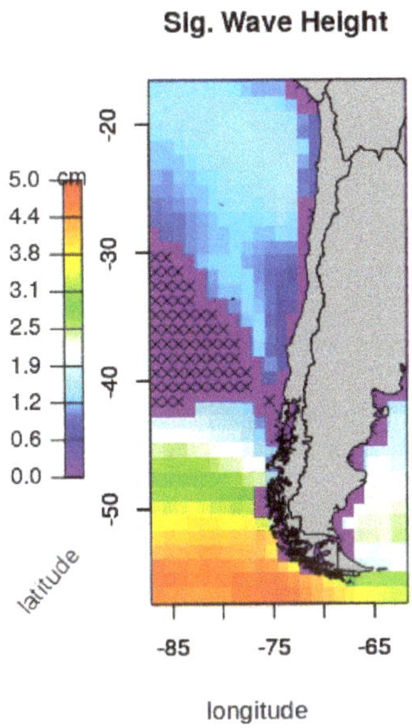

(a) Non-seasonal H_s trends

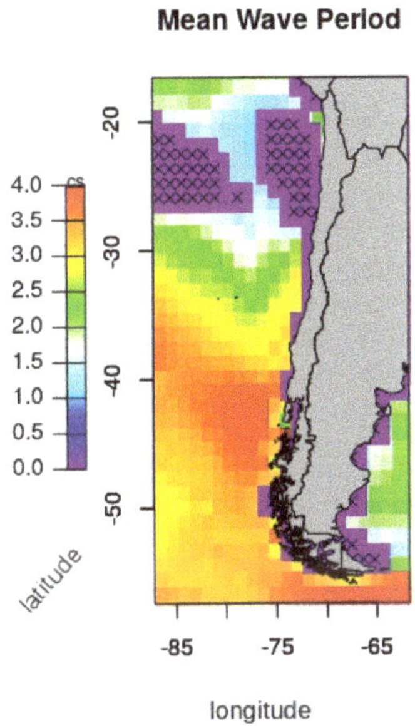

(b) Non-seasonal T_m trends

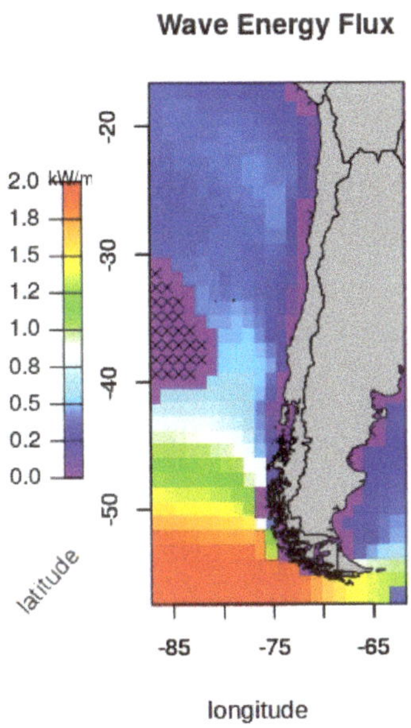

(c) Non-seasonal WEF trends

Figure 6. Trends of H_s (cm/decade) (**a**), T_m (cs/decade) (**b**) and WEF (kW/m/decade) (**c**) in the study area.

4.2.3. Seasonal Wave Energy Trends

The calibration of the ERA20 reanalysis, described in Section 2.2, can also be classified according to the seasonal variations, creating a transfer function for each season, referred to as the seasonally-calibrated ERA20 (scERA20). Figure 7a–d shows the four maps corresponding to the decadal *WEF* trends for autumn, winter, spring and summer, respectively, along the Chilean coast. The results obtained from the scERA20 show a relevant hot spot in the south of Chile, where the wave trend is particularly strong in autumn and winter (up to 2.5 kW/m/decade). This wave trend is still positive in the central and northern latitudes of the Chilean coast, although the wave trend is slightly weaker (about 0.5 kW/m/decade). Wave resource variations are slightly different in spring, where the positive wave trends can be observed in the south (up to 2 kW/m/decade) and north (about 0.5 kW/m/decade) of the Chilean coast. However, no significant variations are observed in the central latitudes, where the resource is consistent all over the 20th century, as illustrated in Figure 7c. Finally, resource variations over the 20th century are negligible in summer, as shown in Figure 7d.

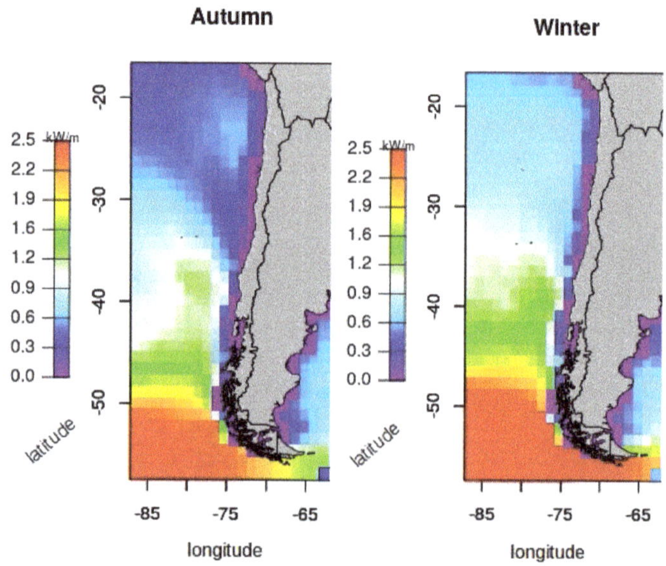

(**a**) *WEF* trends in autumn (**b**) *WEF* trends in winter

Figure 7. *Cont.*

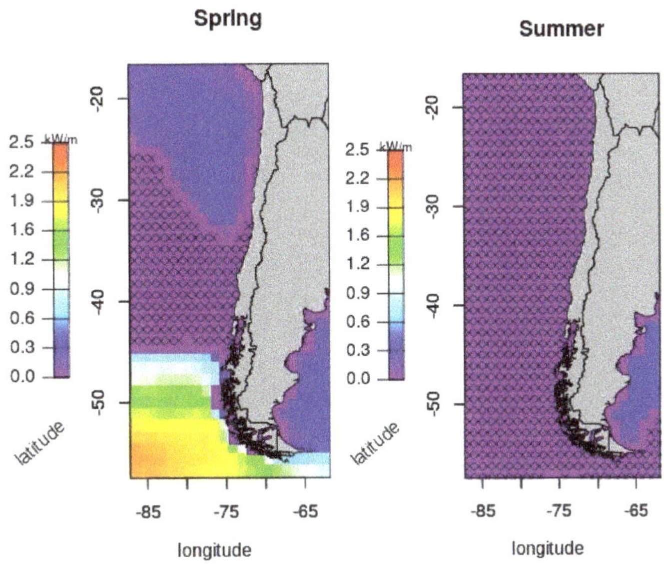

(**c**) WEF trends in spring (**d**) WEF trends in summer

Figure 7. Wave energy trends for autumn (**a**), winter (**b**), spring (**c**) and summer (**d**).

4.3. Wave Trends and Power Production in Valparaiso

4.3.1. Wave Resource Variations

According to Figure 6, which is created using wave data from the dcERA20 model, decadal H_s, T_m and WEF trends are approximately 1.2 cm/decade, 2. 5 cs/decade and 0.4 kW/m/decade, respectively. A more detailed study, analysing each do-decade of the 20th century separately, shows that the mean WEF in the first do-decade, i.e., 1900–1920, was about 19.1 kW/m/decade. Hence, the decadal WEF increase of 0.4 kW/m shown in Figure 6c corresponds to a decadal increase of about 2% ($0.39/19.1 \times 100 = 2\%$).

The scatter-diagram of the resource in this first do-decade of the 20th century is illustrated in Figure 8a, where the most frequent H_s and T_p are shown to be 1.5 m and 10.5 s, respectively. Wave trends of the next do-decades are shown as relative variations (in percentage) with respect to the wave resource in the first do-decade, as shown in Figure 8b, where the decadal increase of 2% is also illustrated. This general trend is consistent with other studies carried out in the same area [24,30]. However, as depicted in Figure 8b, the inter-decadal trends over the 20th century are highly irregular, with significant increases in some do-decades, between 1920 and 1960, for example, and a strong reduction in others, such as between 1960 and 1980.

The bars in Figure 8 show the increase of WEF, H_s and T_p in percentage at each do-decade, with respect to the first do-decade. These bars show a progressive increase of the wave period all over the century, while the WEF and H_s show more irregular patterns, with significant increases in the first two do-decades, and a very strong decrease in the fourth do-decade. The WEF and H_s increase again in the last do-decade. Despite the irregular variations in the different do-decades, wave energy resource variations are significantly more consistent compared to the progressive strong increases detected in the Bay of Biscay and west coast of Ireland in [7] and [8], respectively.

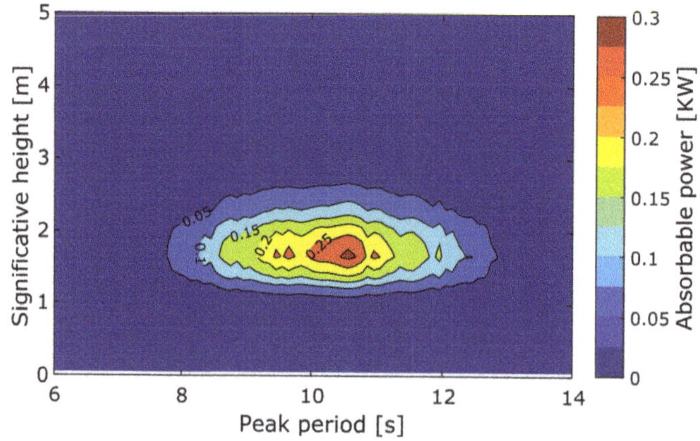

(a)

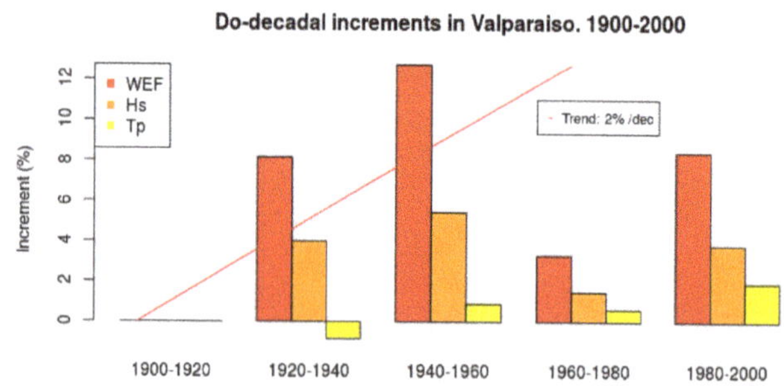

(b)

Figure 8. The scatter-diagram of Valparaiso using the data for the first do-decade of the 20th century (1900–1920) (**a**) and the variations of the resource over the century (**b**).

4.3.2. Impact on Wave Energy Absorption

The power absorbed from ocean waves by the cPA is assessed using the annual mean power production (AMPP) metric. Figure 9 illustrates the variation of the AMPP over the 20th century, if the cPA were deployed in Valparaiso. AMPP variations follow the same pattern as the WEF variations, illustrated in Figure 8. However, AMPP variations are lower than resource variations. An increase of 6% in AMPP is observed in the second do-decade, with respect to the first do-decade, while the resource increases by 8%. This difference is even larger for the third do-decade, with an increase of over 12% in WEF, while the increase in AMPP is almost identical to the second do-decade (6%). The reason the AMPP does not increase with the WEF may be the variation of the mean T_p between the second and third do-decades. While WEF increases in the second do-decade, the variation of the mean T_p is negative, meaning that the resource shows lower wave periods, getting closer to the natural period of the cPA (5.2 s). In contrast, mean T_p variation in the third do-decade is positive, which means that the mean T_p moves away from the natural period of the cPA and, as a consequence, limits the increase in AMPP. Likewise, WEF variation is very similar in the second and fifth do-decades, as shown in Figure 8, while the AMPP increase depicted in Figure 9 is significantly lower in the fifth do-decade. The only difference between resource variations in the second and the fifth do-decades is again the mean T_p, which is significantly larger in the fifth do-decade. Consequently, the same increase in WEF does not involve the same increase in AMPP, illustrating the relevance of wave period variations.

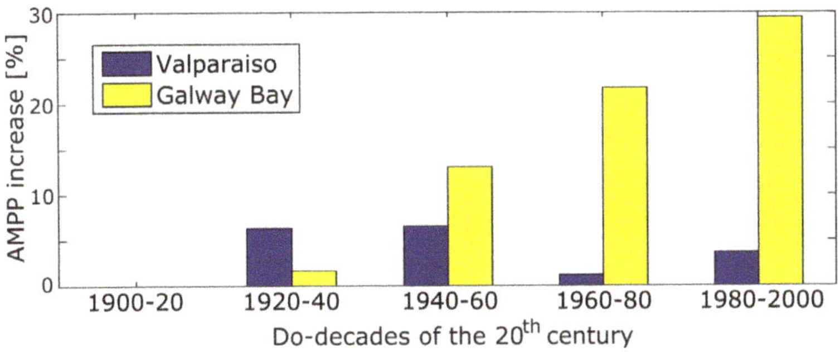

Figure 9. AMPP variation over the 20th century in Valparaiso (Chile) and Galway bay (Ireland).

5. Discussion

The wave trend results of the calibrated wave model presented in this paper for the Chilean coast are consistent with previous studies presented in the literature, either based on data from satellite altimeter [24] or reanalyses [30,31]. The particular case of Valparaiso presented in this study shows an overall WEF increase of 2%/decade over the last century, while previous studies show increments of 0–3.3% from the north to the south of Chile. In any case, this general slope is important only for the validation of the calibration, since the barplot presented in Figure 8 shows that resource variations are quite irregular over the century, including relatively strong decreases between the third and the fourth do-decades. This irregular trend, without a clear uniform variation profile, contrasts the almost linear increment profile observed in the Atlantic Ocean, more specifically, in the Bay of Biscay [7] and the west coast of Ireland [8].

This non-uniform behaviour in trends over the different decades suggests that the causes of the detected variations may be complex and multiple. Although these kind of variations are often attributed to climate change [9], different, relatively unknown global mechanisms may also play an important role.

The comparison of the wave resources in the Chilean coast and the west coast of Ireland is a judicious comparison, due to their similar wave power, as shown in [14]. However, an important difference between Chile and Ireland, also pointed out in [14], is the COV, which is significantly greater in Ireland. In fact, results for the COV presented in [14] are similar to those presented in the present study, where a longer time-period is analysed. Hence, the wave resource in Chile is shown to be significantly more consistent than the wave resource in the west coast of Ireland. However, the impact of a more consistent resource on the AMPP of WECs is not analysed in [14]. Figure 9 compares the variations of the AMPP for the same cPA in the Chilean coast and the west coast of Ireland, where AMPP variations are shown to be significantly lower in Chile, as expected. It should be noted that Valparaiso is located in the centre of Chile, while the variability of the resource (as the WEF) increases most in the southern latitudes [14].

Apart from the differences in resource variations and their impact on WECs' power production capabilities, the frequency of storms is another important factor in the process of selecting the optimum location to install a WEC farm. Storms imply that WECs are shifted into survivability mode in order to protect them from structural damages, which directly affects the cost of the structure. A simple method to quantify the frequency of storms is using a maximum H_s value (H_s^{MAX}) that delimits the operational space of the WEC, above which the WEC shifts into survivability mode. The same method is used in [68] or [8], among others. The $H_s^{MAX} = 4$ m value is found to be quite restrictive in [8], where the AMPP is shown to be drastically reduced, up to 50%, compared to the unlimited case. The events where the wave resource exceeds the H_s^{MAX} limit are referred to as off-limit events, and the frequency of these off-limit events is shown in Figure 10a. These off-limit events represent 20% of the resource off the west coast of Ireland at the beginning of the 20th century, which has significantly increased, up to almost 30%, by the end of the century. In contrast, using the same H_s^{MAX} limitation, the off-limit events represent less than 1% of the resource in Valparaiso, which remains quite constant over the whole century.

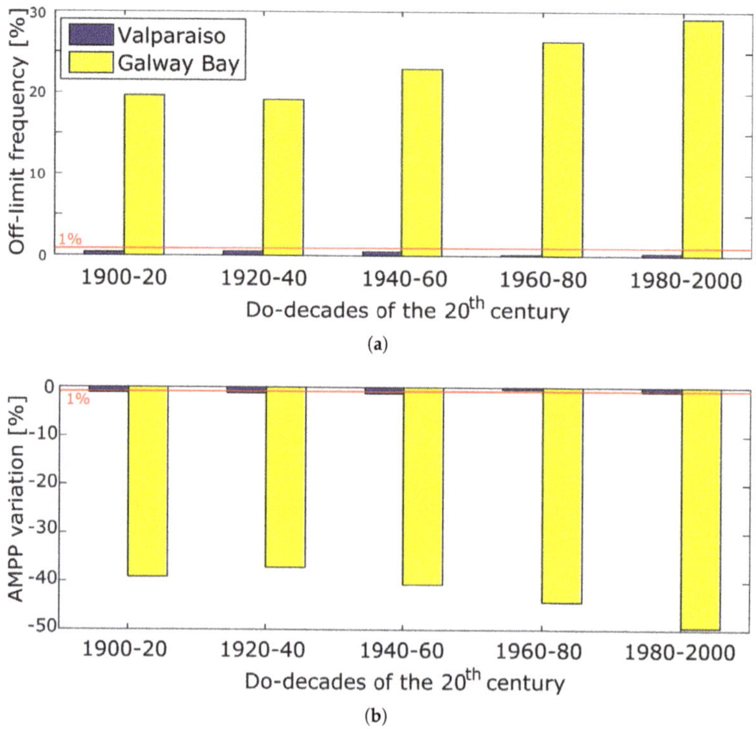

Figure 10. Occurrence of the off-limit sea-states (**a**) and its impact on the AMPP of the Corpower device (**b**) over the 20th century in Valparaiso (Chile) and Galway bay (Ireland).

The impact of this difference in the frequency of off-limit events on the AMPP is shown in Figure 10b, where the AMPP is shown to decrease up to a 50% off the west coast of Ireland, while only decreasing by 1% in Valparaiso. Hence, the significantly more consistent and mild wave resource in Chile represents a more attractive location, compared to Ireland, for the implementation of wave energy projects, allowing for cost minimization and facilitating deployment or maintenance operations.

The Southern areas of the Pacific Ocean lack detailed data coverage for reanalyses during the early parts of the 20th century. Even in periods relatively close in time (1985–2012), the lack of a dense coverage by observations over scarce data regions of the Pacific implies that the computations of surface heat fluxes over the Eastern Pacific show there the highest errors [69]. In long reanalysis spanning back in time to early instrumental periods, such as is the case with ERA-20CM or 20CRA, observations poorly constrain the reanalyses during early 20th Century. This is evident in the higher spread of the forcing by HadISST during 1900 versus 2000 in ERA20-CM or the lack of closure of the surface energy balance during early years over that area even though SSTs are prescribed [70]. In the case of 20CRA, there appear lacking trends of the Pacific Circulation or the Pacific-North American (PNA) pattern [71]. The smaller number of observations over the southern Pacific during early 20th

Century [34] affects the ability of the ERA-20C reanalysis to simulate a realistic PNA index and other atmospheric indices before 1940. Since ocean waves are produced by atmospheric forcing at the surface, the fact that these problems due to data coverage of the early period have already been identified in the literature point that the use of bias correction techniques such as the ones used in this paper might be more important in Southern Pacific areas than in others. However, since different bias correction techniques are available [72,73], it will be interesting in the future to compare the results presented in this paper with the ones from complementary techniques.

6. Conclusions

Results of the ERA20 reanalysis of the European Centre for Medium-Range Weather Forecasts has been shown unreliable when comparing to buoy measurements. However, the directional calibration, based on the ERAI reanalysis, presented in this paper, is shown to significantly improve the results of the ERA20 reanalysis. This calibration provides reasonably reliable wave data for the whole 20th century, which allows the study of long-term wave resource variations off the Chilean coast.

Positive wave trends over the 20th century off the Chilean coast are detected in the present study, with the southern coast showing the most significant variations. However, in contrast to other locations in the Atlantic Ocean, the positive wave trend in the Chilean coast is reasonably irregular, with the strong increases of the wave resource in some do-decades and significant reductions in others. These irregular variations indicate that a straightforward attribution of these changes to climate change may be misleading and, thus, further research is needed to establish the driving mechanisms behind these trends. In addition, although long-term variations are significant, about 2kW/m/decade over the 20th century, inter- and intra-annual variations, represented by the coefficient of variation, are shown to be significantly lower than in other locations in the Atlantic Ocean.

These resource variations over the 20th century also affect the the power absorption of wave energy converters, although variations on annual mean power production do not exactly agree with the variations of the wave energy resource. Similar to the wave energy resource variations, annual mean power production variations are significantly lower in the Chilean coast than off the west coast of Ireland. In addition, the frequency of storms, for which the wave energy converters shift to survivability mode, is significantly lower in the Chilean coast (always lower than 1%), compared to the west cost of Ireland (up to 30%). This directly affects the annual mean power production of wave energy converters, and the design of different aspects of the wave energy converters, such as mooring lines, foundations and the structure.

Hence, the selection of the optimum location for the implementation of a wave energy converter farm should consider both short- and long-term variations of the wave energy resource where the farm is planned to be installed.

Author Contributions: Conceptualization, A.U. and M.P.; Methodology, A.U. and M.P.; Software, A.U., M.P. and A.R.; Validation, A.U. and A.R.; Investigation, A.U., M.P., G. I., J.S., and J. R.; Writing—Original Draft Preparation, A.U.; Writing—Review and Editing, all authors; Supervision, all authors; Project Administration, G.I. and J.R.; and Funding Acquisition, G.I., J.S. and J.R.

Funding: The authors with the Centre for Ocean Energy Research in Maynooth University are supported by Science Foundation Ireland under Grant No. 13/IA/1886. It is also supported by grant CGL2016-76561-R, MINECO/ERDF, UE. Additional funding was received from the University of Basque Country (UPV/EHU, GIU17/002).

Acknowledgments: The buoy data have been kindly provided by the SHOA (Chilean Navy Hydrographic and Oceanographic Service): http://www.shoa.cl/php/inicio.php.

Conflicts of Interest: The authors declare no conflict of interest.

Abbreviations

The following abbreviations are used in this manuscript:

AMPP	Annual mean power production
CF	Capacity factor
COV	Coefficient of variation
cPA	Corpower-like point absorber
dcERA20	directionally-calibrated ERA20
ECMWF	European Centre for Medium-Range Weather Forecasts
ERAI	ERA-Interim
MAPE	Mean absolute percentage error
PA	Point absorber
PTO	Power take-off
RMSE	Root mean square error
scERA20	seasonally-calibrated ERA20
SD	Stadard deviation
WEF	Wave energy flux
WEC	Wave energy converter
WPR	Wave period ratio

References

1. Falcão, A.d.O. Wave energy utilization: A review of the technologies. *Renew. Sustain. Energy Rev.* **2010**, *14*, 899–918. [CrossRef]
2. Ibarra-Berastegi, G.; Sáenz, J.; Ulazia, A.; Serras, P.; Esnaola, G.; Garcia-Soto, C. Electricity production, capacity factor, and plant efficiency index at the Mutriku wave farm (2014–2016). *Ocean Eng.* **2018**, *147*, 20–29. [CrossRef]
3. Rusu, E.; Onea, F. Estimation of the wave energy conversion efficiency in the Atlantic Ocean close to the European islands. *Renew. Energy* **2016**, *85*, 687–703. [CrossRef]
4. Carballo, R.; Sánchez, M.; Ramos, V.; Fraguela, J.; Iglesias, G. The intra-annual variability in the performance of wave energy converters: A comparative study in N Galicia (Spain). *Energy* **2015**, *82*, 138–146. [CrossRef]
5. Reguero, B.; Losada, I.; Méndez, F. A global wave power resource and its seasonal, interannual and long-term variability. *Appl. Energy* **2015**, *148*, 366–380. [CrossRef]
6. Ramos, V.; López, M.; Taveira-Pinto, F.; Rosa-Santos, P. Influence of the wave climate seasonality on the performance of a wave energy converter: A case study. *Energy* **2017**, *135*, 303–316. [CrossRef]
7. Ulazia, A.; Penalba, M.; Ibarra-Berastegi, G.; Saenz, J. Wave energy trends over the Bay of Biscay and the consequences for wave energy converters. *Energy* **2017**, *141*, 624–634. [CrossRef]
8. Penalba, M.; Ulazia, A.; Ibarra-Berastegui, G.; Ringwood, J.; Sáenz, J. Wave energy resource variation off the west coast of Ireland and its impact on realistic wave energy converters' power absorption. *Appl. Energy* **2018**, *224*, 205–219. [CrossRef]
9. Lehmann, M.; Karimpour, F.; Goudey, C.A.; Jacobson, P.T.; Alam, M.R. Ocean wave energy in the United States: Current status and future perspectives. *Renew. Sustain. Energy Rev.* **2017**, *74*, 1300–1313. [CrossRef]
10. Alonso, R.; Jackson, M.; Santoro, P.; Fossati, M.; Solari, S.; Teixeira, L. Wave and tidal energy resource assessment in Uruguayan shelf seas. *Renew. Energy* **2017**, *114*, 18–31. [CrossRef]
11. Lisboa, R.C.; Teixeira, P.R.; Fortes, C.J. Numerical evaluation of wave energy potential in the south of Brazil. *Energy* **2017**, *121*, 176–184. [CrossRef]
12. Cruz, J.; Thomson, M.; Stavrioulia, E. Preliminary Site Selection—Chilean Marine Energy Resources. Technical Report 100513/BR/02, Garrad Hassan and Partners Limited. Available online: http://www.etymol.com/downloads/garrad_hassan_chilean_marine_energy_resources.pdf (accessed on 18 July 2018).
13. Di Lauro, E.; Contestabile, P.; Vicinanza, D. Wave Energy in Chile: A Case Study of the Overtopping Breakwater for Energy Conversion (OBREC). In Proceedings of the 12th European Wave and Tidal Energy Conference, Cork, Ireland, 27 August–1 September 2017.
14. Ringwood, J.; Brandle, G. A new world map for wave power with a focus on variability. In Proceedings of the 11th European Wave and Tidal Energy Conference, Nantes, France, 6–11 September 2015.

15. Lucero, F.; Catalán, P.A.; Ossandón, Á.; Beyá, J.; Puelma, A.; Zamorano, L. Wave energy assessment in the central-south coast of Chile. *Renew. Energy* **2017**, *114*, 120–131. [CrossRef]
16. Mediavilla, D.G.; Sepúlveda, H.H. Nearshore assessment of wave energy resources in central Chile (2009–2010). *Renew. Energy* **2016**, *90*, 136–144. [CrossRef]
17. Monárdez, P.; Acuña, H.; Scott, D. Evaluation of the potential of wave energy in Chile. In Proceedings of the ASME 2008 27th International Conference on Offshore Mechanics and Arctic Engineering, Estoril, Portugal, 15–20 June 2008; pp. 801–809.
18. WMO. *Calculation of Monthly and Annual 30-year Standard Normals*; WCDP-No. 10. WMO-TD/No. 341; World Metheorological Organization: Geneva, Switzerland, 1989.
19. WMO. *The role of Climatological Normals in a Changing Climate*; WCDMP-61, WMO-TD/1377; Word Metheorological Organization: Geneva, Switzerland, 2007.
20. Ruggiero, P.; Komar, P.D.; Allan, J.C. Increasing wave heights and extreme value projections: The wave climate of the US Pacific Northwest. *Coast. Eng.* **2010**, *57*, 539–552. [CrossRef]
21. Gulev, S.K.; Grigorieva, V. Last century changes in ocean wind wave height from global visual wave data. *Geophys. Res. Lett.* **2004**, *31*, doi:10.1029/2004GL021040. [CrossRef]
22. Gulev, S.K.; Grigorieva, V. Variability of the winter wind waves and swell in the North Atlantic and North Pacific as revealed by the voluntary observing ship data. *J. Clim.* **2006**, *19*, 5667–5685. [CrossRef]
23. Woolf, D.K.; Challenor, P.; Cotton, P. Variability and predictability of the North Atlantic wave climate. *J. Geophys. Res. Ocean.* **2002**, *107*.10.1029/2001JC001124. [CrossRef]
24. Young, I.; Zieger, S.; Babanin, A.V. Global trends in wind speed and wave height. *Science* **2011**, *332*, 451–455. [CrossRef] [PubMed]
25. Sterl, A.; Komen, G.; Cotton, P. Fifteen years of global wave hindcasts using winds from the European Centre for Medium-Range Weather Forecasts reanalysis: Validating the reanalyzed winds and assessing the wave climate. *J. Geophys. Res. Ocean.* **1998**, *103*, 5477–5492. [CrossRef]
26. Cox, A.T.; Swail, V.R. A global wave hindcast over the period 1958–1997: Validation and climate assessment. *J. Geophys. Res.* **2001**, *106*, 2313–2329. [CrossRef]
27. Wang, X.L.; Zwiers, F.W.; Swail, V.R. North Atlantic ocean wave climate change scenarios for the twenty-first century. *J. Clim.* **2004**, *17*, 2368–2383. [CrossRef]
28. Wang, X.L.; Swail, V.R. Climate change signal and uncertainty in projections of ocean wave heights. *Clim. Dyn.* **2006**, *26*, 109–126. [CrossRef]
29. Bertin, X.; Prouteau, E.; Letetrel, C. A significant increase in wave height in the North Atlantic Ocean over the 20th century. *Glob. Planet. Chang.* **2013**, *106*, 77–83. [CrossRef]
30. Zheng, C.; Shao, L.; Shi, W.; Su, Q.; Lin, G.; Li, X.; Chen, X. An assessment of global ocean wave energy resources over the last 45 a. *Acta Oceanol. Sin.* **2014**, *33*, 92–101. [CrossRef]
31. Zheng, C.W.; Wang, Q.; Li, C.Y. An overview of medium- to long-term predictions of global wave energy resources. *Renew. Sustain. Energy Rev.* **2017**, *79*, 1492–1502. [CrossRef]
32. Sierra, J.; Casas-Prat, M.; Campins, E. Impact of climate change on wave energy resource: The case of Menorca (Spain). *Renew. Energy* **2017**, *101*, 275–285. [CrossRef]
33. Worley, S.J.; Woodruff, S.D.; Reynolds, R.W.; Lubker, S.J.; Lott, N. ICOADS release 2.1 data and products. *Int. J. Climatol.* **2005**, *25*, 823–842. [CrossRef]
34. Poli, P.; Hersbach, H.; Dee, D.P.; Berrisford, P.; Simmons, A.J.; Vitart, F.; Laloyaux, P.; Tan, D.G.; Peubey, C.; Thépaut, J.N.; et al. ERA-20C: An atmospheric reanalysis of the twentieth century. *J. Clim.* **2016**, *29*, 4083–4097. [CrossRef]
35. Dada, O.A.; Li, G.; Qiao, L.; Ma, Y.; Ding, D.; Xu, J.; Li, P.; Yang, J. Response of waves and coastline evolution to climate variability off the Niger Delta coast during the past 110 years. *J. Mar. Syst.* **2016**, *160*, 64–80. [CrossRef]
36. Patra, A.; Bhaskaran, P.K. Temporal variability in wind–wave climate and its validation with ESSO-NIOT wave atlas for the head Bay of Bengal. *Clim. Dyn.* **2017**, *49*, 1271–1288. [CrossRef]
37. Kumar, P.; Min, S.K.; Weller, E.; Lee, H.; Wang, X.L. Influence of Climate Variability on Extreme Ocean Surface Wave Heights Assessed from ERA-Interim and ERA-20C. *J. Clim.* **2016**, *29*, 4031–4046. [CrossRef]
38. Komen, G.J.; Cavaleri, L.; Donelan, M.; Hasselmann, K.; Hasselmann, S.; Janssen, P. *Dynamics and Modelling of Ocean Waves*; Cambridge University Press: Cambridge, UK, 1996.

39. Bidlot, J.; Janssen, P.; Abdalla, S.; Hersbach, H. *A Revised Formulation of Ocean Wave Dissipation and Its Model Impact;* European Centre for Medium-Range Weather Forecasts: Reading, UK, 2007.
40. Berrisford, P.; Dee, D.; Fielding, K.; Fuentes, M.; Kallberg, P.; Kobayashi, S.; Uppala, S. *The ERA-Interim Archive;* European Centre for Medium-Range Weather Forecasts: Reading, UK, 2009; pp. 1–16.
41. Cahill, B.; Lewis, T. Wave period ratios and the calculation of wave power. In Proceedings of the 2nd Annual Marine Energy Technology Symposium(METS 2014), Seattle, WA, USA, 15–17 April 2014.
42. Contestabile, P.; Ferrante, V.; Vicinanza, D. Wave energy resource along the coast of Santa Catarina (Brazil). *Energies* **2015**, *8*, 14219–14243. [CrossRef]
43. Block, P.; Souza Filho, F.; Sun, L.; Kwon, H.H. A streamflow forecasting framework using multiple climate and hydrological models. *J. Am. Water Resour. Assoc.* **2009**, *45*, 828–843. [CrossRef]
44. Boa, J.; Terray, L.; Habets, F.; Martin, E. Statistical and dynamical downscaling of the Seine basin climate for hydro-meteorological studies. *Int. J. Climatol.* **2007**, *27*, 1643–1655. [CrossRef]
45. Sun, F.; Roderick, M.L.; Lim, W.H.; Farquhar, G.D. Hydroclimatic projections for the Murray-Darling Basin based on an ensemble derived from Intergovernmental Panel on Climate Change AR4 climate models. *Water Resour. Res.* **2011**, *47*, doi:10.1029/2010WR009829. [CrossRef]
46. Piani, C.; Haerter, J.O.; Coppola, E. Statistical bias correction for daily precipitation in regional climate models over Europe. *Theor. Appl. Climatol.* **2010**, *99*, 187–192, doi:10.1007/s00704-009-0134-9. [CrossRef]
47. Rojas, R.; Feyen, L.; Dosio, A.; Bavera, D. Improving pan-European hydrological simulation of extreme events through statistical bias correction of RCM-driven climate simulations. *Hydrol. Earth Sys. Sci.* **2011**, *15*, 2599–2620, doi:10.5194/hess-15-2599-2011. [CrossRef]
48. Teutschbein, C.; Seibert, J. Bias correction of regional climate model simulations for hydrological climate-change impact studies: Review and evaluation of different methods. *J. Hydrol.* **2012**, *456*, 12–29. [CrossRef]
49. Watanabe, S.; Kanae, S.; Seto, S.; Yeh, P.J.F.; Hirabayashi, Y.; Oki, T. Intercomparison of bias-correction methods for monthly temperature and precipitation simulated by multiple climate models. *J. Geophys. Res. Atmos.* **2012**, *117*.10.1029/2012JD018192. [CrossRef]
50. Lafon, T.; Dadson, S.; Buys, G.; Prudhomme, C. Bias correction of daily precipitation simulated by a regional climate model: a comparison of methods. *Int. J. Climatol.* **2013**, *33*, 1367–1381. [CrossRef]
51. Bett, P.E.; Thornton, H.E.; Clark, R.T. Using the Twentieth Century Reanalysis to assess climate variability for the European wind industry. *Theor. Appl. Climatol.* **2015**, 1–20, doi:10.1007/s00704-015-1591-y. [CrossRef]
52. Panofsky, H.; Brier, G. *Some Applications of Statistics to Meteorology;* Pennsylvania State University: University Park, PA, USA, 1958.
53. Applequist, S. Wind Rose Bias Correction. *J. Appl. Meteorol. Climatol.* **2012**, *51*, 1305–1309, doi:10.1175/JAMC-D-11-0193.1. [CrossRef]
54. Taylor, K.E. Summarizing multiple aspects of model performance in a single diagram. *J. Geophys. Res. Atmos.* **2001**, *106*, 7183–7192. [CrossRef]
55. Sen, P.K. Estimates of the regression coefficient based on Kendall's tau. *J. Am. Stat. Assoc.* **1968**, *63*, 1379–1389. [CrossRef]
56. Theil, H. A rank-invariant method of linear and polynomial regression analysis, 3; confidence regions for the parameters of polynomial regression equations. *Indagat. Math.* **1950**, *1*, 467–482.
57. Fiévez, J.; Sawyer, T. Lessons Learned from Building and Operating a Grid Connected Wave Energy Plant. In Proceedings of the 11th European Wave and Tidal Energy Conference, Nantes, France, 6–11 September 2015.
58. Lejerskog, E.; Boström, C.; Hai, L.; Waters, R.; Leijon, M. Experimental results on power absorption from a wave energy converter at the Lysekil wave energy research site. *Renew. Energy* **2015**, *77*, 9–14. [CrossRef]
59. Hals, J.; Ásgeirsson, S.G.; Hjálmarsson, E.; Maillet, J.; Moller, P.; Pires, P.; Guérinel, M.; Lopes, M. Tank testing of an inherently phase controlled Wave Energy Converter. In Proceedings of the 11th European Wave and Tidal Energy Conference, Nantes, France, 6–11 September 2015.
60. Wang, W.; Wu, M.; Palm, J.; Eskilsson, C. Estimation of numerical uncertainty in computational fluid dynamics simulations of a passively controlled wave energy converter. *Proc. Inst. Mech. Eng. Part M J. Eng. Marit. Environ.* **2018**, *232*, 71–84. [CrossRef]

61. Hasselmann, K.; Barnett, T.; Bouws, E.; Carlson, H.; Cartwright, D.; Enke, K.; Ewing, J.; Gienapp, H.; Hasselmann, D.; Kruseman, P.; et al. *Measurements of Wind-wave Growth and Swell Decay During the Joint North Sea Wave Project (JONSWAP)*; Deutsches Hydrographisches Institut: Hamburg, Germany, 1973.
62. Cummins, W. The impulse response function and ship motion. *Schiffstechnik* **2010**, *9*, 101–109.
63. Morison, J.; Johnson, J.; Schaaf, S. The force exerted by surface waves on piles. *J. Petrol. Technol.* **1950**, *2*, 149–154. [CrossRef]
64. Penalba, M.; Giorgi, G.; Ringwood, J.V. Mathematical modelling of wave energy converters: A review of nonlinear approaches. *Renew. Sustain. Energy Rev.* **2017**, *78*, 1188–1207. [CrossRef]
65. Giorgi, G.; Ringwood, J.V. A Compact 6-DoF Nonlinear Wave Energy Device Model for Power Assessment. *IEEE Trans. Sustain. Energy* **2018**, doi:10.1109/TSTE.2018.2826578. [CrossRef]
66. Penalba, M.; Davison, J.; Windt, C.; Ringwood, J. A High-Fidelity Wave-to-Wire Model for Wave Energy Converters: Coupled numerical wave tank and power take-off models. *Appl. Energy* **2018**, *226*, 655–669. [CrossRef]
67. Babarit, A.; Hals, J.; Muliawan, M.; Kurniawan, A.; Moan, T.; Krokstad, J. Numerical benchmarking study of a selection of wave energy converters. *Renew. Energy* **2012**, *41*, 44–63. [CrossRef]
68. Mérigaud, A.; Ringwood, J.V. Power production assessment for wave energy converters: Overcoming the perils of the power matrix. *Proc. Inst. Mech. Eng. Part M J. Eng. Marit. Environ.* **2018**, *232*, 50–70. [CrossRef]
69. Liu, C.; Allan, R.P.; Berrisford, P.; Mayer, M.; Hyder, P.; Loeb, N.; Smith, D.; Vidale, P.L.; Edwards, J.M. Combining satellite observations and reanalysis energy transports to estimate global net surface energy fluxes 1985–2012. *J. Geophys. Res. Atmos.* **2015**, *120*, 9374–9389. [CrossRef]
70. Hersbach, H.; Peubey, C.; Simmons, A.; Berrisford, P.; Poli, P.; Dee, D. ERA-20CM: A twentieth-century atmospheric model ensemble. *Q. J. R. Meteorol. Soc.* **2015**, *141*, 2350–2375. [CrossRef]
71. Compo, G.P.; Whitater, J.S.; Sardeshmukh, P.D.; Matsui, N.; Allan, R.J.; Yin, X.; Gleason, B.E.; Vose, R.S.; Rutledge, G.; Bessemoulin, P.; et al. The Twentieth Century Reanalysis Project. *Q. J. R. Meteorol. Soc.* **2011**, *137*, 1–28. [CrossRef]
72. Kim, K.B.; Kwon, H.H.; Han, D. Bias correction methods for regional climate model simulations considering the distributional parametric uncertainty underlying the observations. *J. Hydrol.* **2015**, *530*, 568–579. [CrossRef]
73. Maraun, D. Bias Correcting Climate Change Simulations—A Critical review. *Curr. Clim. Chang. Rep.* **2016**, *2*, 211–220. [CrossRef]

© 2018 by the authors. Licensee MDPI, Basel, Switzerland. This article is an open access article distributed under the terms and conditions of the Creative Commons Attribution (CC BY) license (http://creativecommons.org/licenses/by/4.0/).

energies

Article

Analysis of the Variability of Wave Energy Due to Climate Changes on the Example of the Black Sea

Yana Saprykina * and Sergey Kuznetsov

Shirshov Institute of Oceanology of Russian Academy of Sciences, Moscow 117997, Russia; kuznetsov@ocean.ru
* Correspondence: saprykina@ocean.ru; Tel.: +7-499-124-6222

Received: 25 June 2018; Accepted: 2 August 2018; Published: 3 August 2018

Abstract: An analysis of the variability of wave climate and energy within the Black Sea for the period 1960–2011 was made using field data from the Voluntary Observing Ship Program. Methods using wavelet analysis were applied. It was determined that the power flux of wave energy in the Black Sea fluctuates: the highest value is 4.2 kW/m, the lowest is 1.4 kW/m. Results indicate significant correlations among the fluctuations of the average annual wave heights, periods, the power flux of wave energy, and teleconnection patterns of the North Atlantic Oscillation (NAO), the Atlantic Multi-decadal Oscillation (AMO), the Pacific Decadal Oscillation (PDO) and the East Atlantic/West Russia (EA/WR). It was revealed that, in positive phases of long-term periods of AMO (50–60 years) as well as PDO, NAO, and AO (40 years), a decrease of wave energy was observed; however, an increase in wave energy was observed in the positive phase of a 15-year period of NAO and AO. The positive phase of changes of EA/WR for periods 50–60, 20–25, and 13 years led to an increase of wave energy. The approximation functions of the oscillations of the average annual wave heights, periods, and the power flux of wave energy for the Black Sea are proposed.

Keywords: wave energy; wave climate variability; wavelet analysis; teleconnection patterns

1. Introduction

Currently, numerous opportunities for renewable natural energy sources, such as wind and wave energy, are being widely discussed. The energy obtained from natural sources is environmentally friendly because it does not require the creation of special production cycles that often can pollute the environment, especially the atmosphere. For example, a unique wave energy plant was constructed in the harbor of a small fishing village Mutriku at the Basque Atlantic coastline.

To convert wave energy into electricity, special wave energy converters (WEC) are under development and are the subjects of significant research. A review of technologies which extract wave energy, an overview of wave energy potential, and the wave energy cost can be found in Rusu and Onea [1]. They have shown that the global wave power resources are 2.11 $\pm$ 0.05 TW [2].

At the same time, the distribution of wave power resources is nonhomogeneous in the World Ocean. Therefore, for the implementation of projects on wave energy use, an analysis of the available local wave energy potential is required. Traditionally, the energy potential estimations of different parts of the World Ocean are performed by modeling waves based on wind reanalysis; for example, studies by Atan and Goggins [3] and Soares et al. [4] discuss the Atlantic.

The Black Sea is an inland sea connected to the Mediterranean Sea by the Bosporus Strait. Its resources, including possible wave energy, have the great economic importance for the countries on its shores, such as Russia, Turkey, Georgia, Ukraine, Bulgaria, and Romania. For the Black Sea, such estimations for the all water area for periods 1900–2010 and 1990–2014 was conducted by Galabov [5], Rusu [6], and Divinsky and Kosyan [7]. The analysis for the wave energy potential along the southeastern coastline for the period 1995–2004 was performed in studies by Akpınar and Kömürcü [8,9].

Variability in the global climate can lead to changes in the wave climate. Recent studies have shown that a variability of climate is reflected in fluctuations in large-scale atmospheric circulation. These fluctuations are determined by so-called teleconnection patterns, for example, the North Atlantic Oscillation (NAO), the Atlantic Multi-decadal Oscillation (AMO), the Pacific Decadal Oscillation (PDO), the Arctic Oscillation (AO), and the East Atlantic-Western Russia pattern (EA/WR). Changes of teleconnection patterns can be described by corresponding climatic indices. Many studies have demonstrated the influence of these climatic indices on the variability of regional storminess, including the region of Black Sea [10–17]. For example, periods of wave climate variability for a 4–10 period were associated with fluctuations of the North Atlantic Oscillation, and a decrease in storms in the 2000s as associated with the low-frequency variability of the East Atlantic-Western Russia teleconnection pattern [11,13]. Long-term fluctuations (40–50 years) of maximum annual wave heights were identified in Polonsky et al. [12], and it was shown that fluctuations can be associated with changes of the Atlantic Multi-decadal and Pacific Decadal Oscillations.

Therefore, for wave energy extraction projects, it is necessary to know the regularities and trends of long-term changes of wave climates as well as their connection with global climate change. Notably, as a result of time-consuming computational costs and insufficient accuracy of wind reanalysis on large time scales, wave energy estimations are often made on time scales of 10–35 years and poorly reproduce fluctuations of long-term periods [18].

Meanwhile, as has been shown in Galabov [5], the modelling data obtained using different wind reanalysis may lead to significantly different estimates of wave energy for the same regions of the Black Sea. Accordingly, estimations for the average annual wind power flux for the entire Black Sea have varied considerably in recent studies, calculated at 3 kW/m in Akpınar and Kömürcü [9], 6–7 kW/m in Aydoğan et al. [19], and about 3–4 kW/m in Divinsky and Kosyan [7]. The maximum decadal average wave power of the Black Sea for 110 years (1900–2010) fluctuated from 2.7 kW/m to 3.3 kW/m during the period 1941–1950 and then decreased to 2.7 kW/m in the 2000s. Notably, wave energy fluctuations can be connected with variability in both the AMO and PDO indices [5].

In Divinsky and Kosyan [7], on the basis of modelling data analyses, the western part of the Black Sea was determined to be more energetic, especially the southwest, where the maximum wave power energy flux can reach 8–10 kW/m; in contrast, the eastern part is less energetic, about 2–4 kW/m. At the same time, for the northwestern Black Sea, average annual wave power energy flux was estimated to be about 4 kW/m [20]. The average annual wave power energy flux estimates made by Galabov [5] for the southwestern part of the Black Sea did not exceed 3 kW/m and fluctuated from 1.5 to 4 for the period 1900–2010.

Obviously, both the accuracy of the wave models and wind reanalysis data and the length of the time interval over which such estimates are made should influence wave energy estimates, especially if there is long-term climate variability of the wave data. To avoid the inaccuracies mentioned above, it would be better to use the data of field observations and to know the main periods of climatic variability of the Black Sea. However, detailed analysis of periods of variability of the wave climate for the Black Sea have not been conducted. Therefore, the main goal of this work is to analyze the wave climate variability of the Black Sea on the basis of long-term field observations data, to identify the main periods of climatic variability, and to make estimates of the possible variability of wave energy in connection with climate change.

2. Data and Methods of Analysis

A significant problem in analysing the variability of the wave climate, especially those of low-frequency, is the wave data of field observations. Wave data are available in a very limited volume, which usually belongs to national hydrometeorological services. The observation stations are located on the shore; therefore, the wave heights are recorded predominantly in the coastal zone of the sea. Concurrently, there are no data of continuous long-term (more than 20 years), full-scale field instrumental observations of the waves. Therefore, for field wave data, we will use the database from the Voluntary Observing Ship Program (VOS). More details about these observations are available at [21]. The database contains meteorological observations, including the visual registration of heights of waves and wave periods carried out by ships on a voluntary basis during sea and ocean routes. On the base of limited collections of VOS data the widly used wave statistics atlas for marine officers and naval engineers were developed [22].

One potential weakness in visual wave observations is that they are less reliable than, for example, satellite or model data because of their low accuracy due to human factor, space-time inhomogeneity, as well as some difficulties with preprocessing and bias corrections. Despite the quantitative discrepancies in estimations of wave heights which have been noted by many researchers (up to 40%), when these estimations are compared with model data and buoys measurements [23–25], these discrepancies largely relate to comparisons of monthly observations. When comparing mean annual data, accuracy is significantly increased, and, as noted in Gulev et al. [25], for the majority of locations, observational uncertainties are within 20% of mean values, which is sufficient for climatological studies. Notably, the accuracy of comparing VOS data with other instrumental measurements depends on the number of measuring stations in a given region: the more measuring stations, the less the discrepancy in the estimation of wave heights. Accuracy also depends on the preprocessing of VOS data [26]. The precision of modern measurements of wave heights and periods is comparable to visual observations, as shown in Grigorieva et al. [27,28] and summarized in Table 1, which is provided by Grigorieva and compiled on the basis of her analysis of wave data of National Data Buoy Center (https://www.ndbc.noaa.gov) and altimetry data of GlobWave project (http://www.globwave.org). As seen in Table 1, VOS data are comparable with the data of direct instrumental measurements and present several advantages. Moreover, in the VOS data, the parameters of wind driven sea waves and swell are recorded separately.

Table 1. Comparisons of resolution of VOS, altimetry, and buoys data (data provided by Grigorieva).

Parameter	VOS	Altimetry	Buoys Data
Wave height	0.5 m	0.4 m	0.2 m
Wave period	1 s	–	≥1 s
Wind speed	1 m/s	1.5 m/s	1 m/s
Direction	10°	17–20°	10°

On the basis of the VOS data in the Shirshov Institute of Oceanology of Russian Academy of Sciences, the Wave Atlas was developed [29]. Because this relies on visual observation data, it is necessary to control a quality of records for the detection of "bugs" and other potential inaccuracies. For the Wave Atlas, the following preprocessing of VOS data was made: artificial errors correction or elimination, correction of small wave heights, examination of extreme wave heights, and inconsistency of wave parameters, etc. Black Sea VOS data from the Wave Atlas for wave heights are available for period 1948–2011. However, records of wave heights and periods are only available from 1960. Accordingly, the wave data of the wind-driven sea with the wave steepness (ratio of height to wavelength) less than 0.1 for the period 1960–2011 years were selected for analysis, which resulted in a total of 85,061 records.

The distribution of observation points along the Black Sea, the density of observation data for different regions, the total number of observations per year, and the dependence of the average annual wave height on the number of observations are shown in Figure 1. The density of observation data for the period 1960–2011 for the five main regions of the sea includes: region 1—18,364 observations, region 2—11,762, region 3—31,132, region 4—15,742, and region 5—7876. Clearly, in recent years, the number of observations has decreased (Figure 1b), but the average annual wave height is practically independent from the number of observations (Figure 1c).

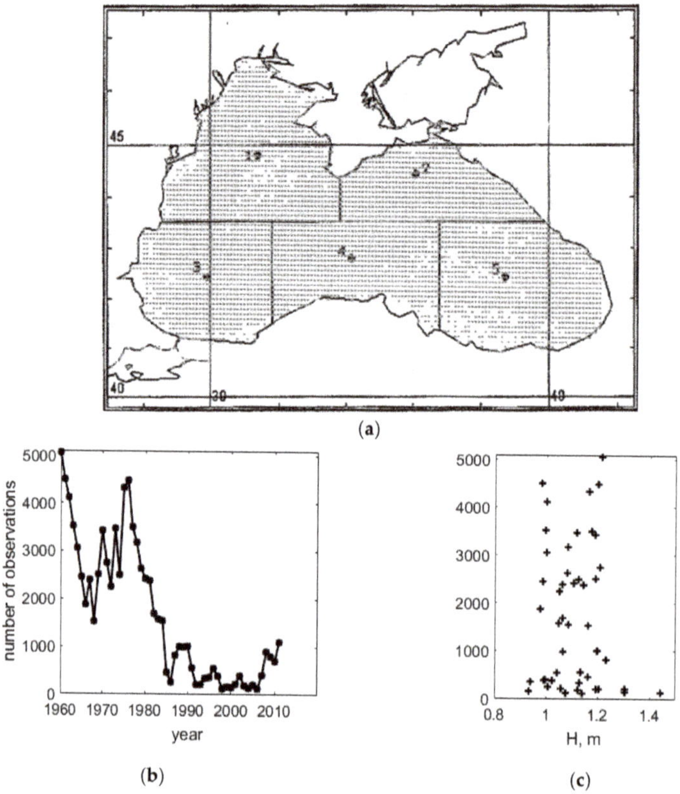

Figure 1. Distribution of observation points for period 1960–2011 along the Black Sea (**a**), the total number of observations per year (**b**), and the dependence of the average annual wave height on the number of observations (**c**).

The impact of global climate variability on the wave climate and wave energy will be estimated on the basis of an analysis of the relationships of the climatic indices (NAO, AO, AMO, PDO, EA/WR) with heights, periods, and energy of average annual waves. The values of dimensionless climatic indices, calculated by certain methods, were taken from the website of the National Oceanic and Atmospheric Administration of the USA [30].

Because the time series of analysed data is rather short (only 52 terms), to reveal the periodicity, the method of continuous wavelet transform with the Morlet wavelet function was used instead of a classical spectral analysis [31]. Wavelet transform is a type of scan of the investigated time series by frequencies that allows us to analyse the structure of non-stationary processes in time and to establish periodicity even in data containing an incomplete period of change. For a mutual correlation analysis, the method of wavelet-correlation developed earlier by the authors was applied instead of a classical correlation analysis [15,18]. Wavelet-correlations are the construction of a correlation function between the wavelet transforms of two signals and between the same wavelet-frequency bands. It is an analog of the correlation analysis, for which the original signals are preliminarily filtered on a multitude of narrow-band signals of characteristic frequency bands; subsequently, the correlations between these narrow-band signals are analyzed. If the number of such narrow-band signals is sufficiently large, then each such narrow-band signal can be considered as a quasi-stationary signal that enables an analysis of the correlation relations between two nonstationary processes.

For a detailed analysis of the structure of variability processes, the spavlet analysis (analysis of the spectra of modules of wavelet coefficients) was used [32]. A spavlet analysis is an analog of spectral analysis of the envelope of the signal. It allows for a determination of amplitude modulations of some frequency scales. To construct the spectra, we used parametric spectral analysis, also known as the Yule-Walker method.

3. Fluctuations of the Average Annual Wave Heights and Periods in Connection with the Changes of Climatic Indices

The changes of average annual wave heights and periods according to VOS data for the Black Sea from 1960 to 2011 are shown in Figure 2.

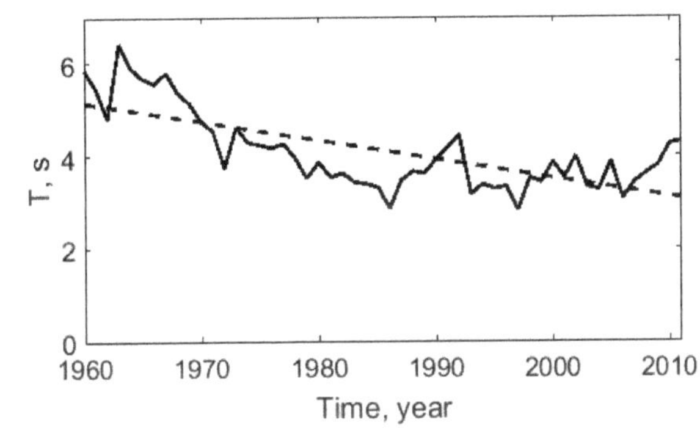

(**a**)

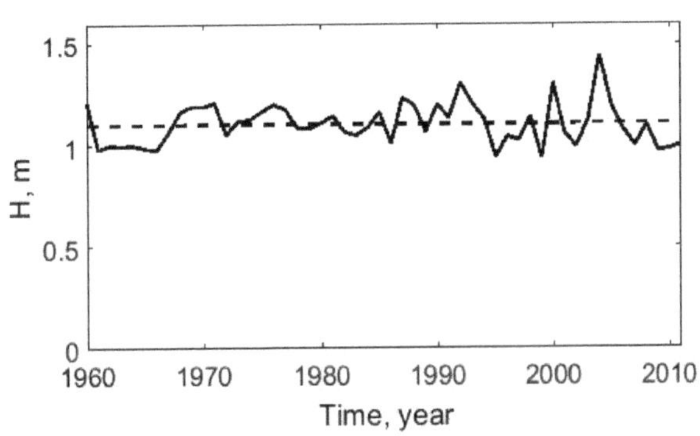

(**b**)

Figure 2. Changes of average annual wave periods (**a**) and heights (**b**), VOS data.

Clearly, there are inter-annual fluctuations of the average annual wave heights and periods. For wave heights, this variability increases in amplitude over the considered time period. However, the average annual wave height remains practically constant, about 1.1 m. For the period 1960–2011, the largest value of average annual waves is 1.4 m, with the smallest at 0.93 m. The average annual wave period decreases 0.04 s·year^{-1}; the longest period is 6.4 s, with the shortest period at 3 s, and the average annual wave period for the considered 51 years is 4.1 s. It can be assumed that the change of the average annual wave period has a low-frequency periodicity: a sharp decrease was observed beginning in the 1960s and then a visible increase, starting approximately in 2005.

As wavelet analysis has shown, for the data of average annual wave periods, there is a periodic change on frequency scales of 0.015–0.02 year^{-1}. Additionally, changes of the average annual period in the frequency range 0.05–0.13 year^{-1} are non-stationary. For example, the characteristic frequency scale of 0.07 year^{-1} increases with time to 0.1 year^{-1} (Figure 3a).

The changes of the average annual wave height are also non-stationary. For example, the characteristic frequency scale 0.05 year^{-1} has an increasing trend to 0.08 year^{-1}. For frequency scales in the range 0.1–0.16 year^{-1}, beginning in 1985, the frequency scale 0.1 year^{-1} increases to 0.16 year^{-1}; after 1985, there are no trends and the frequency 0.16 year^{-1} is stable (Figure 3b).

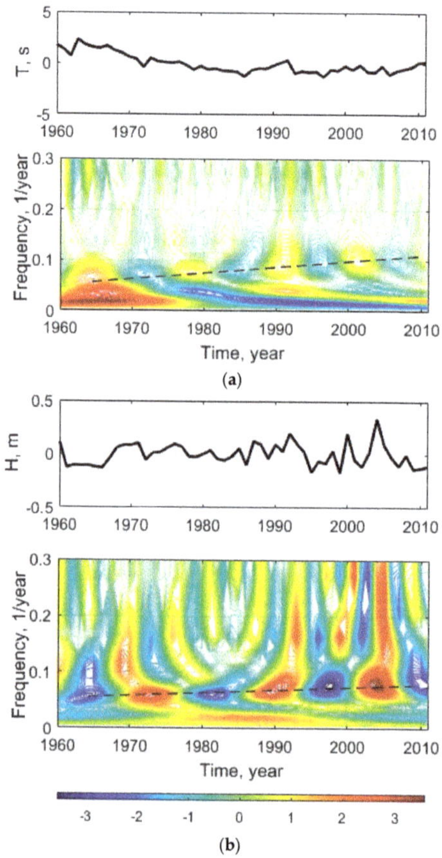

Figure 3. Wavelet transforms of the average annual wave periods (**a**) and heights (**b**). For wavelet analysis, the mean values of the time series were deleted.

All data chosen for analysis climatic indices also changed to non-stationary. Figure 4 shows the wavelet transforms of all indices. According to the wavelet analysis, low-frequency oscillations of all indices can be identified. For AMO, the characteristic low frequencies are 0.022 and 0.05 year^{-1}, for NAO—0.022 and 0.06 year^{-1}, for PDO—0.025 year^{-1}, for AO—0.026 and 0.06 year^{-1}, and for EA-WR—0.035 and 0.1 year^{-1}. Wavelet analysis shows that, for all climatic indices, there are different trends of frequencies changes in the range 0.09–0.13 and 0.2–0.3 year^{-1}.

A visual comparison of the wavelet diagrams shows the similarity between the wavelet transforms of changes of NAO and the average annual wave heights, especially in the frequency range less than 0.16 year^{-1}. Wavelet diagram of the changes of the average period are similar to wavelet diagrams of the changes seen in PDO and AMO. This similarity indicates the occurrence of an identical long-term (decadal and multi-decadal) periodicity of changes of average annual wave heights, periods, and corresponding climatic indices.

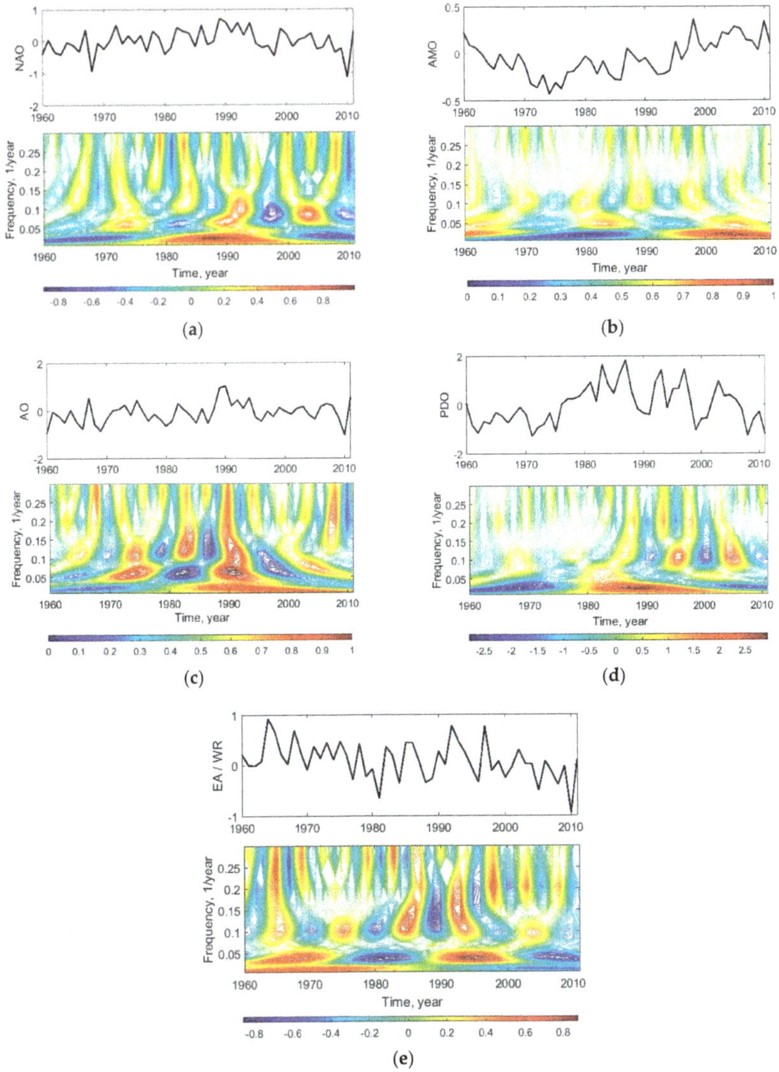

Figure 4. Wavelet transforms of NAO (**a**), AMO (**b**), AO (**c**), PDO (**d**) and EA/WR (**e**). For wavelet analysis, the mean values of the time series were deleted.

A detailed analysis of the structure of non-stationary processes of changes to average annual wave heights, periods, and climatic indices to identify a connection between them can conducted by a spavlet analysis. Spavlet is the set of spectra of the wavelet coefficient modules of each frequency scale of the wavelet transform [32]. The low-frequency components of the wavelet coefficients modules are an analog of the envelope of a narrow frequency band signal centered on the wavelet frequency of a given scale. The spectra were calculated by the Yule-Walker method of parametric spectral analysis for the best estimation of low-frequency components and period that are comparable with the duration of observations.

The spavlet analysis clearly shows the presence of low-frequency modulation, with a frequency of 0.016 year^{-1} and wavelet frequency scales of 0.05, 0.07–0.08, and 0.2–0.25 year^{-1} for average annual wave heights (Figure 5a). As seen in Figure 5a, the figure appears as a ridge parallel to the wavelet frequencies axis at spavlet's frequency of 0.016 year^{-1}. As seen in Figure 5, the ridge on the spavlet diagram at spavlet's frequency equal to the double wavelet's frequency is not significant because it was produced by the doubling of wavelets coefficients frequencies at the creation of its modules. The average annual period of waves is modulated by the same low frequency but only on a wavelet frequency of 0.03 year^{-1} (Figure 5b). On the basis of the structure of the spavlet, one can approximate functions for the process of periodicity of changes of average annual wave heights in the form:

$$F_H(t) = A_1\cos(0.016t) + B_1\cos(0.016t)\cos(0.05t) + C_1\cos(0.016t)\cos(0.075t) + D_1\cos(0.016t)\cos(0.22t), \quad (1)$$

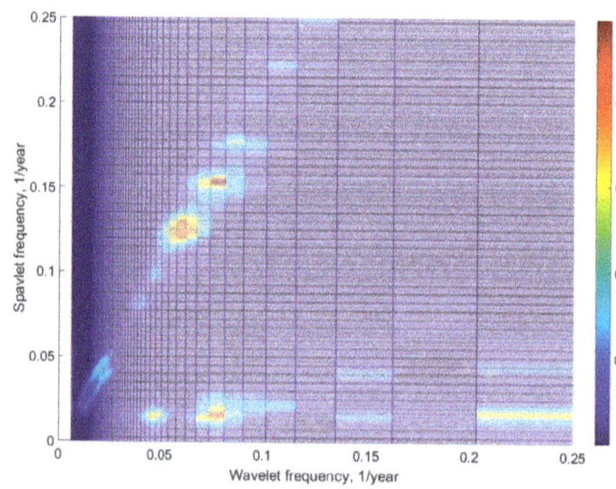

(**a**)

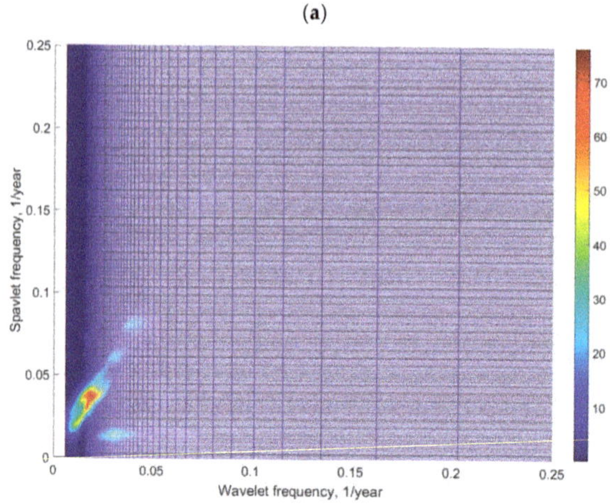

(**b**)

Figure 5. Spavlets of the average annual wave heights (**a**) and periods (**b**).

And for the average annual wave periods as:

$$F_T(t) = A_2\cos(0.016t) + B_2\cos(0.016t)\cos(0.03t), \tag{2}$$

where coefficients A, B, C, and D should be determined empirically.

A spavlet analysis demonstrated that all climatic indices also have low-frequency modulations of wavelet frequencies in the range of 0.015–0.035 year^{-1}. The fluctuations of average annual waves and the NAO index have the similar spavlet structures: the low-frequency modulation at a frequency of 0.016 year^{-1} of the main wavelet frequency scales 0.075 year^{-1} (Figures 5a and 6a). In general, the spavlet structure of the fluctuations of the average annual wave period is similar to the structure of the PDO index: the low-frequency modulation by the frequency of 0.016 year^{-1} of the main wavelet frequency scales 0.03 year^{-1} (Figures 5b and 6b).

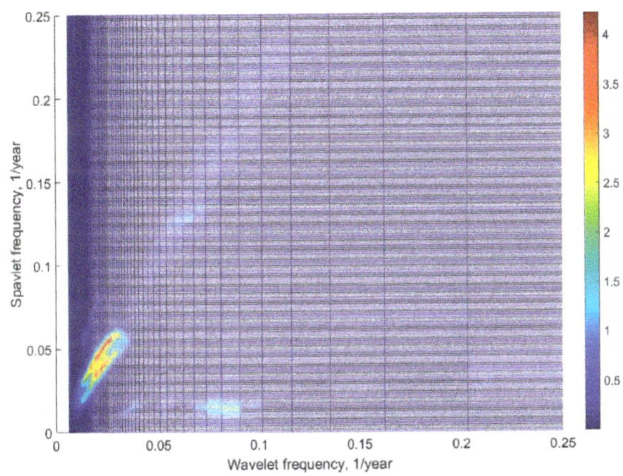

(**a**)

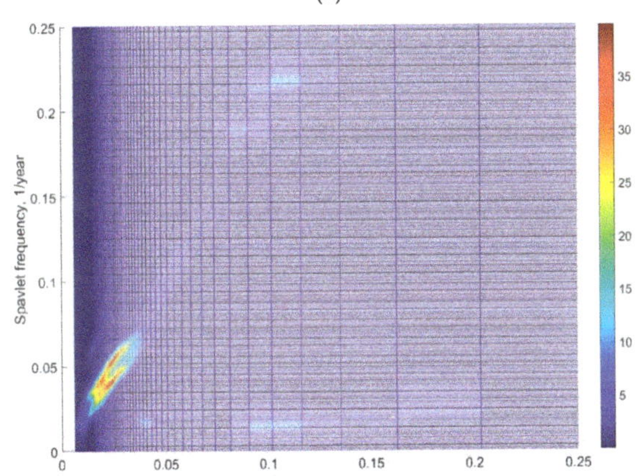

(**b**)

Figure 6. Spavlets of NAO (**a**) and PDO (**b**).

Because the changes in the wave heights, periods, and climatic indices are non-stationary, the classical correlation analysis to identify the correlation between them is inapplicable. Therefore, we propose to apply the wavelet-correlations method we developed in previous research [15,18], in which a correlation function between wavelet frequencies scales of wavelet transforms of two signals is constructed. In Figure 7, the frequency dependencies of the correlation coefficients between climatic indices and average annual wave heights and periods at zero time lags are presented.

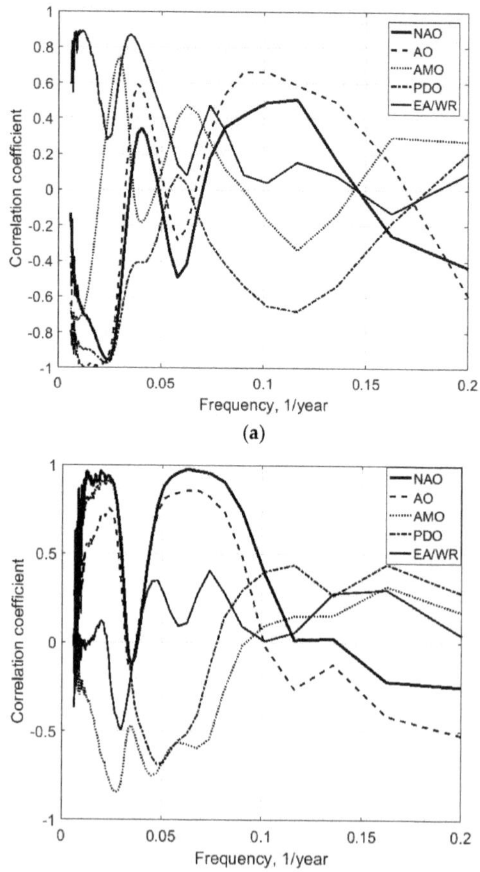

Figure 7. Coefficients of wavelet-correlation functions between climatic indices and average annual wave periods (**a**) and heights (**b**) at zero time lags.

Clearly, the changes of average annual wave heights and periods correlate well with the fluctuations of climatic indices, especially in the low-frequency range (multi-decadal and decadal periods). If we recalculate the frequencies in periods, then it is possible to identify subsequent periods of change to the average annual waves parameters in connection with the changes in climatic indices. The fluctuations of average annual wave heights are correlated with (1) NAO for periods of 40 and approximately 12–15 years; (2) AMO for periods of 30–35 years, 22, and 15 years; (3) PDO for periods of 40, 20, 8, and 6 years; (4) AO for periods of 40, 15, and 5 years; (5) EA/WR for periods of 30 and 13 years. The fluctuations of average annual wave periods are correlated with (1) NAO for periods of 40 and 10–12 years; (2) AMO for periods of 30–35 and 15 years; (3) PDO for periods of 40, 20, and 8 years; (4) AO for periods of 40, 25, 10, and 5 years; (5) EA/WR for periods of 30 and 13 years. In all cases, the values of the modulus of the correlation coefficient are more than 0.5.

According to the sign of correlation coefficients, the low-frequency fluctuations (periods more than 30 years) for the average annual wave heights occur in phase with NAO, PDO, and AO and in anti-phase with the AMO. Simultaneously, for the average annual wave periods, these fluctuations are the reverse: in phase with AMO and in anti-phase with NAO, PDO, and AO.

The largest correlation coefficients of the fluctuations of the average annual wave periods are those with changes in the PDO index and of the average annual wave heights, i.e., with the NAO index, which corresponds to the similarity of the structures of non-stationary processes revealed via a spavlet analysis.

Thus, long-term variability for both the average annual wave heights and periods can be associated with changes in PDO, AO, and NAO indices (for a 40-year period) and with AMO and EA/WR indices (i.e., periods of 30–35 years). Concurrently, there are periods of variability associated with AMO (15–16 years), PDO (8 years), and AO (5 years) indices. In the positive phase of the 40-year period of changes of the NAO, AO, and PDO, the average annual wave period decreases, but the average annual height increases. At the positive 30–35-year phase of AMO and EA/WR, the average annual wave period increases and the average annual wave height decreases. The average annual wave height also decreases in the positive phase of the 20-year period of change within PDO and AMO. The average annual wave height increases in the positive phase of NAO and AO changes during a period of 15 years. The average annual wave period will also increase in this positive phase of changes in the same indices but with a period of 10 years.

4. Fluctuations of the Power of Wave Energy Flux in the Black Sea

The power of the wave energy flux can be calculated by formula [8]:

$$E = \frac{\rho g^2 H_s^2}{64\pi} T_e, \tag{3}$$

where H_s is the significant wave height, $T_e = 0.9T_p$, T_p is the peak period of wave spectrum.

For the Black Sea, $\rho = 1015\ \text{kg/m}^3$ and Formula (3) can be simplified to

$$E = 0.486\ H_s^2\ T_e\ \text{kW/m}. \tag{4}$$

Instead of taking the significant wave height and peak period, we took the average annual wave height and period. The fluctuations of the power flux are shown in Figure 8.

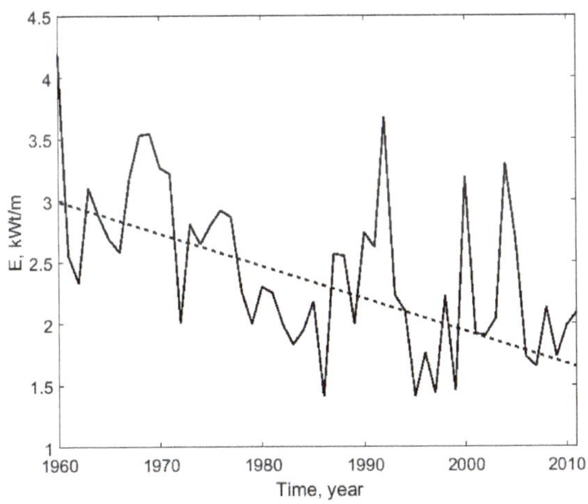

Figure 8. Change of the power of wave energy flux.

Overall, the power of the wave energy flux of the Black Sea is not high and does not exceed 4.2 kW/m. Similar estimates were obtained in Galabov [5], Divinsky and Kosyan [7], Akpınar and Kömürcü [8,9] and Rusu [6] according to numerical modeling. The average power of wave energy flux is 2.43 kW/m, the minimum value is 1.4 kW/m.

If we compare the changes in the average annual wave heights and periods (Figure 2), the changes of the power flux of energy strongly depend on the wave height and wave period. In general, from 1960 to 2011, there is a slight decrease in the power flux of wave energy (0.022 kW/m per year), which can be associated with a decrease of the average annual wave period. The observed increase of the oscillations of the amplitude of the power flux of energy beginning in 1985 is associated with the corresponding fluctuations of the average annual wave heights (Figure 2).

As can be seen from Formula (3), the energy nonlinearly depends on the wave height and linearly on the wave period. A spavlet analysis revealed that the changes of the power flux of wave energy at frequencies 0.03 and 0.075 are modulated by a low frequency of 0.016 year^{-1} (Figure 9).

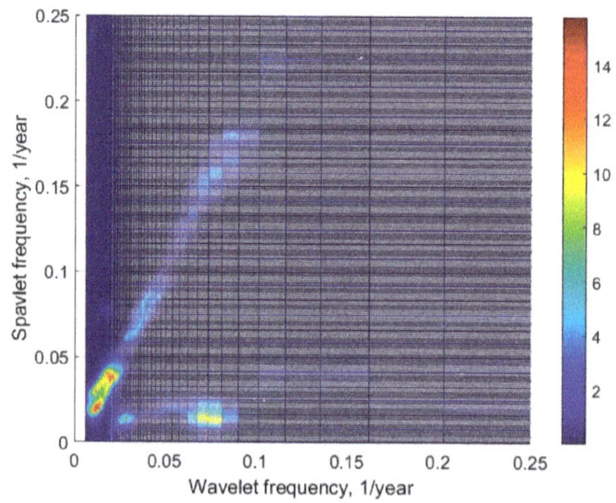

Figure 9. Spavlet of the power flux of wave energy.

From the structure of the spavlets, the fluctuations of power flux of wave energy can be approximately described by a function:

$$F_E\ (t) = A_3\cos(0.016t) + B_3\cos(0.016t)\cos(0.03t) + C_3\cos(0.016t)\cos(0.075t), \quad (5)$$

where coefficients A, B, and C should be determined empirically.

The first term of right part of Equation (5) is similar to Equations (1) and (2). The second term of Equation (5) is determined by the change in the average annual period of waves (Equation (2)), and the third by the change of average annual wave heights (Equation (1)). Note that the frequency 0.016 year^{-1} is also present in the spavlets of the average annual heights, periods, and of the power flux of wave energy; it is also present in spavlets of all considered climatic indices as the fundamental frequency and as the modulating frequency. It is possible that the fluctuations of this period reflect the trends of global climate change.

In Figure 10, the coefficients of wavelet-correlation functions between the power flux of wave energy and climatic indices at zero time lags are shown.

Clearly, long–term changes of the power flux for the periods of 50–60 years is associated with the change in EA/WR and AMO climatic indices: for the 40-year period with the indices of PDO, AO and NAO; for the 30–35-year period with EA/WR and AMO indices; for the 20–25-year periods with EA/WR, AO, and PDO indices. In addition, the variability of the NAO index will influence the changes of the power flux of wave energy with a periodicity about 15 years: for the variability of EA/WR index, a periodicity of about 13 years; for the variability of AO index, a periodicity of about 15 and 5 years. Thus, the short-term variability of the power flux of wave energy is associated with changes to NAO, EA/WR, and AO climatic indices.

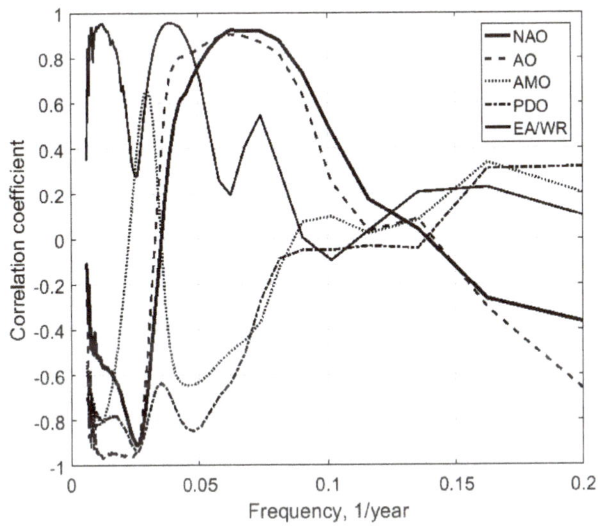

Figure 10. Coefficients of wavelet-correlation functions between the climatic indices and the power flux of wave energy at zero time lags.

Thus, the expected decrease in power flux of wave energy for a periodicity of 50–60 years is associated with the positive phase of AMO, whereas a periodicity of 40 years is associated with the positive phase of NAO, PDO, AO, and the negative phase of AMO. The increase in power flux of wave energy will occur in the positive phases of all periods of the EA/WR change with a frequency of 50–60, 20–25, and 13 years, as well as with a periodicity of 15 years in association with the positive phase of the NAO and AO for this period of variability. After a comparison of the changes in climatic indices and wave energy (Figures 4 and 8) and also on the basis of the wavelet correlation analysis, it can be assumed that a significant decrease in power flux of wave energy in the mid-1980s is associated with a positive phase of PDO and NAO for long-term periods, and the largest increase in the early 1990s is associated with a positive phase of short-period changes of NAO, AO, and EA/WR.

5. Conclusions

As a result of an analysis, which used Voluntary Observing Ship Program wave data, it was found that the average power flux of the wave energy in the Black Sea during the period 1960–2011 was 2.43 kW/m, and the highest value was 4.2 kW/m. These values correspond to the estimations made earlier on the basis of modelling data [5–9].

The fluctuations of the power flux of wave energy are connected with changes of the wave climate, which depend on changes in the global climate and are determined by fluctuations of the teleconnection patterns such as the North Atlantic Oscillation (NAO), the Atlantic Multi-decadal Oscillation (AMO), the Pacific Decadal Oscillation (PDO), and the East Atlantic/West Russia (EA/WR). It has been shown that the changes in the average annual wave heights are primarily affected by NAO and, for the average annual wave periods, by PDO.

On the basis of a spavlet analysis, the approximation functions of the changes of the average annual wave heights, periods, and the power flux of wave energy were proposed.

The wavelet-correlations testified to a significant relationship between the changes of the power flux of wave energy and climatic indices, such as NAO, PDO, AO, AMO, and EA/WR. The main long-term periods of the fluctuations are 40, 30–35, and 20–25 years. The short-term periods of the changes of the power flux of wave energy of approximately 13–15 and 5 years are related to oscillations of NAO, EA/WR, and AO climatic indices.

It has been shown that the positive phase of long-term changes in NAO and AO (40 years) leads to a decrease of power flux of wave energy; the positive phase of short-period (15 years) fluctuation of these indices leads to an increase of power flux of energy. The positive phase of changes of the EA/WR for all periods leads to an increase of power flux of wave energy. Concurrently, the decrease of power flux of wave energy correlates well with the positive phase of long-term changes in PDO (40 years) and AMO (50–60 years).

The obtained results show that, in the implementation of any wave energy extraction projects, it is necessary to make an analysis of estimates of the possible energy potential, taking into account long-term climate variability.

Author Contributions: Conceptualization, data analysis, data preparation, main investigation, and writing of paper was done by Y.S.; Spavlet analysis was conducted by S.K.; All authors read, discussed, and approved the manuscript.

Funding: The work was partly supported by the Russian Foundation for Basic Research (grant No. 16-55-76002 ERA_a) in the frame of BS STEMA project.

Acknowledgments: The work was carried out within the framework of state task (theme No. 0149-2018-0015). The authors are grateful to Vika Grigorieva for personal communications and who kindly provided materials about the accuracy of the VOS data.

Conflicts of Interest: The authors declare no conflict of interest.

References

1. Rusu, E.; Onea, F. A review of the technologies for wave energy extraction. *Clean Energy* **2018**, *2*, 10–19. [CrossRef]
2. Gunn, K.; Stock-Williams, C. Quantifying the global wave power resource. *Renew. Energy* **2012**, *44*, 296–304. [CrossRef]
3. Atan, R.; Goggins, J.; Nash, S. A detailed assessment of the wave energy resource at the Atlantic Marine Energy Test Site. *Energies* **2016**, *9*, 967. [CrossRef]
4. Guedes Soares, C.; Bento, A.R.; Goncalves, M.; Silva, D.; Martinho, P. Numerical evaluation of the wave energy resource along the Atlantic European coast. *Computers Geosci.* **2014**, *71*, 37–49. [CrossRef]
5. Galabov, V. The Black Sea Wave Energy: The Present State and the Twentieth Century Changes. Available online: https://arxiv.org/pdf/1507.01187.pdf (accessed on 21 May 2018).
6. Rusu, L. Assessment of the wave energy in the Black Sea based on a 15-year hindcast with data assimilation. *Energies* **2015**, *8*, 10370–10388. [CrossRef]
7. Divinsky, B.V.; Kosyan, R.D. Spatiotemporal variability of the Black Sea wave climate in the last 37 years. *Cont. Shelf Res.* **2017**, *136*, 1–19. [CrossRef]
8. Akpınar, A.; Kömürcü, M.İ. Wave energy potential along the south-east coasts of the Black Sea. *Energy* **2012**, *42*, 289–302. [CrossRef]
9. Akpınar, A.; Kömürcü, M. Assessment of wave energy resource of the Black Sea based on 15-year numerical hindcast data. *Appl. Energy* **2012**, *101*, 502–512. [CrossRef]
10. Bromirski, P.D.; Cayan, D.R. Wave power variability and trends across the North Atlantic influenced by decadal climate patterns. *J. Geophys. Res. Oceans* **2015**, *120*, 3419–3443. [CrossRef]
11. Vespremeanu-Stroe, A.; Tatui, F. The influence of North Atlantic Oscillation on Romanian Black Sea coast wind regime. *Analele Universitatii din Bucuresti Seria Geografie* **2005**, *64*, 17–26.

12. Polonsky, A.; Evstigneev, V.; Naumova, V.; Voskresenskaya, E. Low-frequency variability of storms in the northern Black Sea and associated processes in the ocean-atmosphere system. *Reg. Environ. Change* **2014**, *14*, 1861–1871. [CrossRef]
13. Zainescu, F.; Tatui, F.; Valchev, N.; Vespremeanu-Stroe, A. Storm climate on the Danube delta coast: evidence of recent storminess change and links with large-scale teleconnection patterns. *Nat. Hazards* **2017**, *87*, 599–621. [CrossRef]
14. Masselink, G.; Austin, M.; Scott, T.; Poate, T.; Russell, P. Role of wave forcing, storms and NAO in outer bar dynamics on a high-energy, micro-tidal beach. *Geomorphology* **2014**, *226*, 76–93. [CrossRef]
15. Saprykina, Y.V.; Kuznetsov, S.Y. Methods of analysis of nonstationary variability of wave climate of Black Sea. *Phys. Oceanogr.* **2018**, *4*, 156–164.
16. Rusu, L.; Bernardino, M.; Guedes Soares, C. Wind and wave modelling in the Black Sea. *J. Op. Oceanogr.* **2014**, *7*, 5–20. [CrossRef]
17. Cherneva, Z.; Guedes Soares, C.; Andreeva, N. Wind and wave climate over the Black Sea. In *Maritime Technology and Engineering*, 2nd ed.; Guedes Soares, C., Santos, T.A., Eds.; Taylor & Francis Group: London, UK, 2015; pp. 1309–1316.
18. Shtremel, M.; Saprykina, Y.; Kuznetsov, S.; Ayat Aydoğan, B.; Aydoğan, B. Wave climate of the Black Sea: Visual Observations and Modelling Data. In Proceedings of the 13th International MEDCOAST Congress on Coastal and Marine Sciences, Engineering, Management & Conservation (MEDCOAST 17), Mellieha, Malta, 30 October–4 November 2017.
19. Aydoğan, B.; Ayat Aydoğan, B.; Yüksel, Y. Black Sea wave energy atlas from 13 years hindcasted wave data. *Renew. Energy* **2013**, *57*, 436–447. [CrossRef]
20. Silva, D.; Rusu, E.; Guedes Soares, C. Evaluation of various technologies for wave energy conversion in the Portuguese nearshore. *Energies* **2013**, *6*, 1344–1364. [CrossRef]
21. World Wide Voluntary Observing Ship Program. Available online: https://www.vos.noaa.gov/ (accessed on 30 July 2018).
22. Hogben, N.; Da Cunha, L.F.; Oliver, H.N. *Global Wave Statistics*; Unwin Brothers Limited: London, UK, 1986.
23. Soares, C.G. Assessment of the uncertainty in visual observations of wave height. *Ocean Eng.* **1986**, *13*, 37–56. [CrossRef]
24. Vettor, R.; Guedes Soares, C. Rough weather avoidance effect on the wave climate experienced by oceangoing vessels. *Appl. Ocean Res.* **2016**, *59*, 606–6015. [CrossRef]
25. Gulev, S.K.; Grigorieva, V.; Sterl, A.; Woolf, D. Assessment for the reliability of wave observations from voluntary observing ships: insights from the validation of a global wind wave climatology based on voluntary observing ship data. *J. Geophys. Res.* **2003**, *108*, 3236. [CrossRef]
26. Grigorieva, V.; (Shirshov Institute of Oceanology, Moscow, Russia). Personal communication, 2018.
27. Grigorieva, V.G.; Gulev, S.K.; Gavrikov, A.V. Global historical archive of wind waves based on voluntary observing ship data. *Oceanology* **2017**, *57*, 229–231. [CrossRef]
28. Grigorieva, V.G.; Badulin, S.I. Wind wave characteristics based on visual Observations and satellite altimetry. *Oceanology* **2016**, *56*, 19–24. [CrossRef]
29. Gulev, S.K.; Grigorieva, V.; Sterl, A. World Ocean Waves (WOW). Available online: http://www.sail.msk.ru/wow/ (accessed on 11 July 2018).
30. National Oceanic & Atmospheric Administration. Available online: https://www.esrl.noaa.gov/psd/data/climateindices/list/ (accessed on 30 July 2018).
31. Torrence, C.; Compo, G.P. A practical guide to wavelet analysis. *Bull. Am. Meteorol. Soc.* **1998**, *79*, 61–78. [CrossRef]
32. Petenko, I.V. Advanced combination of spectral and wavelet analysis ("spavlet analysis"). *Bound.-Layer Meteorol.* **2001**, *100*, 287–299. [CrossRef]

© 2018 by the authors. Licensee MDPI, Basel, Switzerland. This article is an open access article distributed under the terms and conditions of the Creative Commons Attribution (CC BY) license (http://creativecommons.org/licenses/by/4.0/).

energies

MDPI

Article

Modelling Offshore Wave farms for Coastal Process Impact Assessment: Waves, Beach Morphology, and Water Users

Christopher Stokes * and Daniel C. Conley

Plymouth University, School of Biological and Marine Sciences, Drake Circus PL4 8AA, UK; Daniel.Conley@plymouth.ac.uk

* Correspondence: Christopher.stokes@plymouth.ac.uk; Tel.: +44-1752-586177

check for updates

Received: 26 July 2018; Accepted: 19 September 2018; Published: 21 September 2018

Abstract: The emerging global wave energy industry has the potential to contribute to the world's energy needs, but careful consideration of potential impacts to coastal processes in the form of an impact assessment is required for each new wave energy site. Methods for conducting a coastal processes impact assessment for wave energy arrays vary considerably in the scientific literature, particularly with respect to characterising the energy absorption of a wave energy converter (WEC) array in a wave model. In this paper, modelling methods used in the scientific literature to study wave farm impacts on coastal processes are reviewed, with the aim of determining modelling guidance for impact assessments. Effects on wave climate, beach morphology, and the surfing resource for coastal water users are considered. A novel parameterisation for the WEC array transmission coefficient is presented that, for the first time, uses the permitted power rating of the wave farm, which is usually well defined at the impact assessment stage, to estimate the maximum likely absorption of a permitted WEC array. A coastal processes impact assessment case study from a wave farm in south-west Ireland is used to illustrate the application of the reviewed methods, and demonstrates that using the new 'rated power transmission coefficient' rather than a WEC-derived transmission coefficient or complete energy absorption scenario can make the difference between significant and non-significant levels of coastal impacts being predicted.

Keywords: wave energy converter; transmission coefficient; absorption; surfing amenity; resource; impact assessment

1. Introduction

The extraction of wave energy from the world's oceans and seas has the potential to contribute significantly to the global energy mix. In Europe, the emerging wave energy industry could eventually contribute to European Union (EU) renewable energy targets [1–3], if and when full-scale operational wave energy converter (WEC) arrays are deployed. To accommodate these deployments, suitable marine areas for wave energy capture will need to be sought, and for each new WEC test site or operational WEC array, national and international regulations usually require an environmental impact assessment (EIA) to be performed, in order to demonstrate that the project will not have any unreasonable impacts on the ecology and coastal processes in the surrounding environment [3]. In many cases, potential impacts to water users such as surfers will also have to be considered as part of the coastal processes impact assessment, as such groups have a shared interest in the wave resource, are of economic importance to coastal regions [4], and have raised significant concerns and opposition during previous WEC siting proposals, e.g., [5].

In this contribution, methods for conducting a coastal processes impact assessment for wave energy arrays are discussed, with the aim of determining a set of recommended guidelines for a wave farm coastal processes impact assessment. Effects on wave climate, beach morphology, and the surfing resource are considered, while other effects that are more generic to marine engineering, such as those occurring during the installation and decommissioning of a WEC site are omitted, as they are addressed elsewhere in the literature. In Sections 1.1–1.3, methods for investigating the effects of wave energy extraction on wave climate, beach morphology, and water users are discussed, respectively. A novel parameterisation for the WEC array transmission coefficient is presented in Section 1.1.1. In Section 1.4, a coastal processes impact assessment case study for a WEC test site in south-west Ireland is introduced, while the methods and results from the assessment are described in Sections 2 and 3. Sections 4 and 5 conclude the paper by discussing the applied methods and proposing guidance for coastal processes impact assessments based on the methods reviewed throughout the paper.

1.1. *Effects of Wave Energy Extraction on Wave Climate*

The world's first full-scale, grid-connected offshore WEC was tested at the European Marine Energy Centre (EMEC) in Scotland by Pelamis Wave Power in 2004 (www.emec.org.uk/about-us/wave-clients/pelamis-wave-power). However, due to the hostility of the ocean environment and vast costs involved in trialling a WEC in real seas, there have been few prototype-scale deployments like this globally. As a result, the shadow effects from offshore WECs are yet to be fully researched at prototype scale. Instead, scaled physical models and numerical models have been used far more extensively to determine the likely near-field (close to WEC) and far-field (close to shore) effects, respectively, of offshore wave farms. Numerical modelling of such coastal effects has been undertaken for case studies in England [6–12], Scotland [13], Spain [14–16], Portugal [17–19], Mexico [15], Romania [20–22], and the United States [23,24], as well as for generalised cases with idealised bathymetry [25–27].

To quantify the magnitude and characteristics of wave energy capture, the most common approach has been to model WECs as partially transmitting barriers that allow a portion of wave energy to transmit through them, using a numerical wave model. The proportion can be quantified using a transmission coefficient, K_t, which describes the ratio of wave energy (or in some cases wave height) transmitted through a WEC array over that incident upon it. Unsurprisingly, initial modelling studies demonstrated that the near- and far-field attenuation of waves increases with increasing energy absorption (decreasing K_t), and decreases with increasing distance from the WEC array [6,7,10,12,26,27], while array width has been shown to affect the along-coast extent of the resulting wave shadow [18]. Transmission coefficients used in the literature have been determined in various ways, including by estimating device efficiency [6,8], employing breakwater designs [7], or using physical test data [11,14,17–19], and have varied significantly. For example, when modelling the Wave Hub WEC test site in Cornwall, UK, wave height transmission coefficients considered to be 'realistic' ranged from 0 to 0.95 [6,7], and were applied to individual WECs in one study, while being applied over the whole siting area in the other.

Theoretically, the transmission coefficient of an individual WEC can be precisely quantified using the ratio of the device 'capture width' to the spacing [19,27] or width of the devices [28], known as the capture width ratio. Conceptually, capture width describes the length of wave crest completely absorbed by a WEC, and is in the order of 2–21 m, or 12–37% of the device width for the most common types of WEC [28]. In reality, capture width is likely to vary with wave height and period [19], and the absorbed energy may be spread over a greater length of the wave crest than the capture width indicates. Although capture width ratios have now been defined for a wide range of WEC types [28], there is some disparity in how the derived transmission coefficients are applied in modelling studies, with some applying a coefficient across an entire WEC array area, e.g., [17,18], while others have applied transmission coefficients to the precise location of each individual WEC e.g., [10,19]. As spectral wave models do not yet sufficiently simulate device–device interactions, the former approach arguably offers a more conservative method for assessing the impact of a new WEC array site, especially given that WEC array layout (device spacing and alignment) has not yet been optimised for many WEC technologies.

Another key factor that influences the change in wave height at the coast, and the along-coast extent of the impact, is the directional spread, σ_θ, of the waves transmitting past the WEC site [27,29]. A large σ_θ acts to reduce the impact in the lee of a WEC site, as the energy deficit is diffused over a greater area, and wave energy not affected by the WECs can spread into the shadow zone, regenerating some of the lost wave energy. At the same time, a large σ_θ would result in a greater length of coastline being affected by the wave shadow in some way. Swell waves (which are generally preferred by water-users) typically have a small σ_θ, and under such conditions a WEC array would cause a more concentrated reduction in wave height at the coast than during wind-sea or bimodal sea states, which have a larger σ_θ [8,26,27]. Therefore, appropriate characterisation of sea states in the area of interest, and in particular the directional spread of sea states, is essential to the results of a WEC modelling study.

Wave frequencies containing the bulk of energy are logically the most attractive for wave energy capture, and as such, it is likely that WECs will be tuned to resonate optimally at those frequencies [8,25,26]. The resonant frequency of a WEC is likely to be aligned to the average energy period, T_e, of the incident wave climate during the design of the WEC, so as to maximise energy capture [8,26,27,30,31], while frequencies far removed from T_e are likely to be decreasingly affected. The application of a constant transmission coefficient, i.e., one that is applied equally to all wave frequencies, in a number of previous WEC modelling studies [6,7,10,20,21,32] has ignored the fact that devices will naturally resonate at certain frequencies while being less sensitive to others, as demonstrated by published WEC power matrices [28]. However, frequency-dependent modelling of WECs at Wave Hub [8] suggests that the effects would only be slightly lower than those predicted by a comparable frequency-independent study [6]. Therefore, although frequency-dependent modelling is likely to give a more precise representation of energy extraction, it is arguably more conservative to model WECs using a single frequency-independent transmission coefficient in cases where WEC absorption characteristics are not well defined at the impact assessment stage in order to avoid under-prediction of impacts at some frequencies.

WECs have other physical effects on the transmitted wave field that are comparatively less well studied; namely, diffraction of wave energy into the shadow zone, and the transmission of radiated waves caused by the vertical and horizontal motion of the WECs themselves [29,33]. It is likely that for the majority of sea states these effects will not have a significant impact at the coast, however, as numerical modelling by [29,33] indicates that regeneration of lost wave energy by directional spreading is likely to mitigate the effects of diffraction, and any radiated waves should be low in energy and will disperse rapidly with distance. Additionally, modelling by [34] indicates that at distances greater than 500 m from a WEC array, the effects of device–device interaction can be neglected. Ultimately, diffraction and radiation effects are likely to reduce the impact of a WEC array at the coast [34], but the physics are not yet adequately represented in spectral (i.e., phase-averaged) wave models. Such models are currently the only tools capable of efficiently simulating WEC effects at field scales, and have therefore been used widely for this purpose, but the physics of diffraction and wave reflection need to be more accurately parameterised before they will be able to simulate the physics of WEC arrays comprehensively.

Because of the importance of the directional and frequency characteristics of the background wave climate on the potential far-field effects of a WEC array, characterisation of the wave climate is an important step in a WEC impact assessment. The importance of understanding the most common spectral shapes at a potential WEC site have been illustrated by [8], who showed that they can directly affect the degree of wave shadowing predicted (if frequency dependent modelling is applied), especially in the presence of bimodal spectra. Recent work by [35–37] has sought to further characterise wave spectra at a given location for the purposes of wave energy resource assessment, including during bimodal wave conditions. Previous WEC modelling at Wave Hub [6] was critiqued [27] for applying default wave directional spreading values that were not representative of the site under study, or of wave conditions of interest to surfers. Methods for quantifying the directional spread of sea states are discussed by [8], who found that spreading values of around 45° were common at Wave Hub; however, narrow-banded wave conditions of interest to surfers may be less than half of this value, and would be subject to greater wave shadowing in the lee of a WEC array. This demonstrates that although the background wave climate is an important consideration, wave impacts also need to be modelled during conditions that are of importance to specific impact receptors, and that characterisation of the wave climate should include the definition of such conditions.

1.1.1. A New Parameterisation for Wave Energy Converter (WEC) Site Transmission Coefficient, $K_{t,RP}$

A common deficiency of previous WEC impact studies is a lack of consideration for the maximum permitted energy capture of the WEC array. At the impact assessment stage, the permitted array power (i.e., the maximum amount of energy that is allowed to be captured) is usually well defined, and modelling studies therefore need not exceed this level of absorption. Equally, the most conservative approach to modelling a WEC array is to model the maximum amount of energy absorption possible at a site. Therefore, the maximum permitted array power should clearly inform the level of energy absorption, and hence the value of K_T, applied in a WEC modelling study.

To achieve this, one must determine the total amount of power removed from the wave field due to the presence of the WEC array, P_{rem}, based on the array's expected power rating. Assuming the efficiency of the array is equivalent to the device efficiency, γ, this is:

$$P_{rem} = P_a/\gamma \tag{1}$$

where P_a is the permitted power rating of the wave farm (in Watts), which is equivalent to the product of P_{rem} and γ as it is assumed that some energy will be removed from the wave field by the array that cannot be harnessed due to mechanical and electrical inefficiencies represented by γ. In addition, the total wave power transiting the site during design wave conditions, P_S, needs to be determined. Assuming a normally incident wave passing through the site, this is simply:

$$P_S = P_w X \tag{2}$$

where P_w is the power density of the design wave conditions, and X is the length of the permitted array area, as measured parallel to the incident wave crests. The exact array layout is immaterial, as the important considerations are the total power removed by the array and the length of wave crest from which the power is removed; these are the same whether the devices in an array are deployed across multiple lines or a single line (under the assumption of a normally incident wave angle).

The 'rated power' transmission coefficient, $K_{t,RP}$, is then related to the ratio of wave power removed to the total wave power transiting the site as:

$$K_{t,RP} = 1 - \frac{P_{rem}}{P_S} = 1 - (P_a/\gamma)/(P_w X) \tag{3}$$

$K_{t,RP}$, therefore, represents a realistic yet conservative level of wave energy transmission, and is informed by the maximum likely energy absorption (assumed to be achieved during design wave conditions) and area of ocean over which energy extraction is permitted. This parameterisation is analogous to computing the capture width ratio of the entire WEC array by working backwards from the maximum permitted array power. Of the four parameters required to calculate $K_{t,RP}$, X and P_a are likely to be well defined at the impact assessment stage, but γ and P_w may not be known; strategies to estimate these values are presented in Section 2.1.

Table 1 demonstrates the application of the $K_{t,RP}$ parameter to scenarios from previous WEC modelling studies at two WEC array sites where the permitted array power, Pa, has been defined—Wave Hub in Cornwall, England, and the Maritime Pilot Zone in Sao Pedro de Moel, Portugal. As only a selection of wave conditions were modelled in each study, it was not straightforward to determine design wave conditions, so either the highest-occurrence wave scenario or scenario that generated the most power was selected. As device efficiency was not specified in the studies, $\gamma = 1$ was applied in all cases, although in reality the value would be lower than 1. For Wave Hub, the worst-case scenario modelled by [6] far exceeded the maximum power rating of the wave farm (final column in Table 1), representing a level of energy capture that would not be permitted and that would overestimate the level of subsequent coastal impacts (indeed, they refer to this as an 'unachievable scenario'). Conversely, for the Portuguese Maritime Pilot Zone, the selected scenarios [17,18] both heavily under-represent the permitted energy extraction at the site (final column in Table 1) and therefore potentially underestimated the maximum likely wave impacts.

Table 1. The new 'rated power' transmission coefficient, $K_{t,RP}$, compared to existing absorption scenarios from wave energy converter (WEC) impact studies in the literature. Either the wave case that generated the most power, or the highest-occurrence wave case was selected as the design wave condition, depending on the availability of information in the studies. $\gamma = 1$ was applied in all cases, as it was not specified in the studies. All K_t and $K_{t,RP}$ values represent the ratio of transmitted wave energy over incident wave energy.

WEC Site	Study	Wave Scenario			Modelled Array Characteristics			Permitted Array Characteristics			Diff.
		H_s (m)	T_e (s)	P_w (W/m crest)	X (m)	P_a (MW)	K_t	X (m)	P_a (MW)	$K_{t,RP}$	P_a (MW)
Wave Hub, Cornwall, UK	Millar, Smith and Reeve 2007 (worst case)	3.3	14.5	77,576	3000	232.7	0	3000	30	0.87	+202.7
Maritime Pilot Zone, S. Pedro de Moel, Portugal	Palha et al., 2010 (config. A)	2.9	11.1	45,798	13,500	61.8	0.9	17,000	250	0.68	−188.2
	Le Crom et al., 2008 (config. A)	2	9.24	18,133	13,500	12.2	0.95	17,000	250	0.19	−237.8

1.2. *Effects of Wave Energy Extraction on Coastal Sedimentation and Beach Morphology*

In a situation where waves are altered by the presence of a WEC site as they propagate toward the coast, knock-on effects to coastal sedimentation and beach morphology may occur [7,9–12,15,19,26,38]. Before the altered waves reach intermediate waters, their influence on sediments is likely to be negligible, as they will not interact with the seabed or influence wave-driven currents significantly prior to this point [9]. It is therefore assumed here that the potential effects of WECs on sedimentation and morphology are concentrated within shallow water at the coast. Existing studies agree that reduced wave heights in the lee of a wave farm are likely to result in accretion of the beach face, and some have therefore concluded that it is possible for wave farms to provide coastal protection in addition to renewable energy [9–11,15], although the extent and magnitude of the benefits are likely to vary with distance offshore, and with wave directional variation and spreading, as per other forms of coastal protection [39,40]. Wave farms could also affect littoral drift rates, as modelling has shown that longshore current velocity is sensitive to small WEC induced changes in wave conditions [19].

Various approaches have been used to model the effects of WECs on coastal sedimentation. A simple approach is to use the nearshore output from a wave model to indicate potential changes in sedimentation, without running any numerical morphological models [15,16,19,26]. For example, changes in cross-shore and alongshore surf-zone radiation stress gradients were examined by [26] using a spectral wave model to indicate how trends in sediment transport might be affected. A threshold in the alongshore forcing of 0.44 N/m^2 (~0.2 m/s current) was defined in their study to indicate whether a WEC array was likely to induce a significant change in sedimentation. Although this approach has the advantage of lower computational cost compared to running additional numerical models, it ultimately does not quantify the magnitude or location of morphological change that may occur with WECs in place, and may not therefore satisfy the requirements of an impact assessment. Additionally, the threshold they used to define a significant change may be small or large relative to the local variation in forcing and could have instead been informed by the local variability in hydrodynamic or morphodynamic parameters, or sensitivity of local receptors.

Another method involves first modelling the WECs in a numerical wave model (Section 1.1) then propagating the altered wave field into the coast. The altered inshore waves are then used as boundary conditions for one or more additional hydrodynamic and morphodynamic models. For example, process-based morphological models have been used to predict specifically where erosion or accretion of the beach face in the lee of a wave farm may occur, over time scales of hours to months [9–11]. Tidal elevation should be varied during such simulations as it influences wave shoaling and bottom shear stresses [9], and hence, the potential effect of WECs on sedimentation will also vary over tidal cycles and with changes in tidal range. Changes in beach profile in the lee of a hypothetical wave farm were investigated by [10] using the process-based model XBeach [41] to quantify the degree of coastal protection offered by the wave farm at varying distances from the coast. They concluded that significant coastal protection was possible depending on the size, power, and distance of the wave farm. However, this and other similar studies have omitted to allow the measured beach profile to adjust to the background forcing conditions (i.e., without WECs in place) prior to assessing WEC-induced changes. The degree of disequilibrium between the selected beach profile and boundary wave conditions will, therefore, have influenced the degree of impact interpreted from the simulations. In order to truly determine the effect of WECs on profile response, the initial profile should be allowed to equilibrate with the boundary wave condition before any WEC-affected wave conditions are implemented in the model (Section 2.2 presents an example of this), otherwise the choice of measured beach profile can influence the conclusions drawn unduly.

Another potential effect of wave energy extraction is a change in the morphological classification of a beach. Such a change could alter a beach's vulnerability to storms due to the presence or absence of bars [42], the level of bathing hazard via the presence or absence of rip channels [43], or the surfing amenity at the beach via the shape of the bars [44,45]. Changes in beach state are yet to be reliably recreated using process-based numerical models, as the required assumptions and non-linear effects within process models are compounded over large spatial and temporal scales [46,47]. Instead, more traditional sequential beach state models [48–50] that associate relative tide range (RTR) and dimensionless fall velocity (Ω) to different morphological states have been used to predict the beach state likely to develop under wave conditions altered by wave energy extraction [12,38]. Although such models distinguish the key reflective, intermediate, and dissipative states relatively effectively, a number of studies have found that the intermediate sub-states that most affect beach water users are not well distinguished by such models [50–54], and more complex, data-driven behavioral models for bathymetric three-dimensionality have, therefore, been sought for WEC impact studies on intermediate beaches [55,56].

1.3. *Effects of Wave Energy Extraction on Water Users*

Significant opposition was raised by surfers during the Wave Hub consultation in England [5], demonstrating that even a relatively small-scale wave farm can provoke opposition at distant surfing beaches. As this case shows, regions with an optimal wave resource for energy extraction can also be highly valued by water users who have a shared interest in the wave resource. As a result, future interactions between wave energy projects and beach water users are highly likely, and understanding the wave conditions preferred by different groups is, therefore, vital for impact assessment.

The wave conditions most suitable for surfing and other similar activities depend on the level of ability and preferences of individual water users [56,57]. However, globally, very little research has been conducted to determine the preferences of different water-user groups. In general terms, optimum surfing conditions require swell waves with a narrow spread of frequencies and directions [27], effectively approaching monochromatic conditions [7], but this rarely occurs in reality. It was found by [56] that novice to expert water users participating in a variety of surfing-based activities in

Cornwall, UK, shared a common preference for wave periods between $9 \leq T_{1/3} \leq 20$ s, but that different water-user groups preferred different ranges of breaking wave height. For example, women and novice water users preferred smaller waves (significant breaking wave height, $0.8 \leq H_{m0,b} \leq 2.3$ m) than men and expert water users ($1.9 \leq H_{m0,b} \leq 3.7$ m); on average, preferred breaking wave heights ranged between $1.2 \text{ m} \leq H_{m0,b} \leq 3.1$ m. There are no comparable studies to confirm whether the identified preferences apply elsewhere in the world, but it is likely that they are relevant for locations with similar beach morphology and wave climates to the Atlantic coast of Cornwall, where these ranges were determined.

The degree to which waves of interest to water users (for example, those with frequencies of 0.11–0.05 Hz) are affected by wave energy extraction will depend on the frequency response of the WECs deployed. A WEC tuned to extract optimally within the preferred frequency range, or one that simply affects a wide range of frequencies, will reduce the energy of preferred surfing conditions more than a WEC with a narrow frequency response that targets wave frequencies outside the preferred range [56,58]. In cases where the WEC frequency response is known, wave modelling could be used to investigate the precise degree to which preferred surfing waves are likely to be affected, but only if the aforementioned preferences apply to local water users. In other cases, frequency-independent modelling of WECs will provide a more conservative assessment of the likely attenuation of surfing waves.

Additionally, spilling and plunging waves are the only breaker types suitable for most surfing-based activities [59], so potential changes in breaker type should be considered through assessment of the Irribarren number [60]. Similarly, peel angle, which describes how quickly a wave breaks along a bottom contour [61], could be affected by wave energy extraction. Aerial photographs and bathymetric surveys determined that peel angles of 30–70° are suitable for the majority of surfers, while smaller peel angles are only surfable by experts [57]. However, WEC-induced changes to wave peel angle would be impractical to predict without detailed modelling of wave shoaling and breaking within the surf zone.

The effects of wave energy extraction on water users are not limited to changes to inshore wave conditions; knock on effects to beach morphology could also affect the safety and amenity of the surf-zone. Previous research indicates that three-dimensional (3D) 'bar and rip' beach morphology, synonymous with intermediate beach states [48–50], significantly increases the bathing hazard for water users by enhancing rip current circulation [43,50,62,63]. Interestingly, these morphology types also enhance the quality of surfing conditions, as 3D bathymetry can increase the peel angle of breaking waves to within limits suitable for wave riding [44,45,57,64]. As wave height and period are key parameters governing beach state change [48], an alteration to either parameter by wave energy extraction could alter the morphological state, and subsequently the bathing hazard and surfing amenity provided by beaches in the lee of a wave farm [38,55,56,65].

1.4. Case Study: Westwave, South-West Ireland

To explore the application of some of the methods discussed in the previous sections, a WEC impact assessment case study is presented in this paper. 'Westwave' is a WEC test site currently under proposal for installation off the coast of County Clare, south-west Ireland, and would be permitted to operate at 5 MW power capacity. The project aims to demonstrate that a pilot wave energy project can be designed, consented, developed and operated in Ireland using innovative wave energy conversion technologies. As part of the consenting process for the site, a coastal processes impact assessment was undertaken which considered potential impacts from both the operation of WECs at the site (alteration to the wave climate, beach morphology, and surfing amenity), as well as physical impacts that could occur during installation and decommissioning (e.g., disturbance of benthic sediments). Only the aspects of the assessment covering the operation of WECs are discussed in this paper.

The proposed WEC array considered here consists of 6 floating or semi-submerged WECs covering an area of ocean up to 2 km in length, and would be moored at depths of up to 70 m and at distances of up to 12 km from the coast. Commercially operable WEC devices that have been identified as potentially suitable for this array include those in development by Carnegie (CETO 6), Well Oy (Penguin) or Ocean Energy (OE Buoy). However, for the present study these individual devices and their precise spacing or array layout are not simulated, as the focus is on the application of the new generic transmission parameter presented in Equation (3), which is intended for use at proposed WEC sites where only the permitted array power and leased deployment area are known, and for which the exact devices to be deployed may or may not be known a priori.

2. Materials and Methods

2.1. Modelling Effects on Wave Climate

Effects on wave conditions in the lee of Westwave were modelled using the open-source third-generation spectral wave model SWAN [66]. SWAN was designed to simulate the propagation of wind-generated surface gravity waves in the near-shore and accounts for refraction and shoaling due to currents and variations in bathymetry as well as reflection or transmission due to obstacles, making it suited to simulating the effects of WECs, e.g., [6,8,10]. Wave energy dissipation is accounted for in the model by activating processes such as whitecapping, bottom friction, depth-induced wave breaking, and wave–wave interaction. SWAN is based on the spectral action balance equation [66]:

$$\frac{\partial}{\partial_t}N + \frac{\partial}{\partial_x}c_x N + \frac{\partial}{\partial_y}c_y N + \frac{\partial}{\partial_\sigma}c_\sigma N + \frac{\partial}{\partial_\theta}c_\theta N = \frac{S}{\sigma} \tag{4}$$

where the first term represents the rate of change in action density per unit frequency (σ) and direction (θ), $N(\sigma,\theta)$, through time; the second and third terms represent the spatial propagation of N in x and y space, with celerity components c_x and c_y; the fourth term represents the change in relative frequency due to variations in bathymetry and current velocity; and the fifth term represents directional change caused by depth and current induced refraction. The term $S(\sigma,\theta)$ on the right hand side of the action balance equation is the energy density source term that represents wave generation, dissipation, and wave–wave interactions [66].

The wave model configuration used in the Westwave case is summarised in Figure 1, and comprised a nest of three SWAN grids with increasing resolution towards the WEC site, which were forced unilaterally along the western-most boundary by publicly available wave conditions from a single National Oceanic and Atmospheric Administration (NOAA) Wave Watch III model node. The model bathymetric grid was constructed from bathymetry data acquired from the European Marine Observation and Data Network (www.emodnet-bathymetry.eu). Wind forcing and wind induced wave growth were not included in the model, but varying tidal elevation was applied for validation runs using data from Admiralty tide charts. Simulated inshore wave conditions were validated against 2 months of wave data from a directional wave rider buoy (Figure 1), which demonstrated that the model could replicate observed wave conditions at the proposed WEC array location (Figure 2). Root-mean-square error (RMSE) for significant wave height, H_{m0}, peak period, T_p, and peak wave direction, D_p, was 0.63 m, 1.3 s, and 12° respectively.

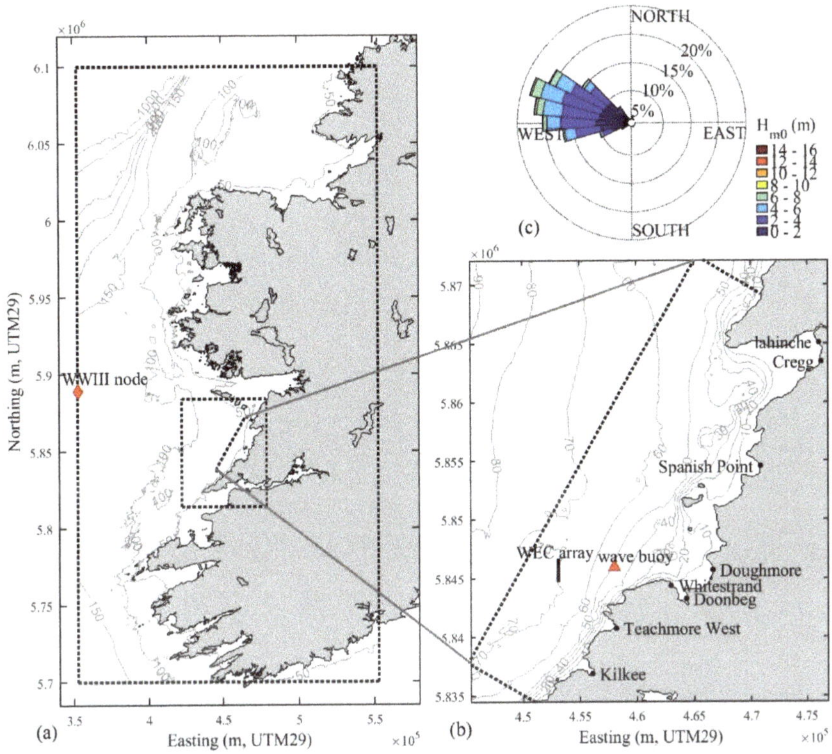

Figure 1. (**a**) Location of Westwave in Co. Clare on the south-west coast of Ireland. The outer, middle, and inner dotted regions represent the extents of the 2000 m, 200 m, and 50 m nested wave model grids, respectively, which were forced by the WWIII node on the western boundary of the coarse model grid (diamond); (**b**) the location of the proposed offshore WEC array at Westwave and known surfing beaches in the region; (**c**) wave height rose demonstrating the energetic wave climate from the Killard wave buoy (triangle in (**b**)).

Three different wave scenarios arriving from three directional sectors (north-west: 285°–360°, west: 255°–284°, south-west: 180°–254°) were simulated, providing a total of nine wave cases (Table 2). Results from the westerly wave scenarios are presented in Section 3:

1. The highest-occurrence wave condition for each directional sector was determined by examining joint probability histograms of H_{m0} vs. T_p, H_{m0} vs. D_p, and D_p vs. σ_θ for each directional sector. These were generated from 10 years of NOAA WWIII hindcast acquired for the study at the model's offshore boundary.
2. 1-year return period wave height for each directional sector was determined using a generalised Pareto distribution fitted to peak wave heights greater than the mean of the data plus one standard deviation, and separated by at least four days. To estimate a suitable wave period and directional spread associated with these wave heights, a one-term power function was fitted to values of H_{m0} vs. T_p and to values of H_{m0} vs. σ_θ from the 10-year NOAA WWIII hindcast time series.
3. The 'optimum' surfing wave condition for each directional sector was determined from the average wave preference expressed by water users in Cornwall (Section 2.3). σ_θ was set at 20°, representing very narrowly spread waves for the region ($\sigma_\theta \leq 20°$ occurred <1% of the time in the 10 year NOAA WWIII hindcast acquired for this study).

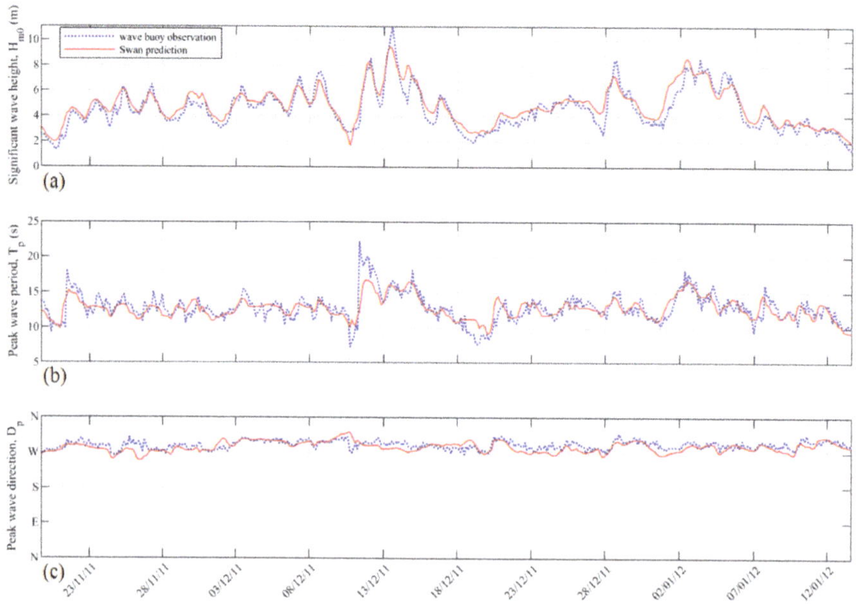

Figure 2. Validation of SWAN model accuracy (solid lines) against measured wave buoy data from the Westwave wave buoy (Figure 1) in ~49 m depth (dotted lines): (**a**) significant wave height, H_{m0} (root-mean-square error (RMSE) 0.63 m); (**b**) peak wave period, T_p (RMSE 1.3 s); (**c**) peak wave direction, D_p (RMSE 12°).

Table 2. Wave scenarios modelled in SWAN. Only results for westerly wave cases are presented in Section 3.

Direction of Wave Approach	Highest-Occurrence Wave				1-Year Return Period Height				Optimum Surf Conditions			
	H_{m0} (m)	T_p (s)	D_p (°)	σ_θ (°)	H_{m0} (m)	T_p (s)	D_p (°)	σ_θ (°)	H_{m0} (m)	T_p (s)	D_p (°)	σ_θ (°)
NW	2.5	10	290	30	10.6	15.0	290	25.6	1.4	14.7	290	20
W	2	10	275	30	11.4	14.5	275	26.4	1.4	14.7	275	20
SW	2.5	10	250	30	8.6	12.6	250	25.5	1.4	14.7	250	20

The proposed WEC array was simulated in the highest resolution (50 m) nested model grid as a partially transmitting obstacle, using a constant (frequency-independent) transmission coefficient, K_t. The standard SWAN model (version 40.72) was used, which allows frequency independent energy absorption from obstacles. However, if frequency dependent absorption was to be simulated, the SNL-SWAN (Sandia National Laboratories—Simulating WAves Nearshore) model could have been applied, which allows this functionality. Three different transmission coefficient cases were tested:

- A #x2018;rated power' scenario, $K_t = K_{t,RP} = 0.9$, calculated with the new parameterisation in Equation (3), using the parameters in Table 3
- A 'WEC derived absorption' scenario that represents the absorption of a single WEC, $K_t = K_{t,WD} = 0.58$, derived from scaled physical model tests [67]
- An extreme and unfeasible 'complete absorption' scenario where $K_t = K_{t,CA} = 0$

It should be noted that the latter two coefficients were modelled for comparison purposes only, as $K_{t,RP}$ is considered the maximum realistic absorption permitted at the site. From the values in Table 3, it can be seen that using the WEC derived transmission coefficient ($K_t = 0.58$) is equivalent to increasing the permitted array power from 5 MW to 12 MW. Using the complete absorption transmission coefficient ($K_t = 0$) is equivalent to increasing the permitted array power from 5 MW to 52 MW.

Table 3. Transmission coefficients applied to three absorption scenarios. The parameters in columns 3, 4, and 7 were used to calculate $K_{t,RP}$ as well as to calculate the equivalent array power modelled under the $K_{t,WD}$ and $K_{t,CA}$ scenarios. As demonstrated by the final column, these two scenarios far exceed the 5 MW power rating of Westwave.

Absorption Scenario	Energy Transmission Coefficient, K_t	Estimated WEC Efficiency, γ	Array Length, X (M)	Design Wave Height, H_{m0} (M)	Design Wave Period, T_e (S)	Design Wave Power, P_w (W/M)	Permitted Array Power (W)	Modelled Array Power, P_a (W)
'Rated power' $K_{t,RP}$	0.90	0.3	2000	5	7	85,874	5,000,000	5,000,000
'WEC derived absorption' $K_{t,WD}$	0.58	0.3	2000	5	7	85,874	5,000,000	12,365,856
'Complete absorption' $K_{t,CA}$	0.00	0.3	2000	5	7	85,874	5,000,000	51,524,400

Assumptions about γ had to be made as no single device had been chosen for deployment at Westwave. Given a notable paucity of device efficiency data for commercial WECs, γ was set at 30% ($\gamma = 0.3$), in line with other studies in the literature [17]. The design wave power, P_w, was estimated using values of wave height and period at which maximum energy absorption occurs for a number of pre-existing commercial WECs (Figure 3). This generic value was determined from published WEC performance data in the form of power matrices [68]; the WECs from which the data were derived may or may not ever be deployed at Westwave and were selected purely on the availability of their performance data. The design wave conditions in Figure 3 range between $4 \leq H_{m0} \leq 6$ m and $6 \leq T_e \leq 8.5$ s and represent a spectrum of device types. Specifically, Aqua Buoy and Pelamis (this WEC is now defunct, but still provides a useful reference for the performance characteristics of a commercial WEC) are examples of offshore devices; Wave Dragon is an intermediate depth device; and Oyster (also now defunct) is an example of a shallow-water device. The values were averaged in order to determine a generic design wave condition for devices of any type: $H_{m0} = 5$ m, $T_e = 7$ s (Figure 3, dashed lines), and these values were used to calculate P_w under the assumptions of linear wave theory. Determining a generic transmission coefficient from such a range of devices was suitable for the Westwave assessment, as both a nearshore and offshore siting option were tested in the original EIA. Although only the offshore WEC array is studied here, the averaged wave conditions are still deemed suitable, as they closely align to the peak performance conditions for the deep water device Aqua Buoy—a point absorber device not dissimilar to the Carnegie CETO 6 device earmarked for Westwave.

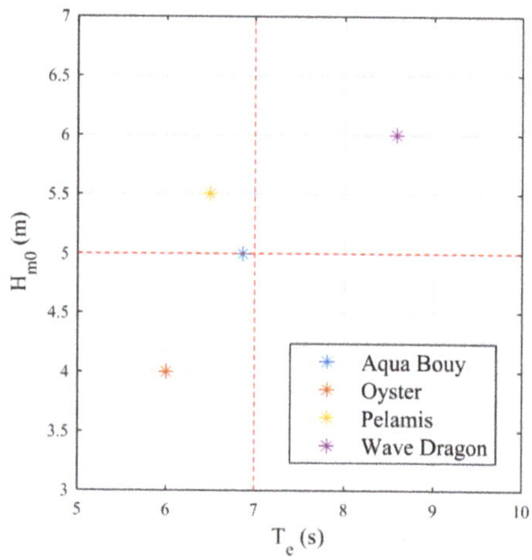

Figure 3. Significant wave height (H_{m0}) and energy period (T_e) at which maximum wave energy absorption occurs for a number of pre-existing wave energy converters, determined from published power matrices [68]. The dashed lines show the average H_{m0} and T_e from these data, used as a generic design wave condition.

2.2. Modelling Effects on Beach Morphology

A morphological modelling study was undertaken which considered the potential for Westwave to alter the morphological classification and profile shape of the beaches in its lee. Morphology results from Doughmore beach (Figure 4), where the largest wave effects were predicted to occur, are presented in Section 3 of this paper. The potential for Westwave to affect the long-term (annual time-scale) three-dimensional shape of the beach was investigated using a conceptual beach state model, while the potential for it to affect the short-term (storm event time-scale) two-dimensional (i.e., '2D-vertical') shape of the beach profile was studied separately using a process-based numerical model. This two-stage approach was taken, rather than modelling the three-dimensional evolution of the beach in a single (i.e., '2D-horizontal') process-based numerical model, as process-based models are not yet capable of reliably recreating measured three-dimensional morphology over the temporal and spatial scales relevant to a WEC impact study [46,47]. Shoreline change modelling was not undertaken as littoral drift was not deemed to be a relevant process for the embayed beaches in the region, and as any alongshore wave height gradients from the WEC shadow are expected to be insignificant at the distance of 10 km, especially given the maximum magnitudes of wave height change that were predicted. However, alongshore varying (three-dimensional) morphology occurs at Doughmore beach (Figure 4), which was accounted for through the beach classification study.

The potential to alter the morphological classification of the beach was investigated with the sequential beach state model of [49]. The model considers nine key beach states and predicts that changes in state are influenced by changes in RTR and Ω which are respectively defined as:

$$\mathrm{RTR} = \mathrm{MSR}/\mathrm{H_b} \quad (5)$$

$$\Omega = \mathrm{H_b}/\mathrm{W_s}\,\mathrm{T} \quad (6)$$

where MSR is the mean spring tide range, here determined from Admiralty tide charts at Galway; H_b is a representative breaking wave height, here calculated using the method of [69]; W_s is the sediment fall velocity, here determined from mid-intertidal surficial sediment samples analysed in a laboratory settling tube; and T is wave period. For this assessment, the highest-occurrence wave height (H_{m0}) and peak period (T_p) were simulated in SWAN and output at the 10 m depth contour adjacent to

Doughmore beach, and were subsequently used to calculate RTR and Ω with and without the influence of WECs at Westwave.

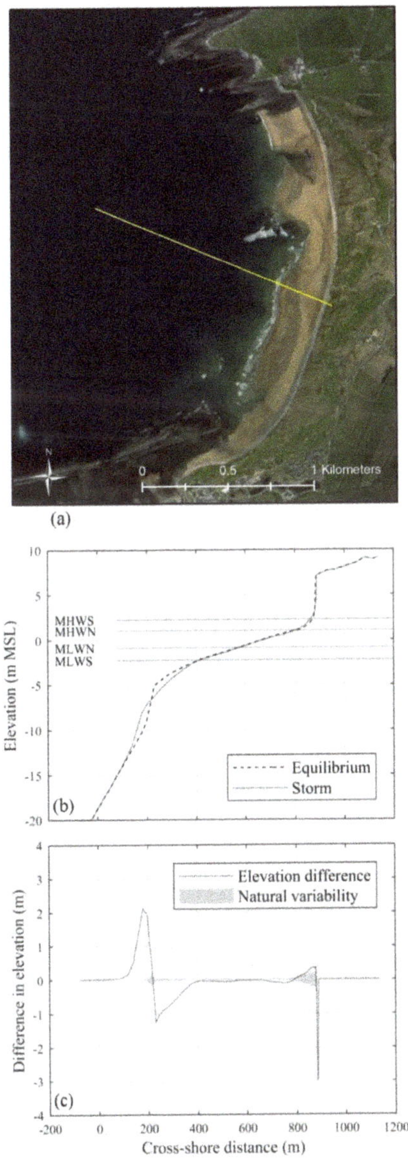

Figure 4. Doughmore beach, Co. Clare, Ireland: (**a**) Location of cross-shore profile used in XBeach (solid line). Imagery courtesy of DigitalGlobe, Data SIO, National Oceanic and Atmospheric Administration (NOAA), U.S. Navy, NGA, and GEBCO; (**b**) equilibrium and storm profiles from XBeach simulation and tidal elevations of mean low water spring (MLWS), mean low water neap (MLWN), mean high water neap (MHWN), and mean high water spring (MHWS); (**c**) difference in elevation between the equilibrium and storm profiles (dotted line) and range of background profile variability (filled area, defined in the text).

Next, the process-based numerical model XBeach was used to investigate whether WECs at Westwave could alter localised patterns of accretion and erosion in the beach profile under either highest-occurrence or storm wave conditions. XBeach can simulate the propagation of incident and infragravity waves, wave induced currents, sediment transport, and morphological changes, solving the time-dependent short wave action-balance equations, roller energy equations, the non-linear shallow water equations of mass and momentum, sediment transport formulations, and bed updating [70]. In the instationary or 'surf beat' mode of XBeach used in this study, short-wave motion is solved using the wave action equation, estimating the variation of the short-wave envelope on the scale of individual wave groups [71]. Wave group dissipation is modelled [72,73], and a roller model [74–76] is used to represent the momentum carried after wave breaking. Radiation stress gradients [77,78] then drive infragravity motion and unsteady currents in the model, which are solved with the non-linear shallow water equations [79]. The surf beat mode of XBeach is valid for dissipative and intermediate beaches, where swash motions are predominantly in the infragravity band, and short waves are mostly dissipated by the time they reach the shore [71]. Since its development, XBeach has proved effective in reproducing measured hydrodynamics and beach profile response under a wide range of scenarios [41,80–83] making it a suitable tool for examining WEC-induced profile changes.

Combined data from a topographic drone survey and hydrographic multibeam survey were interpolated to a 2 m grid using a loess quadratic interpolation method [84], and a single representative beach profile was extracted for use in XBeach (Figure 4). A two-dimensional (i.e., cross-shore profile) non-equidistant grid was employed in XBeach, where the grid resolution was defined as a function of the water depth and offshore wave conditions, with a minimum resolution of 1 m in shallow water. For this purpose, the Courant condition was used to find the optimal grid size in terms of sufficiently resolving physical processes, as well as yielding maximum computational efficiency and grid smoothness. The XBeach morphological acceleration parameter, Morfac, was set to a value of 5, which is within the recommended parameter range. All other model free parameters were set to their default values. Boundary wave conditions for XBeach were provided by SWAN model output at the 20 m depth contour.

Results from two wave cases are presented in Section 3:

1. **The highest-occurrence wave condition**. This case was run continuously in XBeach until the measured profile reached equilibrium with the wave forcing, or in other words, until profile changes no longer varied significantly from one tidal cycle to the next. This occurred after five spring-to-spring tidal cycles (approximately 75 days) at Doughmore. The water level was varied in the simulation using a spring-neap-spring tidal signal, where the tide range varied between 2 m (neap tides) and 4.5 m (spring tides). The profile at the end of the simulation represents the equilibrium profile shape under the most commonly occurring waves.
2. **The 1-year return period wave condition**. This case was run for 24 h with a 4.5 m spring tide range to simulate a single storm event coincident with large tides. This simulation followed immediately from wave case 1 and, therefore, acted upon the equilibrium beach profile. The profile at the end of this simulation represents the storm profile shape.

The background variability of the equilibrium profile (Figure 4c) was quantified as two standard deviations of the profile changes occurring over a spring-to-spring tidal cycle after equilibrium had been reached. This 'baseline profile variation' represents the changes that occur solely due to the action of the tides moving the shoaling, surf, and swash zones up and down the beach profile. WEC induced changes to the beach profile that are smaller than the baseline variation can be considered insignificant, as they are smaller than the level of day-to-day noise in the natural profile. This threshold was used as the context by which to assess the significance of any WEC-induced changes in the profile.

2.3. Modelling Effects on Water Users

To investigate whether WEC operation at Westwave could have any effect on water users at the coast, its potential to change inshore surfing conditions and beach morphological state were considered. Wave model output was extracted at the 10 m depth contour adjacent to eight known surfing locations in the region (Lahinch, Cregg, Spanish Point, Doughmore, Doonbeg, Whitestrand, Teachmore West, and Kilkee, Figure 1) and the nearshore changes predicted by the wave modelling exercise were examined at each site. For brevity, only changes at Doughmore beach are presented in Section 3.

As there is a lack of data to indicate the wave preferences of the local surfing community, wave preferences from water users in the lee of Wave Hub in Cornwall, England (Section 1.3), were used as a proxy for optimal surfing conditions in the Westwave region. Given that the two regions have similar beach morphology types, a similar tidal range, and are both exposed to Atlantic Ocean waves, it is likely that preferred surfing conditions are similar at the two sites. The preferred wave conditions were reverse shoaled using linear wave theory to the depth of the NOAA WWIII data node (143 m) in order to provide SWAN model boundary conditions. At that depth, the average preferred wave conditions for surfing equate to H_{m0} = 1.4 m and T_p = 14.7 s. Breaking wave height was not simulated, as breaker height is influenced locally by wind strength, wind direction, surf-zone currents, and beach morphology, and is therefore impractical to predict. Potential change to the breaker type (spilling, plunging, or surging) was considered by quantification of the Irribarren number [60].

3. Results

3.1. Effects on Waves

The most common wave conditions at the NOAA wave buoy in deep water off the coast of Westwave (Figure 1) are H_{m0} = 1–4 m and T_p = 7–12 s, which predominantly originate from a westerly direction with σ_θ varying between approximately 10–80° and mean $\sigma_\theta = 36°$. At the location of the inshore Westwave wave buoy (~49 m depth) the waves have refracted towards the coast, and approach from predominantly a west north-west direction. The 1-year and 10-year return period westerly wave heights at the deep-water NOAA WWIII node are predicted to be 11.4 m and 15 m, respectively. The wave climate at the Westwave site is, therefore, extremely energetic, and is predominantly driven by Atlantic swell waves. Figure 5 shows the modelled wave field at the Westwave site under highest-occurrence wave conditions (H_{m0} = 2 m, T_p = 10 s, D_p = 275°, σ_θ = 30°), demonstrating that there is along-coast variability in the wave conditions in shallow water without WECs in operation at Westwave.

Figure 6 compares SWAN model simulations of the Westwave WEC array with the value of K_t varied between $K_t = 0$ (complete absorption), $K_t = 0.58$ (WEC derived absorption), and $K_t = 0.9$ (rated power absorption) under highest-occurrence wave conditions. As expected, the selected absorption level has a marked difference on the transmitted wave shadow in the lee of the site. At the 10 m depth contour at the coast (Figure 7), maximum wave height change is predicted to be 44% for $K_t = 0$, 16% for $K_t = 0.58$, and 4% for $K_t = 0.9$ under the three westerly wave scenarios simulated. These maximum changes occur along the rocky coast between Whitestrand beach and Teachmore West (Figure 1), while the maximum predicted changes at Doughmore beach (where the morphological study described in Section 3.2 was undertaken) are predicted to be 10% for $K_t = 0$, 5% for $K_t = 0.58$, and 1.3% for $K_t = 0.9$ (Table 4). Despite being a conservative method for estimating wave energy transmission, the rated power scenario ($K_t = K_{t,RP} = 0.9$) results in the smallest wave impacts of the three approaches.

Table 4. Percentage changes in wave height and morphological parameters under three absorption scenarios (K_t = 0, complete absorption; K_t = 0.58, WEC derived absorption; K_t = 0.9, rated power absorption). Columns 1–3 show changes in wave height for three westerly wave cases at the 10 m depth contour adjacent to Doughmore beach. The last four columns show changes in morphological parameters at Doughmore beach under equilibrium (highest occurrence) and storm (1-year return period) wave conditions.

Absorption Scenario	% Change in Wave Height			% Change in Morphological Parameters			
	Highest-Occurrence Wave	1-Year Return Period	Optimal Surf Conditions	Equilibrium Ω	Equilibrium Relative Tide Range (RTR)	Equilibrium Profile Volume	Storm Profile Volume
$K_t = K_{t,CA} = 0.00$	−10.15	−1.74	−8.11	−8.99	+8.39	+0.12	+0.10
$K_t = K_{t,WD} = 0.58$	−4.54	−0.70	−3.39	−4.01	+3.78	+0.07	+0.05
$K_t = K_{t,RP} = 0.90$	−1.27	−0.17	−0.89	−1.06	+1.07	+0.03	+0.02

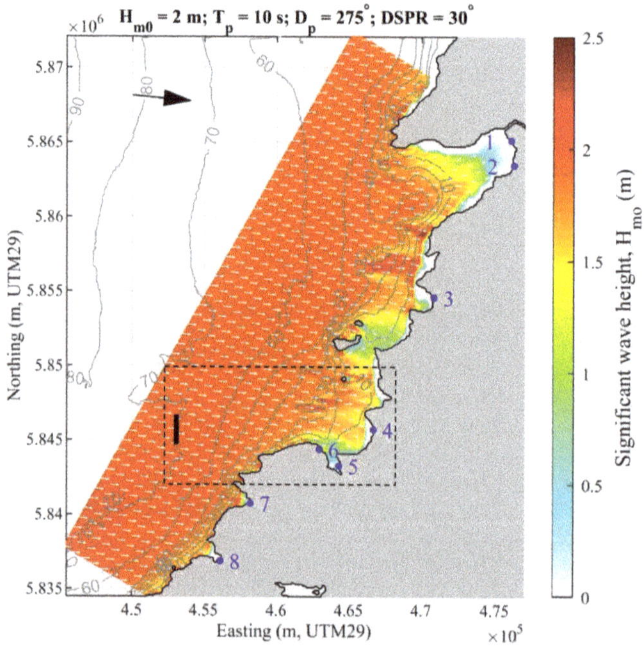

Figure 5. Wave height field for highest-occurrence westerly wave conditions. The location of the proposed WEC site is shown as a solid line for reference, but was not simulated in this model run. The location of the panels in Figure 5 is shown as a dashed line. The locations of known surfing beaches are numbered; 1 = Lahinche, 2 = Cregg, 3 = Spanish Point, 4 = Doughmore, 5 = Doonbeg, 6 = Killard, 7 = Teachmore West, 8 = Kilkee. The numbered contour lines show bathymetric depth from Mean Sea Level (MSL).

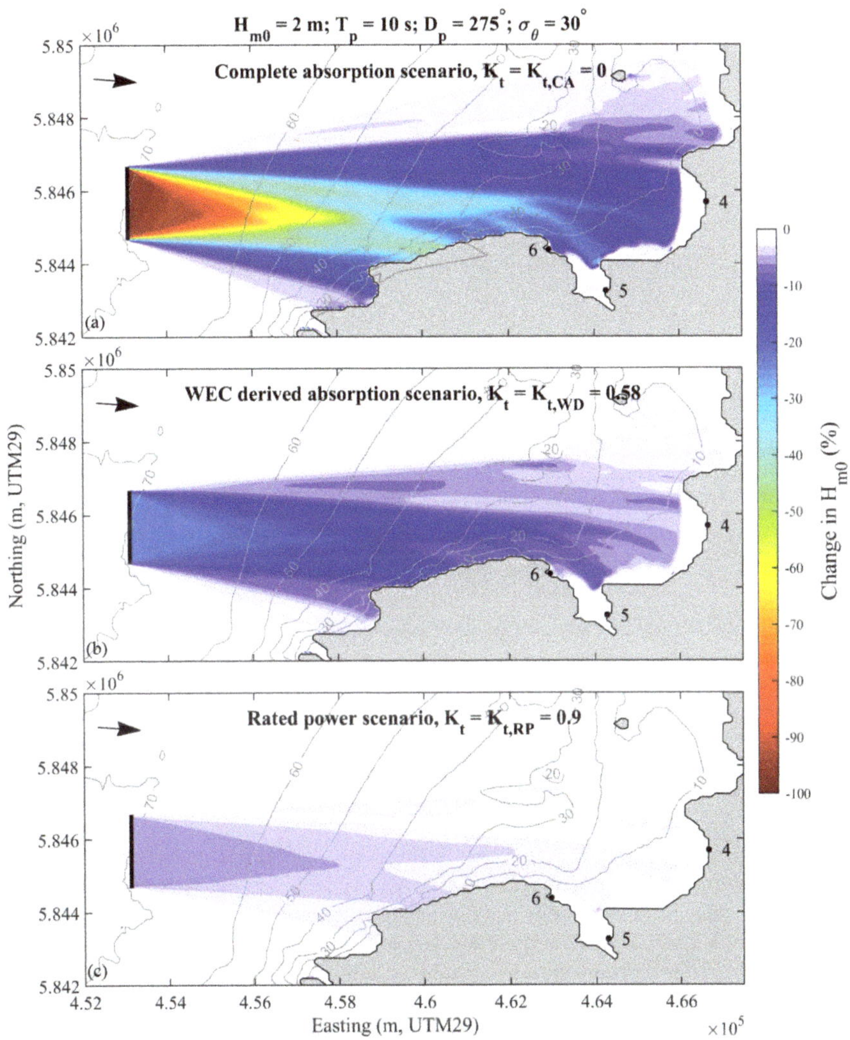

Figure 6. Simulated wave height shadow for highest-occurrence westerly wave conditions, under three different absorption scenarios: (**a**) complete absorption scenario; (**b**) WEC-derived absorption scenario; (**c**) rated power absorption scenario. The position of the Westwave array is shown as a solid line in each panel. The arrows show the angle of wave approach at the offshore model boundary. Points 4, 5 and 6 show the location of beaches in the wave shadow (Doughmore, Doonbeg, and Whitestrand, respectively). The numbered contour lines show bathymetric depth from MSL. The location of the panels is shown in Figure 5 as a dashed box.

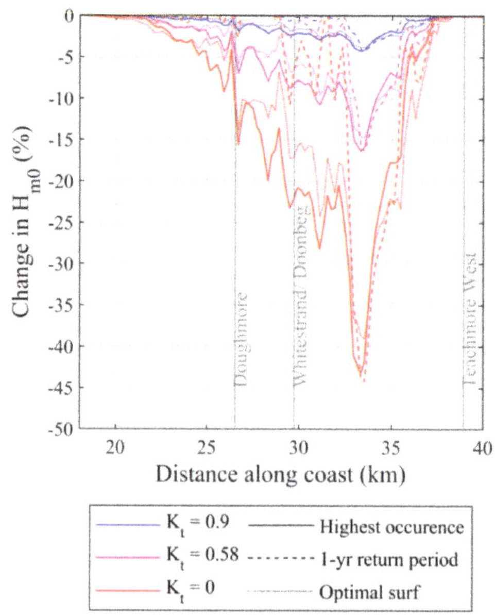

Figure 7. Change in wave height at the 10 m depth contour along the coast in the lee of Westwave for different westerly wave cases (solid, dashed, and dotted lines) under different wave energy extraction scenarios (K_t = 0.9, rated power absorption; K_t = 0.58, WEC derived absorption K_t = 0, complete absorption). The distance along the coast is from north to south, with the along coast distances of Doughmore, Whitestrand, Doonbeg, and Teachmore West beaches shown for reference. The boundary wave conditions for each wave case are detailed in Table 2.

3.2. *Effects on Morphology*

The beach state model of [49] demonstrates that even under the largest predicted changes to inshore wave height caused by the complete absorption scenario (red square in Figure 8), the slight decrease in Ω and increase in RTR (Table 4) are far smaller than the parameter shift required to cause a true change in beach state—the equivalent distance between the centres of two regions in the model. Under the unrealistic complete absorption scenario, the beach could exhibit slightly more alongshore variability, but predicting such subtle changes is beyond the capabilities of a conceptual beach state model. Under the rated power scenario, the changes in Ω and RTR are predicted to be around 3% and 8% less than under the WEC-derived absorption and complete absorption scenarios, respectively.

The XBeach simulations predict that once in equilibrium with the highest-occurrence wave conditions, the background profile variation at Doughmore beach (without WECs in operation) is at most 50 cm, and is greatest around the foot of the dune, with lesser variations also occurring lower on the profile (Figure 9). With WECs in operation, the intertidal area and dune face (250–900 m cross-shore) are predicted to have more sediment, while the subtidal area around the break in slope (150–250 m cross-shore) is predicted to have less sediment during equilibrium and storm wave conditions, representing shoreward transport of sediment and beach steepening. Under all three absorption scenarios the active beach profile is predicted to have more sediment volume than without WECs in operation, indicating that some degree of coastal protection would be provided by the presence of WECs at Westwave.

Changes to the equilibrium and storm profiles well outside the baseline profile variation were predicted by the model under the complete absorption and WEC-derived absorption scenarios. Conversely, under the rated power scenario almost all of the changes in the equilibrium and storm profiles were predicted to be within the baseline profile variation of the beach. At most, a few centimetres of vertical change outside the baseline profile variation was predicted for the rated power scenario, and the overall shape and gradient of the profile was not predicted to change. The predicted profile changes under the rated power scenario are, therefore, not considered significant, while under the WEC-derived and complete absorption scenarios the predicted changes were considered to be potentially significant.

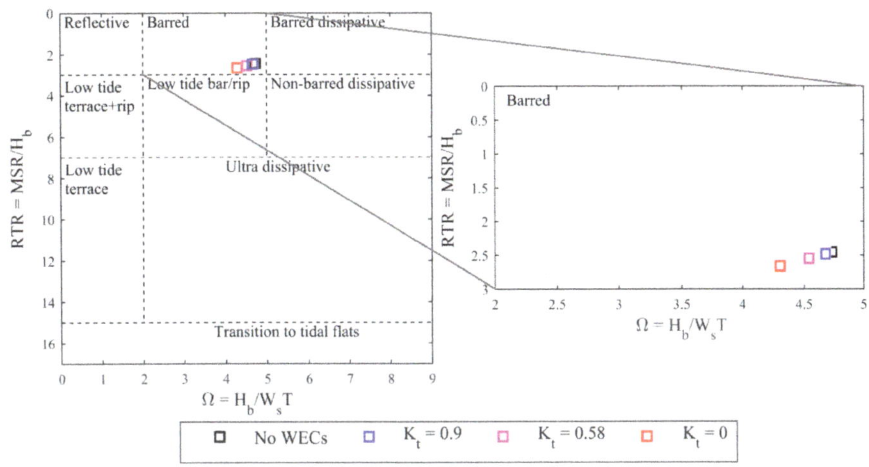

Figure 8. Naturally occurring (without WECs) equilibrium beach morphology state at Doughmore beach (black squares), as well as equilibrium beach states that may occur with WECs in operation at Westwave (coloured squares) under differing energy extraction scenarios (K_t = 0.9, rated power absorption; K_t = 0.58, WEC derived absorption; K_t = 0, complete absorption). The beach state model of [49] is used; for definition of axes parameters, see Section 2.2.

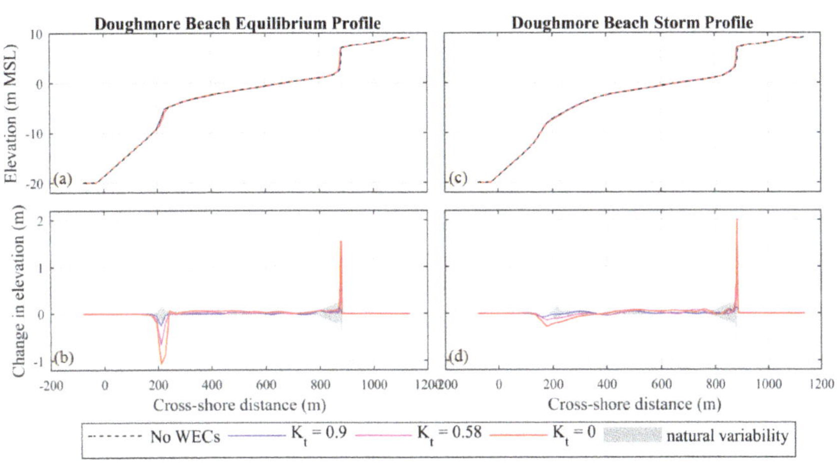

Figure 9. Results of XBeach profile change simulations under different absorption scenarios at Westwave (K_t = 0.9, rated power absorption; K_t = 0.58, WEC-derived absorption; K_t = 0, complete absorption) for equilibrium conditions (panels (**a**,**b**)) and storm conditions (panels (**c**,**d**)): (**a**,**c**) show the profile shape; (**b**,**d**) show the profile change. The shaded areas in panels (**b**,**d**) represent the background profile variability; WEC-induced profile changes that lie outside of the shaded areas are used to indicate where a significant impact may occur.

3.3. Effects on Water Users

In deep water off the coast of Doughmore, suitable wave conditions for surfing are estimated to be $0.5 \leq H_{m0} \leq 2.8$ m, with periods of $9 \leq T_p \leq 20.6$ s. The average 'optimal' conditions for surfers and other water users of all abilities in deep water off the coast of Killard are estimated to be narrow-banded swell waves with height and period of $H_{m0} = 1.4$ m and $T_p = 14.7$ s. From Table 4, the predicted change in wave height at 10 m depth during such optimal surfing conditions is 0.89%, 3.39%, and 8.11% for the rated power scenario, WEC-derived absorption scenario, and complete absorption scenario, respectively. In absolute terms, this equates to reductions in H_{m0} (from 0.94 m at 10 m depth with no WECs in operation) of 1 cm, 3 cm, and 8 cm, respectively. Wave period was predicted to change less than 0.25 s, for all absorption scenarios; however, frequency dependent extraction was not simulated. With these predicted changes, an assessment of the Irribarren number indicates that the natural breaker type would not change from plunging waves. It is assumed that due to the relatively minor predicted changes in the wave height and period, changes in the angle of incidence of waves arriving at the coast will not have a significant influence on wave peel angle for surfing, as this is primarily influenced by localised refraction in the shoaling and breaking zones caused by bathymetric features. However, changes in wave direction caused by wave energy absorption are not yet adequately represented in spectral (phase-averaged) wave models, and the magnitude of such effects therefore remains uncertain.

4. Discussion

Previous WEC impact assessments in the scientific literature have not explicitly accounted for the permitted array power when devising WEC transmission characteristics. Instead, some have modelled a suite of transmission coefficients (including complete energy absorption) in order to capture the full possible range of impacts, e.g., [6,8]; however, as the permitted array power is usually well defined at the impact assessment stage, a conservative maximum absorption value can be estimated for a permitted WEC array, reducing the possibility for over-or under-estimation of the potential coastal impacts. The rated power transmission coefficient, $K_{t,RP}$, presented in this paper offers a conservative maximum absorption level analogous to the capture width ratio of the entire wave farm. It is conservative in that it represents a realistic worst-case scenario (as the absorption level is at full permitted array power capacity) and assumes that WECs operate at design efficiency under all wave conditions, when in reality it is likely that they will operate at a lower absorption efficiency during wave conditions far removed from the design conditions. It also allows the user to be conservative by applying a low efficiency factor (γ), which, for a given permitted array power, has the effect of decreasing the wave transmission due to absorption that does not contribute to the generated power.

Other more precise methods for quantifying and modelling wave energy absorption exist and have been documented in the literature; for example, frequency dependent extraction [8,24] modelled using individual WEC devices [23] offers a more precise estimate of wave shadowing in the lee of a WEC array. However, as frequency response and WEC array layout vary from device to device, modelling these characteristics with great precision will only increase the accuracy of the modelling results if the exact devices to be deployed over the entire lifetime of the permitted array are known with certainty at the impact assessment stage. We propose that this is unlikely to be the case for many WEC sites. Conversely, the permitted array power describes the potential for coastal process impacts without making assumptions about WEC frequency response or array layout. $K_{t,RP}$ therefore offers a simple, generic, conservative and robust approach for estimating coastal process impacts over the lifetime of a permitted WEC array, although it is acknowledged that ignoring WEC array layout and frequency response would be an unnecessary limitation of the presented method, if such parameters were well defined at the impact assessment stage.

The estimation of design wave conditions and the mechanical/electrical efficiency (γ) of energy conversion within WECs are both required in order to calculate $K_{t,RP}$. As these factors require assumptions about the design characteristics of WECs, they are sources of remaining uncertainty. One approach for estimating a generic design wave power value is presented in Section 2.1, based on averaged data from commercially developed WECs presented by [68]; other WEC databases also exist that could be used to inform this value [28]. It is acknowledged, however, that γ should also be informed by published performance data when these are available.

In the Westwave case study presented here, predicted beach profile changes with WECs in operation at Westwave were almost entirely within the magnitude of background profile variation when using $K_{t,RP}$ to quantify the WEC array transmission coefficient. Conversely, cases where the WEC-derived transmission coefficient ($K_{t,WD}$) or complete energy absorption transmission coefficient ($K_{t,CA}$) were used showed potential profile changes well outside the background profile variability. Therefore, the choice of $K_{t,RP}$ over more traditional transmission coefficients entirely determined whether the impacts were predicted to be significant or not in this case study.

Methods for determining impact thresholds have received relatively little attention in the scientific literature, and conclusions about the significance of the predicted effects on, for example, surfing waves or coastal sedimentation have often been subjective, despite being informed by objective modelling, e.g., [6,7]. Using a measure of natural variability offers one objective approach for identifying a significant impact threshold that is relevant to the local dynamics [85]. However, the significance of an impact actually depends on the sensitivity of the local receptors of the impact, and more research is required to provide guidance on identifying these sensitivities for WEC impact assessments. It should also be noted that bias towards erosion or accretion, even when small compared to the natural variability, may have a potentially significant cumulative impact.

In the case of surfing wave impacts, the number of 'surfable' days per year (i.e., days where the wave conditions are within a defined range of preferred wave heights and periods) with and without WECs in operation could be quantified. It is possible for this to increase or decrease in the presence of WECs, as waves previously larger than the preferred range may be reduced to a surfable height, while smaller waves may be made too small to surf. Describing surfing impacts in terms of the number of surfable days may provide a metric that is more comprehensible to water users than the marginal percentage changes in wave height used in the case study presented here and elsewhere in the literature, including [6,7]. However, this approach would require long hindcast wave model runs (at least 1 year in duration) to be conducted at great computational cost, and was, therefore, not undertaken for the Westwave case study.

5. Conclusions

This paper aimed to identify a number of recommended modelling practices for a wave farm coastal processes impact assessment. From a review of studies in the literature, and the subsequent application of modelling techniques for a wave farm case study in southwest Ireland, the following guidelines are proposed:

- Appropriate characterisation of the sea states of interest to the assessment must be made. In particular, attention should be paid to the local directional spread of sea states, as this has a primary effect on the predicted wave shadow in the lee of an offshore wave farm.
- Frequency-dependent modelling of individual WEC devices is likely to offer the most precise representation of WECs for wave model studies, but only in cases where WEC absorption characteristics and array layout are comprehensively defined at the impact assessment stage. In cases where the absorption characteristics are not well defined or where the WECs to be deployed over the whole lifetime of the wave farm are not known with certainty, it is more conservative to model a WEC array using a single frequency-independent transmission coefficient that represents the effects of the entire WEC array.
- To calculate such a transmission coefficient, a novel parameterisation for the frequency-independent WEC array 'rated power' transmission coefficient, $K_{t,RP}$, is presented in this study. This uses the permitted power rating of the WEC array, the area of ocean over which energy extraction is permitted, an estimate of the design wave power, and an estimate of WEC electrical efficiency to inform the maximum likely absorption of a permitted WEC array. As it can be

assumed that the permitted array power will not be exceeded over the lifetime of the array, $K_{t,RP}$ represents a generic method for quantifying a realistic worst-case scenario for wave energy absorption.

- When studying potential morphological impacts from a wave farm, variation in tidal elevation should be simulated, as it influences wave shoaling and bottom shear stresses, and will therefore modulate the potential effects of WECs on sedimentation.
- When using a process-based model to study morphological changes under WEC operation, the measured coastal morphology should initially be allowed to equilibrate to new boundary conditions before implementing WEC-affected boundary conditions, otherwise the degree of disequilibrium between the measured morphology and the boundary wave conditions will influence the degree of WEC induced impact interpreted from the simulations.
- The morphological classification of beaches in the lee of a WEC site should be studied with and without WECs in operation, as beach state can influence coastal vulnerability to storms, the level of bathing hazard, and the surfing amenity at the beach. However, traditional beach classification models do not provide the precision required to predict subtle changes in bathymetric three-dimensionality (i.e., the presence or absence of rip channels), and more complex models may need to be sought where a change in such features would cause an issue.
- Impacts on coastal water-users can range from a change in the available surfing wave resource, to changes in bathing hazards and surfing wave quality as a result of morphological impacts.
- Optimum surfing conditions require swell waves with a narrow spread of frequencies and directions, and preferred wave conditions determined from water-users in Cornwall of $1.2\ \mathrm{m} \leq H_{m0,b} \leq 3.1\ \mathrm{m}$, and $9 \leq T_{1/3} \leq 20$ s may provide a suitable proxy for water-user preferences elsewhere. However, water-users of differing ability levels, activities, and locations are likely to have specific preferences which may also need to be considered.
- Even though some degree of reduction in the height of preferred surfing waves is likely to be predicted under the influence of WECs, a wave farm could actually increase the number of 'surfable' days available per year, depending on the WEC absorption characteristics, wave resource, and wave preferences of local water-users. This may provide a more suitable metric for stakeholder engagement than the traditional use of percentage changes in wave height, but requires a significantly higher computational cost.
- Generic thresholds for 'significant' WEC induced impacts do not yet exist, and are likely to be site specific. Morphological impacts could be set in the context of the natural variability occurring in the region of interest; however, bias towards erosion or accretion, even when small compared to the natural variability, may have a potentially significant cumulative impact. For water-users, opinions on what constitutes a significant change to surfing waves are likely to vary widely. More research is required in order to provide guidance on identifying the sensitivities of local receptors for WEC impact assessments.

Author Contributions: Conceptualization, C.S. and D.C.; Data curation, C.S.; Formal analysis, C.S.; Funding acquisition, C.S. and D.C.; Investigation, C.S.; Methodology, C.S. and D.C.; Project administration, C.S.; Resources, C.S. and D.C.; Supervision, D.C.; Validation, C.S.; Visualization, C.S.; Writing–original draft, C.S.; Writing—review and editing, C.S. and D.C.

Funding: This research was partly funded by the school of Biological and Marine Sciences at Plymouth University, and partly by the Electricity Supply Board of Ireland.

Conflicts of Interest: The authors declare no conflict of interest. The funders had no role in the design of the study; in the collection, analyses, or interpretation of data; in the writing of the manuscript, and in the decision to publish the results.

References

1. Heath, T.; Sarmento, A.; Neumann, F.; Brito E Melo, A.; Prado, M.; Yemm, R.; Tedd, J.; Friis-Madsen, E.; Kofoed, J.P.; Knapp, W. *Ocean Wave Energy: Current Status and Future Perspectives*; Springer: Berlin/Heidelberg, Germany, 2008.
2. de O Falcão, A.F. The development of wave energy utilisation. In *2009 Annual Report, Implementing Agreement on Ocean Energy Systems, (IEA-OES)*; IEAOES Secretary: Lisbon, Portugal, 2009; pp. 30–36.
3. Greaves, D.; Conley, D.; Magagna, D.; Aires, E.; Leitão, J.C.; Witt, M.; Embling, C.B.; Godley, B.J.; Bicknell, A.W.J.; Saulnier, J.-B.; et al. Environmental Impact Assessment: Gathering experiences from wave energy test centres in Europe. *Int. J. Mar. Energy* **2016**, *14*, 68–79. [CrossRef]
4. Guidance on Environmental Impact Assessment of Offshore Renewable Energy Development on Surfing Resources and Recreation. Available online: https://www.sas.org.uk/wp-content/uploads/sas-guidance-on-environmental-impact-assessment.pdf (accessed on 19 May 2018).
5. McLachlan, C. 'You don't do a chemistry experiment in your best china': Symbolic interpretations of place and technology in a wave energy case. *Energy Policy* **2009**, *37*, 5342–5350. [CrossRef]
6. Millar, D.L.; Smith, H.C.M.; Reeve, D.E. Modelling analysis of the sensitivity of shoreline change to a wave farm. *Ocean Eng.* **2007**, *34*, 884–901. [CrossRef]
7. Li, B.; Phillips, M. South West wave energy hub: Coastal impact and wave energy. *Proc. ICE-Energy* **2010**, *163*, 17–29. [CrossRef]
8. Smith, H.C.M.; Pearce, C.; Millar, D.L. Further analysis of change in nearshore wave climate due to an offshore wave farm: An enhanced case study for the Wave Hub site. *Renew. Energy* **2012**, *40*, 51–64. [CrossRef]
9. Gonzalez-Santamaria, R.; Zou, Q.-P.; Pan, S. Impacts of a wave farm on waves, currents and coastal morphology in south west England. *Estuaries Coast.* **2015**, *38*, 159–172. [CrossRef]
10. Abanades, J.; Greaves, D.; Iglesias, G. Coastal defence through wave farms. *Coast. Eng.* **2014**, *91*, 299–307. [CrossRef]
11. Abanades, J.; Greaves, D.; Iglesias, G. Wave farm impact on the beach profile: A case study. *Coast. Eng.* **2014**, *86*, 36–44. [CrossRef]
12. Abanades, J.; Greaves, D.; Iglesias, G. Wave farm impact on beach modal state. *Mar. Geol.* **2015**, *361*, 126–135. [CrossRef]
13. Venugopal, V.; Smith, G.H. Wave climate investigation for an array of wave power devices. In Proceedings of the 7th European Wave and Tidal Energy Conference, Porto, Portugal, 11–14 September 2007; pp. 11–14.
14. Carballo, R.; Iglesias, G. Wave farm impact based on realistic wave-WEC interaction. *Energy* **2013**, *51*, 216–229. [CrossRef]
15. Mendoza, E.; Silva, R.; Zanuttigh, B.; Angelelli, E.; Lykke Andersen, T.; Martinelli, L.; Nørgaard, J.Q.H.; Ruol, P. Beach response to wave energy converter farms acting as coastal defence. *Coast. Eng.* **2014**, *87*, 97–111. [CrossRef]
16. Iglesias, G.; Carballo, R.; Abanades, J.; Greaves, D. Wave farm layout and coastal impacts. *Coast. Eng. Proc.* **2014**, *1*, 36. [CrossRef]
17. Le Crom, I.; Brito-Melo, A.; Sarmento, A. Maritime portuguese pilot zone for wave energy conversion: Modelling analysis of the impact on surfing conditions. In Proceedings of the 2nd International Conference on Ocean Energy (ICOE 2008), Brest, France, 15–17 October 2008.
18. Palha, A.; Mendes, L.; Fortes, C.J.; Brito-Melo, A.; Sarmento, A. The impact of wave energy farms in the shoreline wave climate: Portuguese pilot zone case study using Pelamis energy wave devices. *Renew. Energy* **2010**, *35*, 62–77. [CrossRef]
19. Rusu, E.; Guedes Soares, C. Coastal impact induced by a Pelamis wave farm operating in the Portuguese nearshore. *Renew. Energy* **2013**, *58*, 34–49. [CrossRef]
20. Diaconu, S.; Rusu, E. The environmental impact of a Wave Dragon array operating in the Black Sea. *Sci. World J.* **2013**, *2013*. [CrossRef] [PubMed]
21. Andrei-Tanase, Z.; Florin, O.; Alina, R. Coastal Impact Simulation of a Wave Dragon Farm Operating in the Nearshore of Mangalia. *Univ. Marit. Constanta Analele* **2014**, *15*, 65–70.
22. Zanopol, A.T.; Onea, F.; Rusu, E. Coastal impact assessment of a generic wave farm operating in the Romanian nearshore. *Energy* **2014**, *72*, 652–670. [CrossRef]

23. Chang, G.; Ruehl, K.; Jones, C.A.; Roberts, J.; Chartrand, C. Numerical modeling of the effects of wave energy converter characteristics on nearshore wave conditions. *Renew. Energy* **2016**, *89*, 636–648. [CrossRef]
24. O'Dea, A.; Haller, M.C.; Özkan-Haller, H.T. The impact of wave energy converter arrays on wave-induced forcing in the surf zone. *Ocean Eng.* **2018**, *161*, 322–336. [CrossRef]
25. Alexandre, A.; Stallard, T.; Stansby, P.K. Transformation of wave spectra across a line of wave devices. In Proceedings of the 8th European Wave and Tidal Energy Conference, Uppsala, Sweden, 7–10 September 2009.
26. O'Dea, A.M.; Haller, M.C. Analysis of the impacts of wave energy converter arrays on the nearshore wave climate. In Proceedings of the 2nd Annual Marine Energy Technology Symposium, Seattle, WA, USA, 15–17 April 2014.
27. Black, K.P. Review of Wave Hub Technical Studies: Impacts on Inshore Surfing Beaches. Available online: https://tethys.pnnl.gov/sites/default/files/publications/Impacts_on_Inshore_Surfing_Beaches.pdf (accessed on 9 May 2018).
28. Babarit, A. A database of capture width ratio of wave energy converters. *Renew. Energy* **2015**, *80*, 610–628. [CrossRef]
29. Monk, K.; Zou, Q.; Conley, D. An approximate solution for the wave energy shadow in the lee of an array of overtopping type wave energy converters. *Coast. Eng.* **2013**, *73*, 115–132. [CrossRef]
30. Mollison, D. Assessing the Wave Energy Resource. Available online: http://www.macs.hw.ac.uk/~denis/wave/spruce.pdf (accessed on 9 May 2018).
31. Lenee-Bluhm, P.; Paasch, R.; Özkan-Haller, H.T. Characterizing the wave energy resource of the US Pacific Northwest. *Renew. Energy* **2011**, *36*, 2106–2119. [CrossRef]
32. Bento, A.R.; Rusu, E.; Martinho, P.; Soares, C.G. Assessment of the changes induced by a wave energy farm in the nearshore wave conditions. *Comput. Geosci.* **2014**, *71*, 50–61. [CrossRef]
33. Monk, K.U.; Zou, Q.; Conley, D. The effect of diffraction on the redistribution of wave energy in the lee of an overtopping type wave enegry converter array. *Coast. Eng. Proc.* **2012**. [CrossRef]
34. Babarit, A. Impact of long separating distances on the energy production of two interacting wave energy converters. *Ocean Eng.* **2010**, *37*, 718–729. [CrossRef]
35. Pastor, J.; Liu, Y.C. Hydrokinetic energy: Overview and it's renewable energy potential for the Gulf of Mexico. In Proceedings of the 2012 IEEE Green Technologies Conference, Tulsa, OK, USA, 19–20 April 2012.
36. Pastor, J.; Liu, Y. Wave energy resource analysis for use in wave energy conversion. *J. Offshore Mech. Arct. Eng.* **2015**, *137*, 11901–11903.
37. Pastor, J.; Liu, Y. Wave climate resource analysis based on a revised gamma spectrum for wave energy conversion technology. *Sustainability* **2016**, *8*, 1321. [CrossRef]
38. Poate, T.G. Morphological Response of High-Energy Macrotidal Beaches. Ph.D. Thesis, Plymouth University, Plymouth, UK, 2011.
39. Black, K.P.; Andrews, C.J. Sandy shoreline response to offshore obstacles Part 1: Salient and tombolo geometry and shape. *J. Coast. Res.* **2001**, 82–93.
40. Hsu, J.R.C.; Silvester, R. Accretion behind single offshore breakwater. *J. Waterw. Port Coast. Ocean Eng.* **1990**, *116*, 362–380. [CrossRef]
41. Roelvink, D.; Reniers, A.; van Dongeren, A.; van Thiel de Vries, J.; McCall, R.; Lescinski, J. Modelling storm impacts on beaches, dunes and barrier islands. *Coast. Eng.* **2009**, *56*, 1133–1152. [CrossRef]
42. Thornton, E.B.; MacMahan, J.; Sallenger, A.H., Jr. Rip currents, mega-cusps, and eroding dunes. *Mar. Geol.* **2007**, *240*, 151–167. [CrossRef]
43. Scott, T.M.; Russell, P.E.; Masselink, G.; Wooler, A.; Short, A. High volume sediment transport and its implications for recreational beach risk. *Coast. Eng.* **2008**, 4250–4262.
44. Mead, S.; Black, K. Field studies leading to the bathymetric classification of world-class surfing breaks. *J. Coast. Res.* **2001**, 5–20.
45. Scarfe, B.E.; Healy, T.R.; Rennie, H.G. Research-based surfing literature for coastal management and the science of surfing-a review. *J. Coast. Res.* **2009**, *25*, 539–557. [CrossRef]
46. de Vriend, H.J.; Capobianco, M.; Chesher, T.; de Swart, H.E.; Latteux, B.; Stive, M.J.F. Approaches to long-term modelling of coastal morphology: A review. *Coast. Eng.* **1993**, *21*, 225–269. [CrossRef]
47. Morphodynamic Models: An Overview. Available online: ftp://ftp.geosc.psu.edu/pub/geosc/sling/PUBLICATIONS_SLINGERLAND/2001-2010/Syvtitskietal2009.pdf (accessed on 10 January 2013).

48. Wright, L.D.; Short, A.D. Morphodynamic variability of surf zones and beaches: A synthesis. *Mar. Geol.* **1984**, *56*, 93–118. [CrossRef]
49. Masselink, G.; Short, A.D. The effect of tide range on beach morphodynamics and morphology: A conceptual beach model. *J. Coast. Res.* **1993**, *9*, 785–800.
50. Scott, T.; Masselink, G.; Russell, P. Morphodynamic characteristics and classification of beaches in England and Wales. *Mar. Geol.* **2011**, *286*, 1–20. [CrossRef]
51. Anthony, E.J. Sediment-wave parametric characterization of beaches. *J. Coast. Res.* **1998**, *14*, 347–352.
52. Jackson, D.W.T.; Cooper, J.A.G.; Del Rio, L. Geological control of beach morphodynamic state. *Mar. Geol.* **2005**, *216*, 297–314. [CrossRef]
53. Jiménez, J.A.; Guillén, J.; Falqués, A. Comment on the article "Morphodynamic classification of sandy beaches in low energetic marine environment" by Gómez-Pujol, L, Orfila, A., Cañellas, B., Alvarez-Ellacuria, A., Méndez, F.J., Medina, R. and Tintoré, J. Marine Geology, 242, pp. 235–246, 2007. *Mar. Geol.* **2008**, *255*, 96–101. [CrossRef]
54. Almar, R.; Castelle, B.; Ruessink, B.G.; Sénéchal, N.; Bonneton, P.; Marieu, V. Two-and three-dimensional double-sandbar system behaviour under intense wave forcing and a meso–macro tidal range. *Cont. Shelf Res.* **2010**, *30*, 781–792. [CrossRef]
55. Stokes, C.; Davidson, M.; Russell, P. Observation and prediction of three-dimensional morphology at a high-energy macrotidal beach. *Geomorphology* **2015**, *243*. [CrossRef]
56. Stokes, C. Coastal Impacts in the Lee of a Wave Energy Site: Waves, Beach Morphology and Water-Users (Wave Hub, Cornwall, UK). Ph.D. Thesis, Plymouth University, Plymouth, UK, 2015.
57. Hutt, J.A.; Black, K.P.; Mead, S.T. Classification of surf breaks in relation to surfing skill. *J. Coast. Res.* **2001**, 66–81.
58. Stokes, C.; Beaumont, E.; Russell, P.; Greaves, D. Coastal impacts of marine renewables: Perception of breaker characteristics by beach water users. *J. Coast. Res.* **2014**, *70*. [CrossRef]
59. Mead, S.; Black, K. Functional component combinations controlling surfing wave quality at world-class surfing breaks. *J. Coast. Res.* **2001**, 21–32.
60. Battjes, J.A. Surf similarity. In Proceedings of the 14th International Conference on Coastal Engineering, Copenhagen, Denmark, 24–28 June 1974.
61. Walker, J.R.; Palmer, R.Q. *Surf Parameters; a General Surf Site Concept*; LOOK Laboratory Technical Report; Department of Ocean Engineering, University of Hawaii: Honolulu, HI, USA, 1971; Volume 18.
62. MacMahan, J.; Reniers, A.; Brown, J.; Brander, R.; Thornton, E.; Stanton, T.; Brown, J.; Carey, W. An introduction to rip currents based on field observations. *J. Coast. Res.* **2011**, *27*, 3–6. [CrossRef]
63. Brighton, B.; Sherker, S.; Brander, R.; Thompson, M.; Bradstreet, A. Rip current related drowning deaths and rescues in Australia 2004–2011. *Nat. Hazards Earth Syst. Sci.* **2013**, *13*, 1069–1075. [CrossRef]
64. Scarfe, B.E.; Elwany, M.H.S.; Mead, S.T.; Black, K.P. The Science of Surfing Waves And Surfing Breaks—A Review. Available online: https://escholarship.org/content/qt6h72j1fz/qt6h72j1fz.pdf (accessed on 12 September 2012).
65. Stokes, C.; Russell, P.; Davidson, M. Subtidal and intertidal three-dimensionality at a high energy macrotidal beach. *J. Coast. Res.* **2016**, *75*, 472–476. [CrossRef]
66. Booij, N.; Ris, R.C.; Holthuijsen, L.H. A third-generation wave model for coastal regions: 1. Model description and validation. *J. Geophys. Res. Ocean.* **1999**, *104*, 7649–7666. [CrossRef]
67. Fernandez, H.; Iglesias, G.; Carballo, R.; Castro, A.; Fraguela, J.A.; Taveira-Pinto, F.; Sanchez, M. The new wave energy converter WaveCat: Concept and laboratory tests. *Mar. Struct.* **2012**, *29*, 58–70. [CrossRef]
68. Silva, D.; Rusu, E.; Soares, C.G. Evaluation of various technologies for wave energy conversion in the Portuguese nearshore. *Energies* **2013**, *6*, 1344–1364. [CrossRef]
69. Larson, M.; Hoan, L.X.; Hanson, H. Direct formula to compute wave height and angle at incipient breaking. *J. Waterw. Port Coast. Ocean Eng.* **2010**, *136*, 119–122. [CrossRef]
70. XBeach Model Description and Manual. Available online: https://www.researchgate.net/profile/Arnold_Van_Rooijen/publication/306037851_XBeach_Technical_Reference_Kingsday_Release/links/57ac292f08ae3765c3b88518/XBeach-Technical-Reference-Kingsday-Release.pdf (accessed on 13 June 2016).
71. Roelvink, D.; McCall, R.; Mehvar, S.; Nederhoff, K.; Dastgheib, A. Improving predictions of swash dynamics in XBeach: The role of groupiness and incident-band runup. *Coast. Eng.* **2018**, *134*, 103–123. [CrossRef]

72. Roelvink, D.; Reniers, A. Dissipation in random wave groups incident on a beach. *Adv. Coast. Eng.* **1993**, *19*, 127–150. [CrossRef]
73. Daly, C.; Roelvink, D.; van Dongeren, A.; van Thiel de Vries, J.; McCall, R. Validation of an advective-deterministic approach to short wave breaking in a surf-beat model. *Coast. Eng.* **2012**, *60*, 69–83. [CrossRef]
74. Svendsen, I.A. Wave heights and set-up in a surf zone. *Coast. Eng.* **1984**, *8*, 303–329. [CrossRef]
75. Nairn, R.B.; Roelvink, J.A.; Southgate, H.N. Transition zone width and implications for modelling surfzone hydrodynamics. In Proceedings of the 22th International Conference on Coastal Engineering, Delft, The Netherlands, 2–6 June 1990; pp. 68–81.
76. Dano, R.; Reniers, A. *A Guide to Modeling Coastal Morphology. Advances in Coastal Engineering*; World Scientific: Singapore, 2011; ISBN 978-981-4304-25-2.
77. Longuet-Higgins, M.S.; Stewart, R.W. Radiation stress and mass transport in gravity waves, with application to 'surf beats'. *J. Fluid Mech.* **1962**, *13*, 481–504. [CrossRef]
78. Longuet-Higgins, M.S.; Stewart, R.W. Radiation stresses in water waves; a physical discussion, with applications. *Deep Sea Res. Oceanogr. Abstr.* **1964**, *11*, 529–562. [CrossRef]
79. Phillips, O.M. *The Dynamics of the Upper Ocean*, 2nd ed.; Cambridge University Press: New York, NY, USA, 1977; ISBN 0-521-29801-6.
80. Van Dongeren, A.; Bolle, A.; Vousdoukas, M.I.; Plomaritis, T.; Eftimova, P.; Williams, J.; Armaroli, C.; Idier, D.; Van Geer, P.; van Thiel de Vries, J.; et al. MICORE: Dune erosion and overwash model validation with data from nine European field sites. In *Proceedings of Coastal Dynamics 2009*; World Scientific: Singapore, 2009; Volume 9, pp. 1–15.
81. McCall, R.T.; van Thiel de Vries, J.S.M.; Plant, N.G.; Van Dongeren, A.R.; Roelvink, J.A.; Thompson, D.M.; Reniers, A.J.H.M. Two-dimensional time dependent hurricane overwash and erosion modeling at Santa Rosa Island. *Coast. Eng.* **2010**, *57*, 668–683. [CrossRef]
82. Bolle, A.; Mercelis, P.; Roelvink, D.; Haerens, P.; Trouw, K. Application and validation of XBeach for three different field sites. *Coast. Eng. Proc.* **2011**, *1*, 40. [CrossRef]
83. Splinter, K.D.; Palmsten, M.L. Modeling dune response to an East Coast Low. *Mar. Geol.* **2012**, *329*, 46–57. [CrossRef]
84. Plant, N.G.; Holland, K.T.; Puleo, J.A. Analysis of the scale of errors in nearshore bathymetric data. *Mar. Geol.* **2002**, *191*, 71–86. [CrossRef]
85. Robins, P.E.; Neill, S.P.; Lewis, M.J. Impact of tidal-stream arrays in relation to the natural variability of sedimentary processes. *Renew. Energy* **2014**, *72*, 311–321. [CrossRef]

© 2018 by the authors. Licensee MDPI, Basel, Switzerland. This article is an open access article distributed under the terms and conditions of the Creative Commons Attribution (CC BY) license (http://creativecommons.org/licenses/by/4.0/).

energies

MDPI

Article

Spatial Environmental Assessment Tool (SEAT): A Modeling Tool to Evaluate Potential Environmental Risks Associated with Wave Energy Converter Deployments

Craig Jones [1,*], Grace Chang [1,*], Kaustubha Raghukumar [1], Samuel McWilliams [1], Ann Dallman [2] and Jesse Roberts [2]

[1] Integral Consulting Inc., Santa Cruz, CA 95060, USA; kraghukumar@integral-corp.com (K.R.); smcwilliams@integral-corp.com (S.M.)

[2] Sandia National Laboratories, Albuquerque, NM 87123, USA; ardallm@sandia.gov (A.D.); jdrober@sandia.gov (J.R.)

* Correspondences: cjones@integral-corp.com (C.J.); gchang@integral-corp.com (G.C.); Tel.: +1-831-466-9630 (C.J.); +1-805-967-5640 (G.C.)

Received: 28 June 2018; Accepted: 30 July 2018; Published: 6 August 2018

check for updates

Abstract: Wave energy converter (WEC) arrays deployed in coastal regions may create physical disturbances, potentially resulting in environmental stresses. Presently, limited information is available on the nature of these physical disturbance or the resultant effects. A quantitative Spatial Environmental Assessment Tool (SEAT) for evaluating the potential effects of wave energy converter (WEC) arrays on nearshore hydrodynamics and sediment transport is presented for the central Oregon coast (USA) through coupled numerical model simulations of an array of WECs. Derived climatological wave conditions were used as inputs to the model to allow for the calculation of risk metrics associated with various hydrodynamic and sediment transport variables such as maximum shear stress, bottom velocity, and change in bed elevation. The risk maps provided simple, quantitative, and spatially-resolved means of evaluating physical changes in the vicinity of a hypothetical WEC array in response to varying wave conditions. The near-field risk of sediment mobility was determined to be moderate in the lee of the densely spaced array, where the potential for increased sediment deposition could result in benthic habitat alteration. Modifications to the nearshore sediment deposition and erosion patterns were observed near headlands and topographic features, which could have implications for littoral sediment transport. The results illustrate the benefits of a risk evaluation tool for facilitating coastal resource management at early market marine renewable energy sites.

Keywords: marine renewable energy; ocean energy; wave energy; environmental effects; wave modeling; wave propagation; numerical modeling; sediment dynamics; risk assessment

1. Introduction

Wave energy has enormous potential for supporting energy security and reducing greenhouse gas emissions. For example, it has been estimated that the recoverable wave resource potential along the United States outer continental shelf is 1170 TWh/year [1]. This amounts to over a quarter of the United States' total annual electricity consumption. Wave energy converter (WEC) technology development over the past few decades has produced a wide array of wave energy technologies. The International Energy Agency Ocean Energy Systems has developed categories to describe these technologies, which include oscillating water columns, wave activated bodies (floating point absorbers), and overtopping device archetypes [2]. Within these categories, the European Marine Energy Center has counted over 250 developers to date, showing the significant interest in wave energy conversion [3].

Scaled versions of commercial devices are currently being tested in dedicated wave tanks as well as at oceanic test sites [4–6]. While present-day testing has generally involved laboratory-scale devices or single devices, arrays of multiple ocean-deployed WEC devices will be necessary for commercial-scale wave energy conversion to onshore electrical power. These WEC arrays, or wave farms, may conflict with other ocean uses, e.g., shipping and navigation, commercial fishing, and recreation [7]. Wave farms consisting of tens to hundreds of devices may have combinations of beneficial and adverse near- and far-field physical effects on the marine environment, such as changes to wave characteristics, circulation patterns, and sediment dynamics (e.g., [8–15]).

Characterization of the physical environment and associated potential alteration of an environment due to a single WEC or arrays of WECs must be ascertained in order to make informed predictions of device performance, system design requirements for hydrodynamic loads, and resource management decisions based on environmental responses. The survivability and maintainability of wave energy devices is directly influenced by nearshore circulation and waves, primarily where WEC infrastructure (e.g., anchors, piles) is exposed to ocean hydrodynamic forces. The WEC-associated infrastructure has the potential to modify the circulation in the water column and, depending on the location, may alter important environmental processes. For example, nutrient delivery, light availability, benthic habitat, larval motility, and many other environmental parameters may be changed by WEC-induced alteration of circulation and sediment transport patterns, thereby affecting marine and shoreline ecosystems [16].

Multiple numerical case studies have quantified the effects of wave farms on coastal processes and assessed the influences of WECs on hydrodynamics and device performance. The optimum wave farm layout for the Wave Dragon WEC was determined by Beels et al. [9] by investigating wake effects using a mild-slope wave propagation model. O'Boyle et al. [17] performed wave tank experiments to evaluate near-field wave disturbances from wave scattering and radiation around different array layouts of five oscillating water column WECs. These studies suggest that the effects of the wave farm layout on hydrodynamics due to wave scattering can be controlled by device spacing relative to wavelengths typical to the wave energy site. They also determined that wave radiation effects were found to be significant and dependent on WEC device performance characteristics and stated the importance of considering these characteristics when optimizing power capture or evaluating WEC effects on nearshore processes. Similar to O'Boyle et al. [17], Özkan-Haller et al. [15] performed wave tank experiments and numerical modeling to investigate wave field modifications in the presence of WECs. Their results suggested that the downstream environmental effects of WEC arrays may be diminished through device operational changes, specifically indicating that WECs should be designed to operate in a wave climate that is equal to or longer than the period of peak energy extraction.

Gonzales-Santamaria et al. [18] discovered significant wave farm impacts on morphological changes at the Wave Hub site (England). The authors quantified littoral transport in the lee of the WEC array and concluded that longshore currents could lead to enhanced deposition in the lee of the wave farm as well as local erosion and a northward shift in erosion and deposition patterns due to wave farm-induced wave diffraction. Abanades et al. [12] used wave and morphodynamic numerical models to investigate the effectiveness of a wave farm as a coastal defense mechanism in Cornwall, England. Their results showed substantial reduction in the erosion of beach profiles and the authors concluded that a wave farm could be considered a viable method of coastal protection. Similarly, Zanopol et al. [19] performed numerical wave and circulation models and ascertained that the northern Romanian coast of the Black Sea may benefit from certain configurations of WEC arrays, which could induce the transport of sediment from the north and enhance shoreline accretion.

Many studies have shown that the deployment of a WEC array has the potential to alter waves and nearshore currents that govern sediment mobility. It is therefore critical to evaluate the potential effects of WEC array deployments and address key physical stressors at a study site, including changes to sediment transport. The primary goal of this study is to describe the Spatial Environmental Assessment Tool (SEAT), a quantitative tool that can be used to provide input to the environmental assessment process by evaluating the potential effects of WECs on waves, circulation, and sediment transport, and thus understand potential environmental risks associated with WEC deployments. The SEAT is not intended to replace the traditional environmental assessment process, but it can provide information on a subset of potential environmental risks. As a baseline example, the SEAT was used to quantitatively assess seabed changes as the result of WEC deployments for a case study off the coast of Newport, Oregon, USA. Two stressors, bottom shear stress and seabed elevation, and metrics specific to each are presented to establish the basis of the SEAT framework for use in evaluating other environmental metrics.

2. Materials and Methods

Nonlinear combinations of winds, tides, and waves typically control nearshore circulation and mixing, with waves being the dominant driver of nearshore circulation in energetic coastlines such as the Oregon coast. Therefore, in order to capture complex wave-, tide-, and wind-induced currents and mixing, the SEAT incorporates a coupled wave, circulation, and sediment transport model with quantitative spatial risk assessment for investigating simulated WECs in nearshore environments. The risk metrics developed for the SEAT relate changes in a stressor (e.g., bottom shear stress) to values considered critical for benthic habitats (e.g., changes in the bottom substrate).

2.1. Study Site

The case study site is located approximately 3.7 km to 5.5 km offshore of Newport, Oregon, USA, north of Yaquina Head, at a water depth of approximately 50 m (Figure 1). The site is known as the North Energy Test Site (NETS) of the Pacific Marine Energy Center (PMEC), where WECs have been deployed and future deployments are planned. This 1.8-km^2 site was chosen as the potential location of a WEC array due to the available energy resources and limited potential environmental effects. The availability of data necessary for model input boundary conditions, calibration, and validation near NETS is an additional benefit to the evaluation of this site. Long-term wave characterization and current velocity measurements are available through Oregon State University's (OSU) Acoustic Wave and Current (AWAC; Nortek, Boston, MA, USA) data collected in the summer of 2005 near NETS. Additionally, daily-averaged surface velocities encompassing the model domain are available from two Coastal Ocean Dynamics Applications Radar (CODAR) systems deployed in the area by OSU researchers (http://bragg.oce.orst.edu/ORCoast/).

The waters off Newport, Oregon are characterized by a year-round energetic wave field. Swell heights range from zero to 7 m, swell directions vary seasonally from 180° to 315°, and wave periods range from 4 s to 20 s. The site coastline is defined by pocket beaches up to 16 km long with headlands of large cliffs. Sandy sediment ranging from 100 µm to 500 µm in diameter (fine to coarse sand) compose much of the beach material along the site's coastline [20,21] and extend into the offshore region [22]. Seasonal variations in storm intensity move sediment between the offshore and nearshore zones. Variations in sediment transport are primarily driven by waves whose bulk statistics vary with the seasons. Winter storms are predominantly from the south. Southerly waves transport sands northward along the beaches where headlands trap sediment and promote deposition [23]. Less energetic summer swells move sediment to the south. Over long periods, this process results in little net sediment transport out of the system due to alongshore transport.

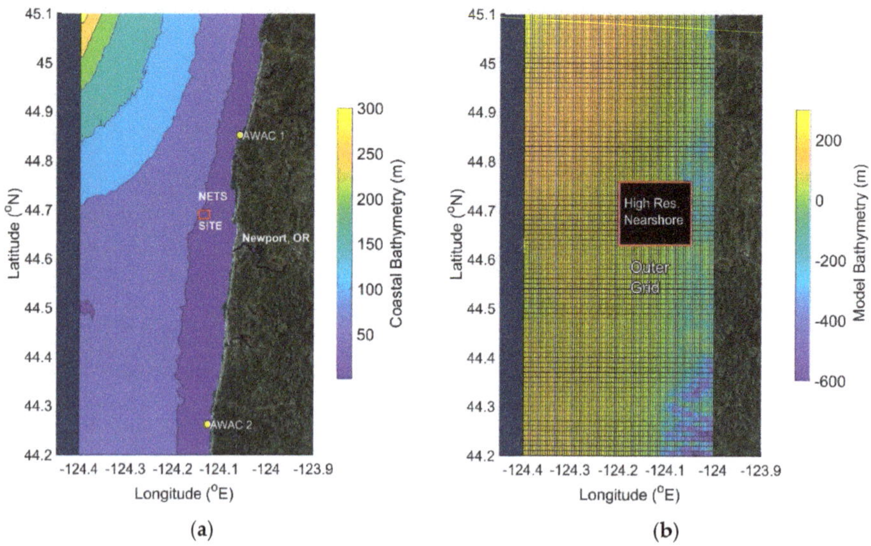

Figure 1. (**a**) Coupled Delft3D-FLOW-SNL-SWAN model domain and locations of available measurements; (**b**) nested model grids. The star indicates the location of the National Oceanic and Atmospheric Administration National Data Buoy Center (NOAA NDBC) buoy station 46089. NETS: North Energy Test Site; AWAC: Acoustic Wave and Current; OR: Oregon.

2.2. Spatial Environmental Assessment Tool (SEAT) Coupled Numerical Model

A coupled wave, circulation, and sediment transport model was developed using a hybrid modeling framework (Figure 2). Waves were modeled using a modified version of Simulating WAves Nearshore (SWAN) [24,25], named SNL-SWAN (Sandia National Laboratories-SWAN), which is a module developed by SNL and incorporated into the open-source Delft3D framework. SNL-SWAN incorporates a WEC module that accounts for device-specific WEC power take-off characteristics to more accurately evaluate each device's effects on wave propagation and ultimately nearshore hydrodynamics [26–29]. Delft3D-FLOW is a multidimensional hydrodynamic and sediment transport model that is capable of quantifying circulation (non-steady), waves, and sediment transport phenomena as a result of forcing by tides and meteorological processes [30,31]. The coupled model allows for the evaluation of tidal- and wave-driven circulation and sediment transport in the model domain. Within the model framework, WECs are treated as obstacles that allow wave energy to be propagated and to be absorbed by the WECs.

The Newport, Oregon case study model configuration consisted of a three-dimensional (latitude, longitude, depth) nested grid over which circulation and waves were modeled (Figure 2). In Delft3D-FLOW, circulation in a nested grid was modeled as a two-way coupling between grids, while the coupled wave module was run as one-way nesting, meaning information only passed from the outer to the inner grid. The coupled Delft3D-FLOW-SNL-SWAN model simulated the hydrodynamics of tidal- and wave-driven circulation at a 3-s time step for model stability, then coupled the circulation to SNL-SWAN every 3 h to capture variations in the temporal wave field (Figure 2). This two-way coupling of waves and currents allowed for the model to capture the non-linear interactions between currents and waves and the subsequent effects on the system (Case study model set-up files are available in Supplementary Materials).

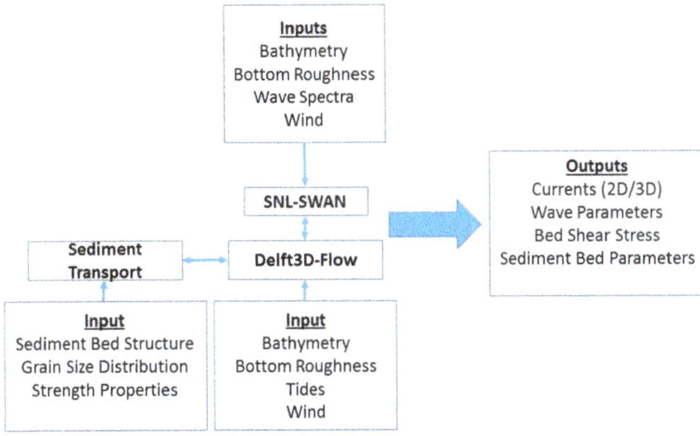

Figure 2. Flow chart illustrating the coupled Delft3D-FLOW-SNL-SWAN wave, circulation, and sediment transport models. SNL-SWAN: Sandia National Laboratories-Simulating WAves Nearshore.

2.2.1. Model Inputs

Wave propagation and circulation in the coastal zone are fundamentally controlled by bathymetry; therefore, the coupled numerical model required accurate bottom bathymetry data as input. For the Newport, Oregon case study, model depths were defined from the National Oceanic and Atmospheric Administration (NOAA) National Geophysical Data Center high-resolution (approximately 10 m) digital elevation model of the Oregon coast. The bathymetry data was mapped to the model grid using bi-linear interpolation (Figure 1). Generally, the digital elevation model has higher spatial resolution than the computational model. Water depths within the model domain varied from less than 5 m nearshore to over 200 m in the northwestern corner of the computational grid.

2.2.1.1. Circulation Model

Surface currents in the model domain were driven by the application of time varying, spatially uniform winds. Hourly winds were derived from measurements by NOAA National Data Buoy Center (NDBC) buoy station 46089 at the outer edge of the largest-scale model domain (Figure 1). Tidal boundary conditions were applied using the TOPEX/POSEIDON global tide model (TPXO09) interpolated onto the model grid [32]. Water density was assumed constant at 1025 kg/m^3 and horizontal eddy viscosity was held constant at 1.0 m^2/s throughout the computational domain. Variations in density due to temperature and salinity gradients were not included in these screening level coastal models, but can be readily incorporated into future studies.

2.2.1.2. Wave Model

A comprehensive analysis of expected wave conditions on the Oregon coast was conducted using seven years (2005–2011) of modeled wave conditions [33]. This analysis provided approximately 300 discrete sea state bins, characterized by significant wave height (H_s), peak period (T_p), and mean wave direction (D_p) [34], along with the probability of occurrence of each sea state. In order to incorporate directionality into the characterization of the sea states, the data were split into four directional bins (200°–230°, 230°–260°, 260°–290°, 290°–320°) over which the analyzed wave directions were distributed. A joint probability distribution (JPD) of H_s and T_p was calculated for each of the four directional bins (Figure 3).

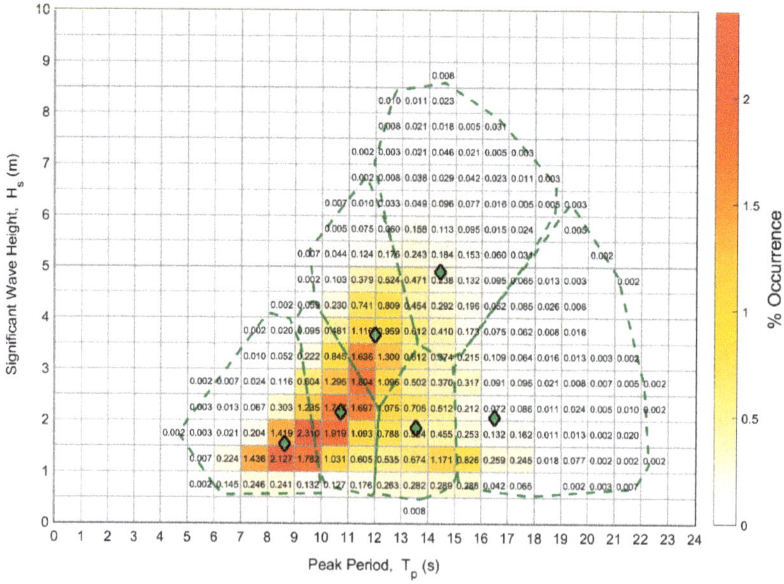

Figure 3. An example joint probability distribution for significant wave height and direction when applied to swell directions arriving from between 260° and 290°. The green diamonds indicate cluster centroids and the green lines indicate cluster boundaries.

Clustering analysis (k-means) was used to reduce the number of sea states from 300 events to 24 events for model input following Bull and Dallman [35] (Table 1). Briefly, k-means clustering analysis determined the sea states, or cluster centroids, via the minimization of the squared Euclidian distance between each point and a cluster centroid. The number of clusters within each directional bin was defined to be six, and an iterative partitioning method optimized the centroid of each cluster and the cluster size by minimizing the sum of point-to-centroid distances, summed over all six clusters. The resulting sea states generated by k-means are represented by the green diamonds in Figure 3.

The probability of occurrence of each sea state within each direction bin was defined as the number of individual (H_s, T_p) points assigned to a cluster, divided by the total number of (H_s, T_p) data points within a direction sector. Similarly, the probability of occurrence over all direction bins was calculated by dividing by the total number of (H_s, T_p) data points within the entire model data record. The direction, D_p, assigned to the sea states in Table 1 was determined by taking the average of the individual D_p values over the cluster. This process provided a reproducible and adjustable method to characterize the occurrence of site conditions. Clustering analysis enabled the modeling of discrete events based on probability analysis and the development of risk potential based on modeling results. This approach has the benefit of reducing run times by modeling discrete events rather than simulating long-term time series, but becomes limited by the lack of continuity between events. Despite this limitation, pairing the probability of an event with results from the modeled conditions can be used to inform stakeholders of potential long-term effects of a WEC array on a specific environment and vice versa.

Table 1. Operational sea states selected for each of the four direction vectors. The sea states are the cluster centers, identified using k-means clustering [35].

Directional Bin	H_s (m)	T_p (s)	D_p (deg)	% of Direction Bin	Total % Occurrence of the Cluster
200°–230°	1.76	6.60	221.8	15.41	0.39
	2.67	8.62	220.8	40.68	1.03
	4.06	10.16	221.3	23.47	0.59
	1.37	15.33	224.0	8.06	0.20
	7.05	12.60	223.6	3.42	0.09
	2.11	11.63	223.8	8.96	0.23
230°–260°	4.91	13.62	251.2	11.99	1.76
	1.70	7.73	244.8	15.03	2.21
	2.69	9.80	247.6	26.75	3.94
	1.23	14.62	248.8	18.54	2.73
	2.31	17.54	249.4	3.97	0.58
	2.94	11.77	250.6	23.72	3.49
260°–290°	4.90	14.43	275.8	8.78	4.70
	1.54	8.62	278.0	20.90	11.19
	3.66	12.00	277.2	20.95	11.21
	2.16	10.71	277.5	25.39	13.59
	1.85	13.54	277.2	16.21	8.67
	2.05	16.51	276.4	7.77	4.159
290°–320°	1.81	9.11	298.6	28.46	8.297
	2.16	13.35	295.7	10.70	3.120
	1.49	7.12	304.6	22.86	6.664
	2.66	11.02	297.2	26.48	7.720
	2.08	16.53	295.6	5.28	1.540
	4.65	13.23	296.2	6.22	1.813

The results of clustering analysis provided 24 discrete wave model inputs, characterized by bulk wave parameters (H_s, T_p, and D_p), for the Newport, Oregon case study. Significant wave height ranged from 1.23 m to 7.05 m, peak period ranged from 6.6 s to 17.5 s, and the directional window ranged from 220° to 305° (Table 1). Uniform wave boundary conditions with a JONSWAP spectrum centered on the 24 sets of bulk wave parameters were applied to the SNL-SWAN wave model. While the use of bulk parameter wave conditions (converted to a JONSWAP spectrum) seemingly represents a lower resolution wave model compared to the full spectral specification of boundary conditions, it avoids the contamination of model results by wave conditions that fall outside the pertinent events determined though the cluster analysis described above and will generally represent the peak wave-associated stressors being evaluated.

2.2.1.3. Sediment Transport Module

The Delft3D-FLOW-SNL-SWAN sediment transport module uses vertical layers in the sediment bed to define sediment characteristics and the initial thickness of each initial bed layer defines the amount of sediment available for transport [36]. Sediment layers are defined as fractions of cohesive and noncohesive sediment types, with size classes of various fractions defined in terms of mean diameter (noncohesive) and settling velocity (cohesive sediment). A unique critical shear stress for erosion is associated with each sediment type, and the critical shear stress of the sediment layer is calculated as the weighted fraction of critical shear stresses associated with each size class. Removal of non-cohesive sediment from a parent bed layer (i.e., erosion) occurs if the modeled shear stress exceeds the critical shear stress. Sediment can be advected as a suspended load or a bedload, with the only difference relating to the treatment of transport properties within the water column versus the sediment bed. Erosion rates of the sediment bed for sandy non-cohesive sediment in Delft3D-FLOW-SNL-SWAN are based on empirically defined values [37].

Based on the Coastal and Marine Ecological Classification Standard (CMECS, [22]), the majority of the coastal region within the Newport, Oregon study site is comprised of sandy sediment. However, fine cohesive sediment does exist in the Oregon coastal environment; therefore, three non-cohesive sediment size classes (sands) and one cohesive size class (silt) were incorporated into the calibrated and validated Delft3D-FLOW-SNL-SWAN wave and circulation model. The sandy sediment was represented with grain sizes of 250 μm, 200 μm, and 100 μm in equal fractions and a specific bulk density of 2650 kg/m^3, representative of quartz sand. The three grain size classes highlighted differing behaviors of sediment in the environment. Non-cohesive sediment such as mud was included with a settling velocity of 2.5×10^{-4} m/s.

2.2.2. Model Validation

Prior to the introduction of simulated WECs to the coupled numerical model, it was necessary to validate the Delft3D-FLOW-SNL-SWAN model. The model wave and current validation data were provided by in situ CODAR and AWAC measurements within the model domain, collected by OSU researchers. Due to disparate time periods of wave and current data availability, multiple model runs over different time periods were used to compare the flow and wave model results to measurements. The SNL-SWAN wave model was initiated with offshore NOAA NDBC data and validated with in situ directional wave spectra collected by an AWAC deployed at a nearshore location within the model grid (−124.058, 44.8514; AWAC1; Figure 1) at a water depth of 12 m. Modeled versus measured comparisons were made over a 28-day period from 1 to 28 August 2005 during a period of multiple swell conditions, which provided an excellent comparison period for the model (Figure 4). CODAR daily-averaged surface current velocity measurements during a four-week period between 1 and 28 October 2009 were used to validate the circulation model (not shown). Two measurement locations, C1 and C2, were chosen for model validation due to their proximity to the NETS site (Figure 1).

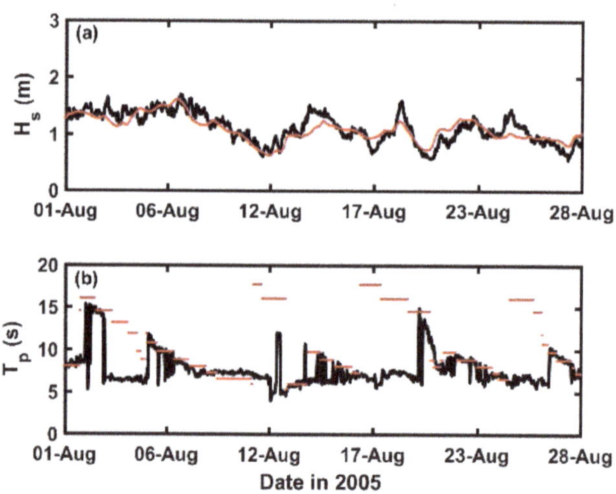

Figure 4. SNL-SWAN model validation results comparing (**a**) significant wave height, H_s, and (**b**) peak wave period, T_p. Measured data from the Acoustic Wave and Current meter (AWAC), operated by Oregon State University, are shown in black and SNL-SWAN model data are shown in red.

Statistical metrics were determined for modeled versus measured wave parameters and current velocities over the time periods of model validation (see Section 2.2.2) (Table 2). The metrics included model skill (skill), root mean square error (RMSE), bias, and mean percent difference (Avg % diff), and were computed as follows:

$$\text{Skill} = 1 - \frac{\sum|\text{modeled} - \text{measured}|^2}{\left(\sum\left|\text{modeled} - \overline{\text{measured}}\right| + \left|\text{measured} - \overline{\text{modeled}}\right|\right)^2}, \quad (1)$$

$$\text{RMSE} = \sqrt{\overline{(\text{modeled} - \text{measured})^2}}, \quad (2)$$

$$\text{Bias} = \overline{\text{modeled} - \text{measured}}, \quad (3)$$

$$\text{Avg \% diff} = \overline{100 \times (\text{modeled} - \text{measured}) / \text{measured}}, \quad (4)$$

where overbars indicate means and summation is over the entire model domain. Coefficients of determination (correlation) were determined through least-squares linear regression analysis between modeled and measured quantities. The results demonstrate the excellent ability of the numerical model to reproduce measurements (Table 2; Figure 4).

Table 2. Comparison of modeled versus measured H_s and T_p and east-west (u) and north-south (v) current velocities. In situ wave data were measured at location AWAC1 (Figure 1). Coastal Ocean Dynamics Applications Radar (CODAR) data were compared for locations south (C1) and west (C2) of the wave energy converter (WEC) array (Figure 1). See Equations (1)–(4) for definitions of Skill, root mean square error (RMSE), Bias, Avg % diff, and Correlation.

Metric	H_s	T_p	u (C1, C2)	v (C1, C2)
Skill	0.90 m	0.90 s	0.7, 0.7 cm/s	0.90, 0.9 cm/s
RMSE	0.18 m	1.90 s	5.6, 6 cm/s	5.5, 6.3 cm/s
Bias	0.09 m	−0.71 s	3.5, 4.1 cm/s	−0.8, 1.4 cm/s
Avg % diff	12.42%	15.51%	18.9%, 13.9%	9%, 7%
Correlation	0.84	0.71	0.8, 0.8	0.9, 0.8

2.2.3. Case Study Model Runs

Model runs for the Newport, Oregon case study were conducted for baseline scenarios (no WECs) and WEC scenarios to study the changes within the system due to the presence of WECs. A single type of WEC device was considered for this study, the floating oscillating water column (F-OWC) inspired by the Ocean Energy (OE) Buoy developed by Ocean Energy Ltd., which has a modeled power matrix computed according to Babarit et al. [38] (device characteristics described by Babarit et al. [38]). The power matrix represents the energy extracted from the environment by the device, given a specific sea state prescribed by the significant wave height and peak period. One 18-device array configuration was evaluated at the Newport, Oregon NETS case study site, where each of the 18 modeled devices had a diameter of 50 m spaced 200 m (four diameters) apart in both the east–west and north–south directions (Figure 5).

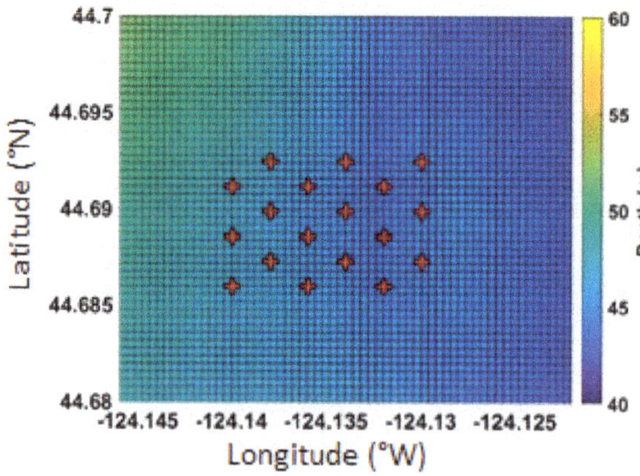

Figure 5. The simulated wave energy converter (WEC) array configuration.

The fully coupled Delft3D-FLOW-SNL-SWAN model was run for each of the 24 discrete wave events as returned by the cluster analysis (Table 1) with and without a simulated WEC array for a total of 48 model runs. Each model run was conducted for a period of 24 h (approximately two tidal cycles), allowing for variability in tidal conditions to affect sediment transport. Results were compiled at the end of the 24-h period, allowing for 12 h over which sediment could potentially be mobilized. The spin-up period of 12 h allowed the flow model to equilibrate and reduce the potential for numerical instabilities to cause spurious and unwanted changes to the sediment bed. The sediment transport module was run to update the bed thickness with every model time step to simulate the two-way feedbacks between the changing water depth, flow, and wave field and erosion and deposition processes. Following each 24-h model run, the sediment bed at the last time step was chosen for further evaluation. Parameters of interest such as bottom shear stress and bed elevation were extracted at the end of the model run period, and risk metrics were computed.

2.3. SEAT: Risk Assessment

The Newport, Oregon case study focused on two SEAT stressors (near-bed shear stress and seabed elevation) that are considered critical for benthic habitat assessment. The ratio of the shear stress to critical shear stress plays a key role in assessing the risk of sediment mobility associated with the introduction of WECs in an environment. For example, WECs may have the effect of lowering shear stress in the lee of the WEC array. Should the WEC-induced decrease in shear stress result in shear stresses below critical that were once greater than critical prior to the WEC deployment, then an erosional environment would be transformed into a depositional environment. The SEAT therefore defines a sediment transport parameter, T, for each model grid point as:

$$T(x,y) = \frac{\tau(x,y)}{\tau_c(x,y)}, \tag{5}$$

where τ and τ_c represent near-bed shear stress and sediment bed critical shear stress and (x,y) denotes model grid points in the longitudinal and latitudinal directions. The sediment transport parameter provides an indication of the ability of sediment to be mobilized. For example, a value of $T(x,y) > 1$ occurs when shear stresses exceed the critical shear stress of the underlying sediment, as determined by the median particle size of underlying sediment, allowing sediment to be mobilized.

When summed over the 24 cases of wave conditions as returned from the cluster analysis (see Table 1) and dropping the explicit dependence on (x,y), the resulting transport parameter for the baseline (no WECs) scenarios can be expressed as:

$$T_b = \sum_i P_i \frac{\tau}{\tau_c}, \tag{6}$$

where T_b represents the baseline transport parameter and P_i is the probability of the occurrence of the wave event. Similarly, a transport parameter, T_w, was defined for the model scenario with WECs present using the analog of Equation (6). Possible scenarios associated with sediment mobility, based on the relationship between T_b and T_w, and the critical shear stress of the underlying sediment are shown in Figure 6.

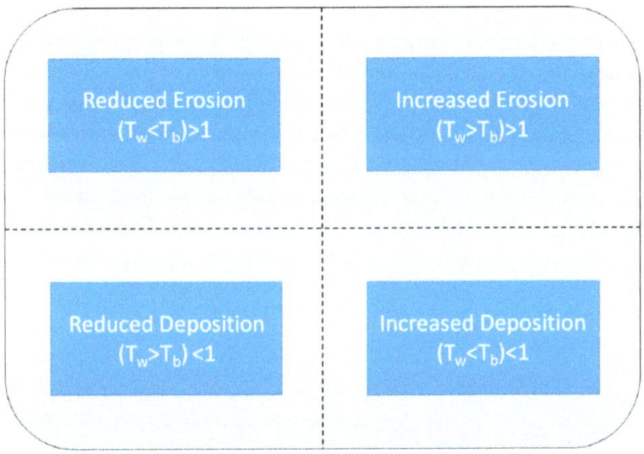

Figure 6. Sediment mobilization regimes associated with the relationship between the transport parameter for baseline (no WECs) and WEC scenarios, T_b and T_w (Equations (5) and (6)), respectively.

Mathematically, the risk of sediment mobility, R_τ, can be defined in terms of a non-dimensional number:

$$R_\tau = \text{int}\left(\frac{T_w - T_b}{|T_w - T_b|} T_w\right) + [T_w - T_b], \tag{7}$$

where the int( ) operation represents the integer value of the term. The first term in Equation (7) can be thought of as constraining the risk to the correct quadrant ($R < -1$, $-1 \leq R < 0$, $0 \leq R < 1$, $R > 1$; see Figure 6). The second term quantifies the risk value within the quadrant. The resulting risk metrics for sediment mobility, Equation (7), therefore consists of four regimes. The first, $R_\tau \leq -1$, represents a reduction in erosion. The second, $-1 < R_\tau \leq 0$, represents an increase in deposition. The third, $0 \leq R_\tau < 1$, represents a reduction in deposition, and the fourth, $R_\tau \geq 1$ represents an increase in erosion.

Bottom critical shear stress (and therefore erosion and deposition) is closely related to water depth through the dependences between depth and water column velocities. Physical processes such as tides, waves, and sediment transport result in a change in effective seabed elevation relative to that of a quiescent ocean. The presence of WECs can influence variability in seabed elevation and affect benthic habitats. Therefore, seabed elevation changes were incorporated into the case study risk framework associated with the SEAT. The risk associated with seabed elevation changes associated with WEC deployments was defined as:

$$R_\eta = \sum_i P_i (\eta_w - \eta_b) \tag{8}$$

where η_w and η_b are the seabed elevations in the absence and presence of WECs at each model grid point. Here, R_η has units of elevation (meters) unlike the non-dimensional risk metrics for sediment mobility (Equations (7) and (8)). This difference is primarily due to the difficulty in defining a critical depth of importance to benthic habitat. For example, one could define a critical depth below which eelgrass cannot grow. However, the growth of eelgrass is governed not only by depth, but also by other parameters such as water clarity and photosynthetically available radiation, water column nutrient concentrations, ambient currents, and a host of other environmental parameters. However, in the event that a critical depth for eelgrass growth can be determined for a site, a revised metric similar to Equation (7) can be defined for seabed elevation changes.

3. Results and Discussion

3.1. Wave Energy Converter (WEC) Effects on Sediment Transport

The fully coupled Delft3D-FLOW-SNL-SWAN model of the SEAT framework enables the assessment of changes in hydrodynamic and sediment transport parameters simulated with and without WECs in the environment. For the Newport, Oregon case study, maximum shear stress and bed elevation change were identified as two key parameters that influence variability in nearshore morphological characteristics. Two wave events with the highest annual probability of occurrence out of a total of 24 events were simulated within the SEAT framework and normalized changes in shear stress and change in bed elevation with and without an array of WECs for these two wave events were evaluated (Figures 7–10), where normalized change = $\frac{N_W - N_b}{|N_b|}$ and change = $N_W - N_b$. The parameters N_W and N_b represent shear stress or bed elevation in the presence and absence of WECs, respectively. Therefore, negative values of normalized change indicate decreases in values as the result of the WEC array and vice versa. The two wave events shown are typical of summer (Figures 7 and 9) and winter (Figures 8 and 10) conditions on the Oregon coast.

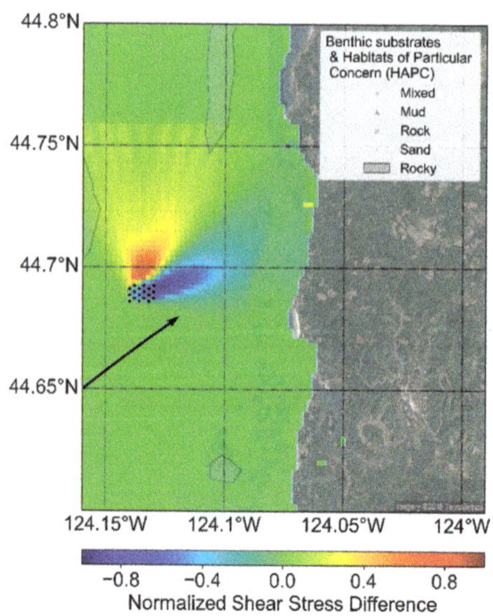

Figure 7. Spatial Environmental Assessment Tool (SEAT) coupled numerical model results showing normalized changes in bottom shear stress for input waves typical of summer conditions, where Max. Shear Stress Diff. = $\frac{\tau_W - \tau_b}{|\tau_b|}$. Input wave model parameters were H_s = 1.37 m, T_p = 15 s, and D_p = 224°. The black arrow indicates peak wave direction and the location of the WEC array is show by red crosses. Also shown are various benthic substrates and habitats of particular concern (HAPC, [22]).

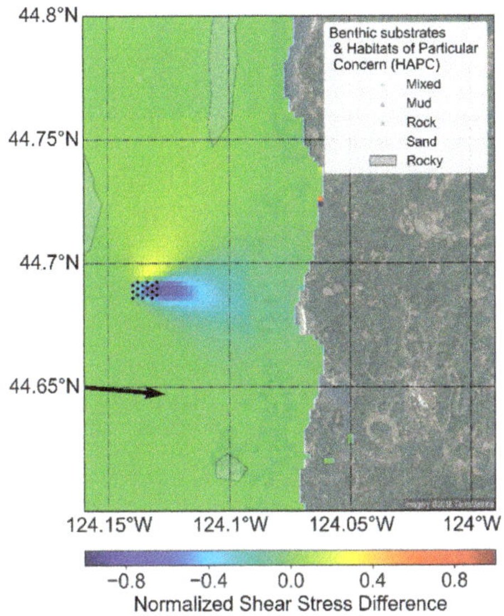

Figure 8. Same as Figure 7 but showing model results for winter conditions, where input wave parameters are H_s = 2.16 m, T_p = 10 s, D_p = 277°.

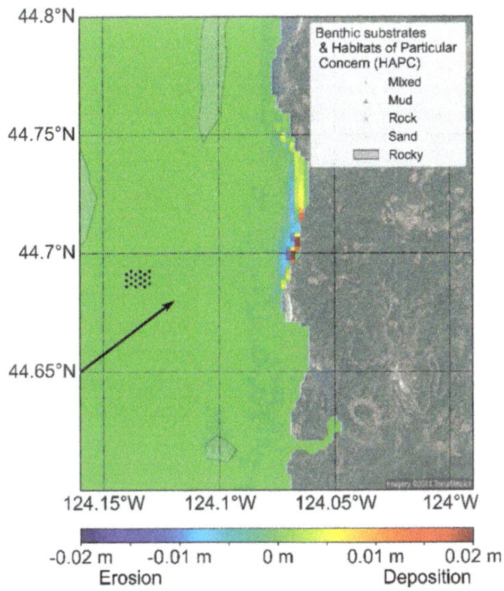

Figure 9. Same as Figure 7 but showing changes in bed elevation in the presence and absence of WECs, $\eta_W - \eta_b$.

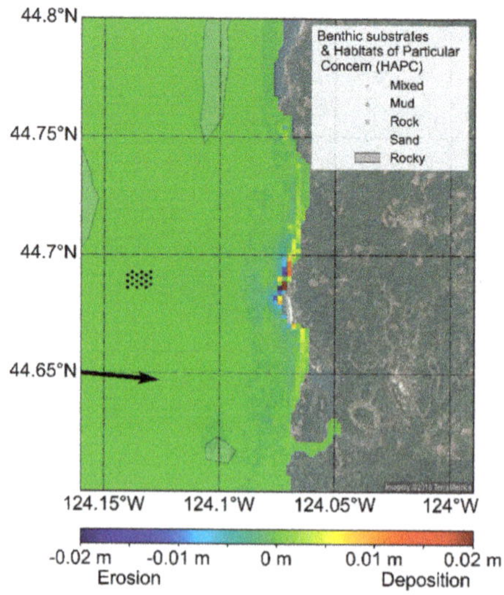

Figure 10. Same as Figure 8 but showing changes in bed elevation in the presence and absence of WECs, $\eta_W - \eta_b$.

The SEAT numerical model results indicate that variations in normalized shear stress are primarily observed in the vicinity of the WECs. During conditions typical of summer and winter, decreases in shear stress are observed to occur at the WECs and in the lee of the WEC array along an approximately 15-km swath oriented in the same direction as peak wave direction. Shear stress increased to the north of the WEC array in both cases in the presence of WECs. In winter conditions, reductions in normalized shear stress exceeded increases to the north of the array. It is important to note that while there are noticeable changes in normalized shear stress in the vicinity of the WEC array, absolute changes in shear stress are generally insufficient to cause significant changes in coastal sediment transport as the magnitudes are much lower than the critical shear stress for sediment mobility; however, even small changes can have implications on the nearshore circulation where sediment is mobile. For example, if shear stresses are above the critical shear stress for erosion in the absence of WECs, then the reduction in velocity in the lee of the WEC can change an erosional environment to a depositional environment.

Wave shadowing and coastal transport processes resulted in changes in modeled bed elevation between the presence and absence of WECs, which were expected [20]. While minimal, the wave shadow in the lee of the WEC array, in simulations in the presence of WECs, resulted in enhanced deposition in the nearshore regions (<20 m water depth). A corresponding reduction of deposition was observed in adjacent offshore cells. These deviations were in the order of 0.5% relative to the simulations in the absence of WECs. Interestingly, greater changes in the nearshore sediment bed were observed during summer as compared to winter conditions. It is likely that this is due to a combination of Yaquina Head acting as a wave shadow for the peak wave direction of 224° during summer conditions as compared to 277° during winter and the shift in the peak wave period from 15 s (summer) to 10 s (winter), (Figures 9 and 10).

3.2. SEAT Risk Analysis

The SEAT enables the application of the probabilistic framework to compute the risk associated with key potential stressors to the marine environment. Figure 11 shows the risk of sediment mobility (R_τ) (Equation (7)) for the Newport, Oregon case study in the context of habitats of particular concern (HAPC), as defined by CMECS [22]. As can be expected, there are moderate changes to sediment mobility in the lee of a WEC array, with expected increased deposition. However, all of these changes are in a region where the potential for changes in deposition rates are naturally limited by sediment supply and are therefore not significant. Most importantly, little to no risk is seen in HAPC, such as rocky reefs and kelp patches.

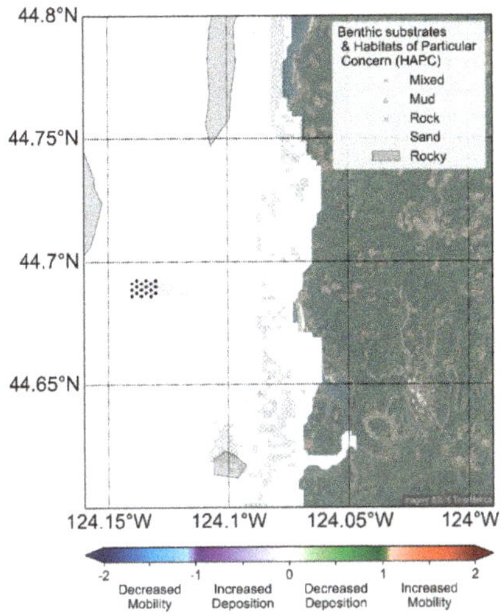

Figure 11. Risk metrics computed for changes in sediment mobility.

Figure 12 is a map of risk metrics when applied to bottom elevation changes. As discussed in Section 2.3, bottom elevation risk is expressed in changes in bed elevation in units of meters. Changes less than zero meters reflect an erosional environment and changes greater than zero meters reflect a depositional environment. SEAT results indicate that there is generally no change in bed elevation in the majority of the model domain. In regions where bottom elevation risk is non-zero, e.g., in shallow coastal areas, changes in the order of 1 cm are observed. These changes are largely insignificant (<0.5%) when considering that coastal bed elevations are in the order of 5 m.

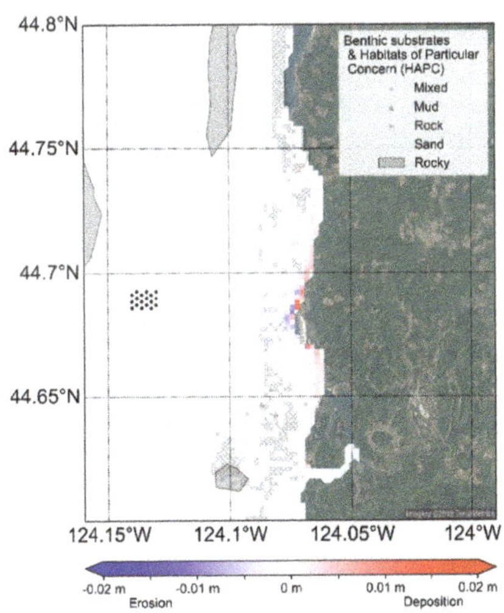

Figure 12. Risk metrics computed for changes in bed elevation.

A quantitative assessment of risk by habitat type is presented in Table 3, which shows the maximum risk of sediment mobility and bed elevation for different habitats found within the model domain. The low numbers suggest that maximum risk values for both sediment mobility and bed elevation are negligible in the presence of WECs. Sediment mobility risk values are very close to zero, noting that regime shifts in sediment mobility occur at ±1 (Equation (7)) and modeled bed changes are generally considerably less than 1 cm. The maximum risk values are in the most dynamic rocky and sandy environments that have high baseline sediment mobility potential.

Table 3. Change in SEAT risk for seabed parameters by habitat type (Equations (7) and (8)).

Habitat	Sediment Mobility Risk (max, non-dim)	Bed Elevation Risk (max, meters)
Estuary	9.38×10^{-7}	0.0008
Kelp	0.0009	0.0006
Seagrass	0	0
Rocky	0.0050	1.83×10^{-5}
Mixed	5.12×10^{-5}	1.04×10^{-5}
Mud	−0.0004	8.71×10^{-5}
Rock	0.0015	0.0013
Sand	0.0355	0.0132

4. Summary and Conclusions

The present study accomplishes three goals:

1. To develop a suitable case study that exemplifies tools and techniques for supporting marine environmental assessments,
2. To outline the application of the SEAT at the case study site, and
3. To develop maps and tables of WEC-induced stressors and the relationship to sediment transport and seabed changes as determined in the assessment.

The inclusion of sediment parameters in the coupled Delft3D-FLOW-SNL-SWAN model allows for a probabilistic assessment of the effect of WECs on the benthic and coastal environments, forming the basis of the SEAT. The methodology was examined using a subset of wave events, based on long-term site evaluations, which form the typical annual wave climate off the coast of Oregon. The data and results generated by a coupled wave, circulation, and sediment transport model can provide a wealth of information to assess the potential nearshore effects of WECs on hydrodynamics and sediment characteristics. However, the interpretation of the model results may be challenging and subjective. Therefore, the development of a risk assessment tool that utilizes model results, such as the SEAT, may provide a more effective means for coastal resource management.

The SEAT utilizes a holistic approach to site assessments that incorporates a joint probability distribution to capture wave dynamics over the course of a climatological year. Using the occurrence distribution of various wave events based on the analysis of a seven-year wave record, representative discrete wave events were chosen using clustering methodology to reduce the number of discrete wave events to a reasonable number. The clustering method is grounded in the specific wave dynamics in the region of interest, such as typical directional ranges, directional spreading associated with typical swell events, and the variance in significant wave height and period. When taken together, the cumulative sum of discrete wave events formed a climatological set that enabled an evaluation of the spatial risk potential. This probabilistic approach allowed for the calculation of expected changes to parameters such as annual changes in critical shear stress and bed elevation.

The risk metrics presented in this study allow for a spatial characterization of physical risk for several key parameters such as sediment mobility and bed elevation change. These risk metrics for sediment mobility are based on the concept of critical thresholds above which sediment mobilization is likely. The critical threshold for sediment mobility was defined as the critical shear stress of the underlying sediment. The spatial visualization of risk allows for the rapid identification of potential changes to the system and percent changes for parameters in each habitat type allow for a comprehensive evaluation of change. The overall WEC-induced physical stressors risk is very low for the Newport, Oregon case study, with no significant changes in risk that affect sediment mobility or bed elevation change.

The results of the SEAT illustrate the benefits of a site evaluation tool in facilitating coastal resource management at early market WEC sites. Though the SEAT is not intended to replace the traditional environmental assessment process, it can provide important information on potential components of environmental risk. Furthermore, whereas the methodologies shown here were applied to a hypothetical WEC array, these analysis techniques can be applied to other sectors of the renewable energy market such as tidal and offshore wind installations. In addition, risk parameters can be extended to include the effects of marine hydrokinetic energy installations on larval motility or light levels in the context of seagrass sustainability.

Supplementary Materials: The following are available online at http://www.mdpi.com/1996-1073/11/8/2036/s1. Model boundary files; Title: Model.zip.

Author Contributions: Conceptualization, C.J., G.C., K.R., S.M., A.D., and J.R.; Data curation, C.J., K.R., and S.M.; Formal analysis, C.J., K.R., S.M., and A.D.; Funding acquisition, C.J. and J.R.; Investigation, C.J., G.C., K.R., S.M., and A.D.; Methodology, C.J., K.R., S.M., and A.D.; Project administration, C.J., S.M., and J.R.; Resources, C.J. and J.R.; Software, K.R., S.M., and A.D.; Supervision, C.J. and J.R.; Validation, G.C., K.R., S.M., and A.D.; Visualization, C.J., G.C., K.R., S.M., and A.D.; Writing—original draft, G.C. and K.R.; Writing—review and editing, C.J., S.M., and A.D.

Funding: This research was made possible by support from the Water Power Technologies Office, within the Office of Energy Efficiency and Renewable Energy at the U.S. Department of Energy. The views expressed in the article do not necessarily represent the views of the U.S. Department of Energy or the United States Government. Sandia National Laboratories is a multi-mission laboratory managed and operated by National Technology and Engineering Solutions of Sandia, LLC, a wholly owned subsidiary of Honeywell International, Inc., for the U.S. Department of Energy's National Nuclear Security Administration under contract DE-NA0003525.

Acknowledgments: Wave hindcast data, wave model setup files, and wave measurement validation data for the area offshore of Newport, Oregon were kindly provided by Tuba Özkan-Haller and Gabriel García-Medina from Oregon State University.

Conflicts of Interest: The authors declare no conflict of interest. The funders had no role in the design of the study; in the collection, analyses, or interpretation of data; in the writing of the manuscript, and in the decision to publish the results.

References

1. Jacobson, P.; Hagerman, G.; Scott, G. *Mapping and Assessment of the United States Ocean Wave Energy Resource*; Technical Report for Electric Power Research Institute: Palo Alto, CA, USA, December 2011.
2. *Handbook of Ocean Wave Energy*; Pecher, A.; Kofoed, J.P. (Eds.) Springer International Publishing: Basel, Switzerland, 2017; pp. 1–287.
3. EMEC: European Marine Energy Centre. Available online: http://www.emec.org.uk/marine-energy/wave-developers/ (accessed on 23 July 2018).
4. de Sousa Prado, M.G.; Gardner, F.; Damen, M.; Polinder, H. Modelling and test results of the Archimedes wave swing. *Proc. IMechE Part A: J. Power Energy*, **2006**, *220*, 855–868. [CrossRef]
5. Contestabile, P.; Iuppa, C.; Di Lauro, E.; Cavallaro, L.; Andersen, T.L.; Vicinanza, D. Wave loadings acting on innovative rubble mound breakwater for overtopping wave energy conversion. *Coast. Eng.* **2017**, *122*, 60–74. [CrossRef]
6. Cameron, L.; Doherty, R.; Henry, A.; Doherty, K.; Van't Hoff, J.; Kaye, D.; Naylor, D.; Bourdier, S.; Whittaker, T. Design of the next generation of the Oyster wave energy converter. In Proceedings of the 3rd International Conference on Ocean Energy, Bilbao, Spain, 6–8 October 2010; pp. 1–12.

7. Nelson, P.A.; Behrens, D.; Castle, J.; Crawford, G.; Gaddam, R.N.; Hackett, S.C.; Largier, J.; Lohse, D.P.; Mills, K.L.; Raimondi, P.T.; Robart, M.; Sydeman, W.J.; Thompson, S.A.; Woo, S. Developing wave energy in coastal California: potential socio-economic and environmental effects. 2008; California Energy Commission. Available online: https://www.energy.ca.gov/2008publications/CEC-500-2008-083/CEC-500-2008-083.PDF (accessed on 5 April 2018).
8. Millar, D.; Smith, H.C.M.; Reeve, D.E. Modelling analysis of the sensitivity of shoreline change to a wave farm. *Ocean Eng.* **2007**, *34*, 884–901. [CrossRef]
9. Beels, C.; Troch, P.; De Visch, K.; Kofoed, J.P.; De Backer, G. Application of the time-dependent mild-slope equations for the simulation of wake effects in the lee of a farm of Wave Dragon wave energy converters. *Renew. Energy* **2010**, *35*, 1644–1661. [CrossRef]
10. Haller, M.C.; Özkan-Haller, T. Physical-environmental effects of wave and offshore wind energy extraction: a synthesis of recent oceanographic research. Available online: https://core.ac.uk/download/pdf/10196785.pdf (accessed on 20 April 2018).
11. Smith, H.C.M.; Pearce, C.; Millar, D.L. Further analysis of change in nearshore wave climate due to an offshore wave farm: An enhanced case study for the Wave Hub site. *Renew. Energy* **2010**, *40*, 51–64. [CrossRef]
12. Abanades, J.; Greaves, D.; Iglesias, G. Wave farm impact on the beach profile: A case study. *Coast. Eng.* **2014**, *86*, 36–44. [CrossRef]
13. O'Dea, A.M.; Haller, M.C. Analysis of the impacts of wave energy converter arrays on the nearshore wave climate. In Proceedings of the 2nd Marine Energy Technology Symposium, METS2014, Seattle, WA, USA, 15–17 April 2014; pp. 36–44.
14. Riefolo, L.; Lanfredi, C.; Azzellino, A.; Vicinanza, D. Environmental impact assessment of wave energy converters: A review. In Proceedings of the International Conference on Applied Coastal Research, SCACR 2015, Florence, Italy, 28 September–1 October 2015.
15. Özkan-Haller, T.; Haller, M.C.; McNatt, J.C.; Porter, A.; Lenee-Bluhm, P. Analyses of wave scattering and absorption produced by WEC arrays: Physical/numerical experiments and model assessment. *Mar. Renew. Energy* **2017**, 71–97.
16. Watson, J.R.; Hays, C.G.; Raimondi, P.T.; Mirarai, S.; Dong, C.; McWilliams, J.C.; Blanchette, C.A.; Caselle, J.E.; Siegel, D.A. Currents connecting communities: nearshore community similarity and ocean circulation. *Ecology* **2011**, *92*, 1193–1200. [CrossRef] [PubMed]
17. O'Boyle, L.; Elsabet, B.; Whittaker, T. Experimental measurement of wave field variations around wave energy converter arrays. *Sustainability* **2017**, *9*, 70. [CrossRef]
18. Gonzalez-Santamaria, R.; Zou, Q.; Pan, S. Modelling of the impact of a wave farm on nearshore sediment transport. *Coast. Eng. Proc.* **2012**, *1*, 66. [CrossRef]
19. Zanopol, A.T.; Onea, F.; Rusu, E. Coastal impact assessment of a generic wave farm operating in the Romanian nearshore. *Energy* **2014**, *72*, 652–670. [CrossRef]
20. Shih, S.M.; Komar, P.D. Sediments, beach Morphology, and Sea Cliff Erosion within and Oregon Coast Littoral Cell. *J. Coast. Res.* **1994**, *19*, 144–157.
21. Twenhofel, W.H. *Minerological and Physical Composition of the Sands of the Oregon Coast from Coos Bay to the Mouth of the Columbia River*; Bulletin No. 30; Department of Geology and Mineral Industries: Portland, OR, USA, 1946.
22. Federal Geographic Data Committee. Coastal and Marine Ecological Classification Standards (CMECS); June 2016. Available online: https://iocm.noaa.gov/cmecs/ (accessed on 19 June 2018).
23. Harlett, J.C.; Kulm, L.D. Suspended sediment transport on the northern Oregon continental shelf. *Geol. Soc. Amer. Bull.* **1973**, *84*, 3815–3826. [CrossRef]
24. Booij, N.; Ris, R.C.; Holthuijsen, L.H. A third-generation wave model for coastal regions. Part I: Model description and validation. *J. Geophys. Res.* **1999**, *104*, 7649–7666. [CrossRef]
25. Delft University of Technology. *SWAN User Manual*, SWAN Cycle III version 41.01; Delft University of Technology: Delft, The Netherlands, 1993.
26. Ruehl, K.; Porter, A.; Posner, A.; Roberts, J. Development of SNL-SWAN, a Validated Wave Energy Converter Array Modeling Tool. In Proceedings of the 10th European Wave and Tidal Energy Conference, Aalborg, Denmark, 2–5 September 2013.
27. Ruehl, K.M.; Chartrand, C.; Porter, A. *SNL-SWAN Manual, Version 1.0*; SAND Report; Sandia National Laboratories: Albuquerque, NM, USA, 2014.

28. Ruehl, K.M.; Porter, A.; Chartrand, C.; Smith, H.; Chang, G.; Roberts, J. Development, Verification and Application of the SNL-SWAN Open Source Wave Farm Code. In Proceedings of the 11th European Wave and Tidal Energy Conference, Nantes, France, 6–11 September 2015; pp. 1–8.
29. Chang, G.; Ruehl, K.; Jones, C.A.; Roberts, J.; Chartrand, C. Numerical modeling of the effects of wave energy converter characteristics on nearshore wave conditions. *Renew. Energy* **2016**, *89*, 636–648. [CrossRef]
30. Roelvink, J.A.; Van Banning, G.K.F.M. Design and development of DELFT3D and application to coastal morphodynamics. *Oceanogr. Lit. Rev.* **1995**, *42*, 925–938.
31. Gerritsen, H.; deGoede, E.D.; Platze, F.W.; van Kester, J.A.Th.M.; Gensenberger, M.; Uittenbogaard, R.E. *Validation Document Delft3D-FLOW, A Software System for 3D Flow Simulations*; Deltares: Delft, The Netherlands, 2008.
32. Egbert, G.D.; Erofeeva, S.Y. Efficient inverse modeling of barotropic ocean tides. *J. Atmos. Ocean. Technol.* **2002**, *19*, 183–204. [CrossRef]
33. García-Medina, G.; Özkan-Haller, H.T.; Ruggiero, P. Wave resource assessment in Oregon and southwest Washington, USA. *Renew. Energy* **2014**, *64*, 203–214. [CrossRef]
34. Kuik, A.J.; vanVledder, G.P.; Holthuijsen, L.H. A method for the routine analysis of pitch-and-roll buoy wave data. *J. Phys. Oceanogr.* **1988**, *18*, 1020–1034. [CrossRef]
35. Bull, D.; Dallman, A. Wave Energy Prize experimental sea state selection. In Proceedings of the OMAE 36th International Conference on Ocean, Offshore and Arctic Engineering, Trondheim, Norway, 25–30 June 2017. [CrossRef]
36. Lesser, G.R.; Roelvink, J.A.; van Kester, J.A.T.M.; Stelling, G.S. Development and validation of a three-dimensional morphological model. *Coast. Eng.* **2004**, *51*, 883–915. [CrossRef]
37. Soulsby, R.L. *Dynamics of Marine Sands—A Manual for Practical Applications*; Thomas Telford Publications: London, UK, 1997.
38. Babarit, A.; Hals, J.; Muliawan, M.J.; Kurniawan, A.; Moan, T.; Krokstad, J. Numerical benchmarking study of a selection of wave energy converters. *Renew. Energy* **2012**, *41*, 44–63. [CrossRef]

© 2018 by the authors. Licensee MDPI, Basel, Switzerland. This article is an open access article distributed under the terms and conditions of the Creative Commons Attribution (CC BY) license (http://creativecommons.org/licenses/by/4.0/).

energies

Article

Energy Efficiency Analysis of Multi-Type Floating Bodies for a Novel Heaving Point Absorber with Application to Low-Power Unmanned Ocean Device

Dongsheng Cong [1,2], Jianzhong Shang [1], Zirong Luo [1,*], Chongfei Sun [1] and Wei Wu [1]

1 School of Intelligent Science, National University of Defense Technology, Changsha 410073, China; cdssdjz@sdjzu.edu.cn (D.C.); shangjianzhong@nudt.edu.cn (J.S.); sunchongfei@yeah.net (C.S.); wuwei14@nudt.edu.cn (W.W.)
2 School of Mechanical and Electronic Engineering, Shandong Jianzhu University, Jinan 250101, China
* Correspondence: luozirong@nudt.edu.cn; Tel.: +86-731-8457-4932

Received: 2 September 2018; Accepted: 21 November 2018; Published: 25 November 2018

Abstract: Long-term energy supplies hinder the application of the low-power unmanned ocean devices to the deep sea. Ocean wave energy is a renewable resource with amount stores of enormous and high density. The wave energy converter (WEC) could be miniaturized so that it can be integrated into the devices to make up the power module. In this paper, a small novel heaving point absorber of energy supply for low-power unmanned ocean devices is developed based on the counter-rotating self-adaptive mechanism. The floating body as an important part of the heaving point absorber, the geometric parameters is optimized to increase the efficiency of power production. Through constructing the constitutive relation between the geometric parameters, the wave force, the motion displacement, the motion velocity, and the capture width ratio of the floating body, the energy efficiency characteristics of the multi-type floating bodies are calculated, and the optimal shape is selected. On the other hand, in the calculation process of the wave force, the Froude-Krylov method is an effective method to accurately calculate the wave excitation force. Meanwhile, nonlinear static and dynamic Froude-Krylov force effectively overcomes the inaccuracy of the linear models and reduces the time consumed to simulate. Finally, the wave force, heaving velocity, heaving displacement, and capture width ratio of the three floating bodies are compared and analyzed, and the results show that the cylindrical floater that is vertically placed on the wave surface is more suitable for the novel heaving wave energy point absorber.

Keywords: unmanned ocean device; wave energy; multi-type floating bodies; nonlinear Froude-Krylov force; energy efficiency

1. Introduction

Low-power unmanned ocean device, such as unmanned surface/under water vehicles, ocean robots, and ocean buoys are widely used in unmanned combat, deep-sea exploration, and marine communications, and so on [1–5]. With the continuous development of deep sea and offshore strategies, the scope of work of low-power unmanned ocean devices continues to expand [6]. Long-term, stable, and reliable energy supply is the basis for the rapid development of low-power unmanned ocean device [7]. Currently, the common energy supply technology is a battery. However, its energy density is low and it is necessary to carry multiple batteries to meet the requirements. The device space is limited, which limits the amount of battery that is carried [8]. Therefore, it is difficult for the battery to meet the requirements of the long-range and long-life capability of the current low-power unmanned ocean device. Therefore, studying the energy supply technology of high energy density and improving the long range and long battery life of the ocean device is the fundamental guarantee for effectively solving the application of low-power unmanned ocean device to the deep sea [9].

The ocean covers about 70% of the Earth's surface, making it the world's largest solar collector and energy storage system [10]. There are many types of renewable resources for ocean storage, such as Wave Power, Tidal Power and Tidal Current Energy, Ocean Thermal Energy Conversion (OTEC), and Salinity Gradient Energy. We can convert them into electricity by means of a power generator [11]. Wind blowing over the surface of the ocean creates waves that can be harvested for energy and creates uninterrupted, continuous wave energy in the Ocean's surface [12]. The amount of wave power that is stored by the ocean is enormous and high density. Theoretically, the global wave power resource is 2.11 ± 0.05 TW, of which 4.6% is extractable with the chosen WEC configuration [13,14]. Therefore, wave energy is the most ideal resource for a low-power unmanned ocean device.

The utilization of wave energy goes far back in time—people using wave push water mill taken place in the 1200 s. The world's first wave energy technology patent is filed by Frenchman Girard and his son in 1799 [15]. In 1910, the Frenchman Pocec-Plesic built the world's first wave energy converter (WEC) on the coast, namely air turbine private power station with a capacity of 1kW [10]. Salter of the University of Edinburgh successfully developed a duck-type WEC, and first published an article entitled "Wave Power" in Nature in 1974 [16]. However, due to the unsolved technical problems, such as the economy and stability of WEC, the research progress of wave energy technology was slower in the 1980s and early 1990s. In the last 20 years, with the emergence of problems, such as energy depletion, environmental pollution, and the greenhouse effect, the development of green and renewable energy has become the mainstream [7]. Wave energy has regained rapid growth under the guidance of the government, especially in some mariner countries, such as Ireland, Denmark, Norway, and UK [17–20].

WEC's are generally categorized by the method that is used to capture the energy of the waves, by location and by the power take-off (PTO) system [10,21]. Locations are shoreline, near shore, and offshore. Types of power take-off include: hydraulic ram, elastomeric hose pump, pump-to-shore, hydroelectric turbine, air turbine, and linear electrical generator. In addition, WEC'S can also be categorized by the energy transfer method, namely point absorber, surface attenuators, oscillating water columns, and overtopping devices.

Comparative and analytic research on the working principle of current WEC's, the point absorber is the best choice for the power supply of low-power unmanned ocean device [7]. However, the research on point absorber is mainly focused on large-scale grid-connected generation in shore and near shore, and the dimension of the device is larger, and the power generation operation is more complicated [10]. At present, typical point absorber includes Power Buoy [22], Wave bob [23], Sea Based [24], Fred Olsens Lifesaver [25], and Carnegies CETO [26]. The shape dimension of them is very large, such as Power Buoy, where the diameter of the buoy is about 3 m and the overall height about 14 m. As the characteristics, such as concealment, camouflage, small dimension, wide and far working area of low-power unmanned ocean device, and the huge dimension of the above point absorbers, makes it still difficult to provide a satisfactory power source for it.

Considering the technical requirements of the power supply for the low-power unmanned ocean device and the current research state of power generation technology, we developed a small novel heaving point absorber that is based on the counter-rotating self-adaptive mechanism, with the advantages of small space device and a stable and reliable energy conversion process [7]. It is an ideal WEC of power supply for the low-power unmanned ocean device, so as to increase their working hours and improve independent operation ability.

Previous work shown that the influence law of the blade angel on the impeller speed was obtained and the model prototype was trial produced base on the study of the blade angle and the relative speed of the upper and lower impellers for the Underwater PTO [7].The adjustment of the blade angle in impellers is passive action. The move displacement of the floating body and the Underwater PTO affects the deflection angle. The locking devices of the impellers limit the angle. In addition, previous works show that the maximum blade deflection angles between 35° and 55° are the best efficiency characteristics [7]. In order to increase the power production of the novel heaving point absorber, it is necessary to increase the elevation of the Underwater PTO in the vertical direction, that is, increase the heave amplitude of the floating body. Because the displacement of the body is performed by its immersion depth, shape, and dimension, so the geometric parameters of the body are optimized, which is of great significance for increasing the power production of the novel heaving point absorber.

The floating body dynamics of the WEC have been investigated by many researchers. For example, Black et al. [27] comparatively analyzed the wave radiation forces and scattering forces for horizontal rectangular and vertical circular cylinders using the Haskind's theorem. Mohapatra and Guedes Soares [28] studied the wave forces on a two-dimensional rectangular floating structure based on linearized Boussinesq. Rodriguez et al. [29] investigated the numerical nonlinear heave response of a rectangular box concerning the importance of the relative body dimensions. Islam et al. [30] analyzed the wave radiation of a heaving box-type floating structure based on CFD simulations with a volume of fluid method. Yeung et al. [31] and Sabuncu et al. [32] discussed the added mass and damping of a vertical cylinder in finite depth water. Calisal et al. [33] presented an efficient method of hydrodynamic coefficients calculation for vertical composite cylinders at finite depth. Mansour et al. [34] analyzed the diffraction of linear waves by a uniform vertical cylinder with cosine-type radial perturbations. Bhata et al. [35] studied scattering and radiation problems for a cylinder on nonlinear wave loading at finite depth. Kim [36] researched the hydrodynamic coefficients of the floater with elliptical cylinder and ellipsoid on a free surface. Bihs et al. [37] simulated a horizontal cylinder in heave motion and the motion of a freely floating rectangular barge in waves using the CFD model and compared the results with experimental data. Koh et al. [38] used Matched Eigenfunction Expansion Method (MEEM) to solving the radiation problem of the heaving circular cylinder in the context of linear potential theory. Wang sheng [39] discussed the added mass and damping of an ellipsoid in infinite and finite depth water. Finnegan et al. [40] determined an analytical approximation for the wave excitation forces on a floating truncated vertical cylinder in water of infinite depth and solved the appropriate boundary value using the method of separation of variables. Ghadimi et al. [41] presented a detailed analytical solution for the boundary value problem to evaluate the wave loads for the cylinder with heave and pitch motions in water of finite depth in the presence of an incident wave. Although the floating body dynamics of the WEC are more researched, the energy efficiency of multi-type floating bodies in the heaving wave energy point absorber is poorly studied. Furthermore, in the above paper, the separation variable method and the eigenfunction expansion method is usually used for the calculation of the radiation force and diffraction force of the buoy base on the potential flow theory. However, it is difficult to solve their analytical solutions and is often time consuming to calculate.

Under the Airy's wave theory, the Froude-Krylov approximation method is implemented to solve the wave force when the device dimension is considerably smaller than the wave length [42]. Froude-Krylov approximation method [33–45] is assumed that the original wave pressure distribution of the incident wave does not change due to the presence of the floating body. Therefore, wave excitation force is the product of the force of the undisturbed incident wave pressure on the floating body (dynamic Froude-Krylov force) and the diffraction correction coefficient. The diffraction coefficients reflect the attached mass effect and diffraction effects, which is determined by the model test. Although this method is an approximation method, it is simple to calculate and it is a very practical method for estimating the wave force. Moreover, because the method is built on model tests, accurate calculations can be obtained [42]. Falnes and Perlin [46,47] analyze the oscillating bodies in low-amplitude waves and obtain that the diffraction is negligible when the device dimension is considerably smaller than the wave length. Clement and Ferrant [48] described a method for the computation of free surface flows generated by submerged bodies, and obtain that radiation nonlinearities are negligible for floating bodies that are small as compared to the wave length. Merigaud et al. [43] added specific nonlinear terms to hydrodynamic models for wave energy devices, to improve the validity of such models across the full operational spectrum, showing that the response of the device is mainly affected by nonlinear FK forces, while nonlinear radiation and diffraction forces have minor effects on the system dynamics.

The focuses of the paper is on the novel heaving point absorber operating in the power production region, using the Froude-Krylov method to optimal shape design and maximize energy capture of the floating body for the power supply of the low-power unmanned ocean device. The remainder of this paper is organized as follows: Section 2 introduces the structure model of the novel heaving point absorber. Section 3 gives the mathematical model of the floating body of the novel absorber. Section 4 presents the algebraic solution of the wave excitation force in the vertical direction of floating bodies and nonlinear Froude-Krylov force integral. The numerical simulation and simulation analysis for the wave force, heaving velocity, heaving displacement, and capture width ratio of the multi-type floating bodies are in Sections 5 and 6. Some conclusions and final remarks are presented in Section 7.

2. Structure Model

The waves are generally generated by wind blowing across the surface of the ocean. The motion of waves is regular and periodic on the water surface, with the amplitude decreasing exponentially with depth. When the depth to be greater than half the wavelength, the wave-induced motions is only approximately 4% of those at the surface and thus could be considered to be insignificant [10]. Therefore, this depth range is defined as the hydrostatic layer, in which the motion of waves is hardly perceived as the depth increases. Based on the above characteristics of wave motions, a novel heaving wave energy point absorber is designed, which can be used as a power module for the low-power unmanned ocean device [7]. Figure 1 shows the working scene of this point absorber. First, when the low-power unmanned ocean device is working on the ocean surface, the value of the battery energy is decreasing from the maximum to the lowest safe. Secondly, utilizing the release mechanism of the low-power unmanned ocean device, the novel heaving point absorber is released. The Underwater PTO subsystem of this absorber converts the captured wave energy into electricity. The value of the battery energy reaches the highest energy. Finally, utilizing the recovery mechanism of the ocean device, the absorber is recovered. The above process cycles back and forth, of which ensures the energy supply of unmanned ocean device for long-range and long-term work.

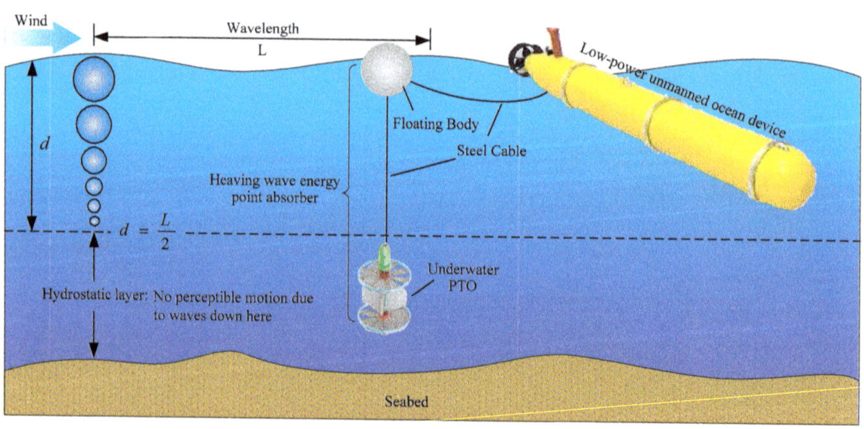

Figure 1. The working scene of the novel heaving wave energy point absorber.

The system configuration and working principle of the novel heaving point absorber are illustrated in Figure 2. The absorber mainly includes two parts, a floating body and an Underwater PTO, which are connected with each other by a steel cable. The floating body floats on the ocean surface and the Underwater PTO suspends the hydrostatic layer at a depth of about 40 m. The Underwater PTO mainly consists of a power generator, upper impeller, lower impeller, steady blade, and transmission shaft and planet-gear increaser. The impellers are mainly composed of blades, connecting rods, locking devices, center wheels, and external fixation rings. The center wheel and the external fixation ring are connected by eight radials arranged in connecting rods. Eight fan-shaped blades are fixed on corresponding connecting rods with locking devices and are arranged in a centrally symmetric circumferential array. In addition, the steel cable not only acts as dragging the Underwater PTO, but also transmissions electricity and control signals.

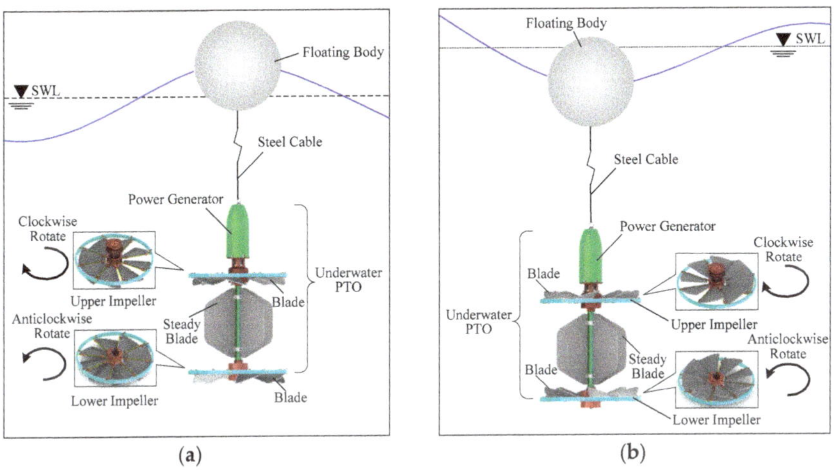

Figure 2. The system configuration and working principle of the novel heaving point absorber. (**a**) Rising state; and, (**b**) Sinking state.

The working principle of the novel heaving point absorber is shown in Figure 2. (1) During the movement of the floating body from the wave trough to the wave crest, the Underwater PTO is pulled up by the steel cable. The upper surface of the blade is impacted by the water flow and the blade adaptively swings downward, as shown in Figure 2a. Due to the limitation of the locking devices, the blade then stops swinging and is in a slanted state after reaching the maximum angle of inclination. The water flow continues to impact the slanted blades and propel the blade forward. The circumferential array of the blades enables the impeller to be subjected to circumferential thrust. Since the blades of the upper and lower impellers are arranged in opposite directions, the upper impeller is clockwise rotated by the water flow and the lower impeller is anticlockwise rotated by the water flow, and they are relatively reversed. The upper and lower impellers are fixedly connected with the stator and rotor of the generator, respectively, and then drive the generator to generate electricity. (2) During the movement of the floating body from the wave crest to the wave trough, the Underwater PTO sinks under the influence of gravity. The lower surface of the blade is impacted by the water flow and the blade adaptively swings upward, as shown in Figure 2b. Due to the limitation of the locking devices, the blade then stops swinging and it is in a slanted state after reaching the maximum angle of inclination. Since the direction of the water flow impinging on the blade does not change, the water flow continues to impact the slanted blades and propel the blade forward. The upper impeller is continuously clockwise rotated by the water flow and the lower impeller is continuously anticlockwise rotated by the water flow. Therefore, the direction of rotation of the generator does not change and it continues to generate electricity. According to the different impact directions of water flow, the impellers' blades adaptively adjust the blade deflection. The upper and lower impellers act as components that interact directly with the water flow and they provide continuous rotational motion to the generator during rising and sinking of the Underwater PTO.

3. Mathematical Model

The floating body is an important part of the novel heaving wave energy point absorber and is in direct contact with the waves on the ocean surface. In order to increase the energy harvesting of floating body, it is necessary to increase the heave amplitude, heave velocity, and wave force in the vertical direction. It can increase the rising and sinking amplitude of the Underwater PTO and the relative rotational speed of the upper and lower impellers. Moreover, the power production of the novel heaving point absorber can also be increased. This section analysis of the force of the floating body, establishes its motion equation, and solves its wave excitation force and the Underwater PTO damping force. Also, this section derives the model of energy conversion efficiency between the floating body and the wave energy, and calculates its energy conversion efficiency.

The hydrodynamics model and energy efficiency model of the floating body is established to calculate the wave force and the conversion efficiency: Section 3.1 considers nonlinear Froude-Krylov force in the vertical direction and Underwater PTO damping force for the force analysis of the floating body in micro-wave amplitude. Section 3.2 presents the capture width ratio of the floating body for the energy efficiency analysis.

3.1. Hydrodynamics Model of the Floating Body

We consider that the floating body of the novel heaving point absorber floats freely in water of uniform depth. Under the action of linear regular waves, the floating body does micro amplitude heave motion. The fluid is assumed inviscid and the incident flow irrotational and incompressible. Figure 3 is the coordinates and force analysis of the floating body. The right-handed inertial reference frame is centered at the hydrostatic equilibrium position of the body. X is the vertical distance from the reference fluid surfaces to the body waterline (positive downwards). Y is the vertical distance from the reference fluid surface to the fluid surface (positive downwards). $Y = r\cos\varphi t$, where r is the wave amplitude and φ is the wave angular frequency. Z is the vertical distance from the reference fluid surface to any point in the fluid (positive downwards). D is the waterline depth of the floating body. Newton's second law can be used to describe the system dynamics, as follows:

$$m\ddot{\mathbf{X}}(t) = G - \iint\limits_{S(t)} P(t)\mathbf{n}dS - Q - I + \mathbf{F}_{\mathrm{PTO}}(t) \tag{1}$$

where, m is the mass of the novel heaving point absorber, $\mathbf{X}$ is the heaving displacement of the body from its hydrostatic equilibrium position, $\ddot{\mathbf{X}}$ is the heaving acceleration of the body, G the gravity force, S the submerged surface, P the pressure, $\mathbf{n}$ a vector normal to the surface, Q the viscous damping force, I the inertia force of attached mass effect, and $\mathbf{F}_{\mathrm{PTO}}$ the Underwater PTO damping force.

The pressure P can be derived from the incident flow applying Bernoulli's equation:

$$P(t) = -\rho g z(t) - \rho\frac{\partial\phi(t)}{\partial t} - \rho\frac{|\nabla\phi(t)|^2}{2} \tag{2}$$

where, ρ the water density, g the acceleration of gravity, $P_{\mathrm{st}} = -\rho g z$, hydrostatic pressure and ϕ the potential flow, which can be decomposed as the sum of the undisturbed incident flow potential ϕ_{I}, the diffraction potential ϕ_{D}, and the radiation potential ϕ_{R}:

$$\phi = \phi_{\mathrm{I}} + \phi_{\mathrm{D}} + \phi_{\mathrm{R}} \tag{3}$$

The Airy's wave theory assumes that the motion of the floating body is a small amplitude. The solution of Equation (1) is solved around the equilibrium position of the buoy. Under the linear assumption, the wetted surface is constant. However, the nonlinear of Froude-Krylov forces to be considered in the actual calculation process. Thus, the wetted surface is exactly instantaneous, namely integrating the fluid pressure over the actual submerged portion of the buoy, as it moves through the water. Froude-Krylov forces include the static and dynamic forces. They depend on the instantaneous wetted surface, which depends both on the incident wave elevation and the displacement of the buoy. Froude-Frylov force can be written as:

$$\mathbf{F}_{FK} = \mathbf{F}_{FK_{st}} + \mathbf{F}_{FK_{dy}} = \iint\limits_{S(t)} \left(P_{st}(t) + P_{dy}(t)\right)\mathbf{n}dS \tag{4}$$

where, $\mathbf{F}_{FK_{st}}$ is the static Froude-Krylov force, given as the balance between the gravity force and the Archimedes force, and $\mathbf{F}_{FK_{dy}}$ is the dynamic Froude-Krylov force.

$$\mathbf{F}_{FK_{st}} = \iint\limits_{S(t)} P_{st}(t)\mathbf{n}dS = -\iint\limits_{S(t)} \rho gz dS \tag{5}$$

$$\mathbf{F}_{FK_{dy}} = \iint\limits_{S(t)} P_{dy}(t)\mathbf{n}dS \tag{6}$$

where, $P_{\text{dy}} = -\rho\frac{\partial\phi_I}{\partial t} - \rho\frac{|\nabla\phi_I|^2}{2}$ is the dynamic pressure.

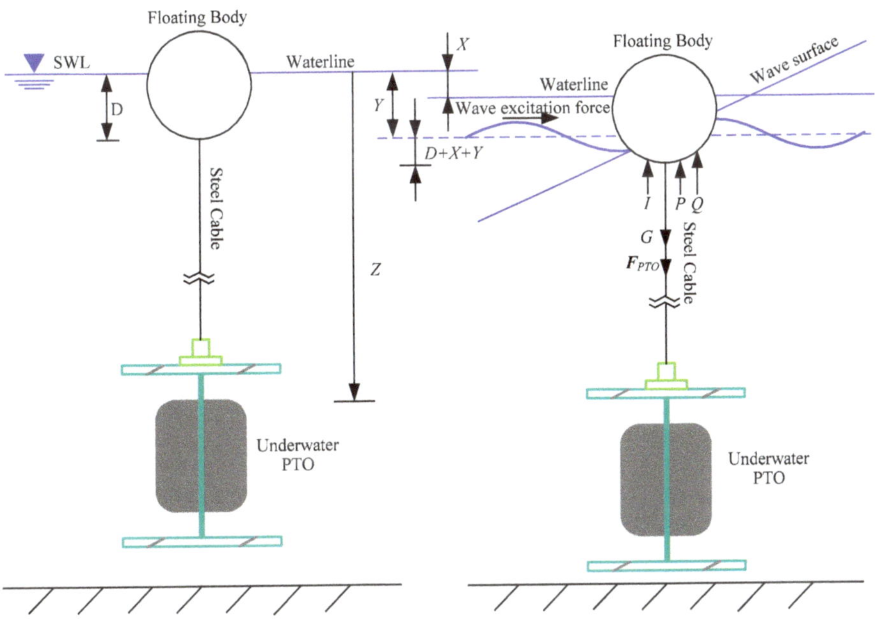

Figure 3. The coordinates and force analysis of floating body.

The time-dependence annotation will be omitted for brevity hereafter. Equation (1) can be rewritten as [42]:

$$m\ddot{\mathbf{X}} = G + \mathbf{F}_{FK} + \mathbf{F}_D + \mathbf{F}_R + \mathbf{F}_{\text{PTO}} \tag{7}$$

where, $\mathbf{F}_D$ is the diffraction force and $\mathbf{F}_R$ is the radiation force.

Note that, since the fluid is assumed to be inviscid, irrotational, and incompressible, no viscous force and inertia force appears in Equation (7). In addition, due to Froude-Krylov force already includes the inertia force that is caused by the attached mass effect, the inertia force also does not appear in this Equation.

Wave excitation force is the force of wind blows the waves to disturb the motion of the floating body on the ocean surface. This force can be written as:

$$\mathbf{F}_{\mathrm{ex}} = C\mathbf{F}_{FK_{\mathrm{dy}}} + \mathbf{F}_D \tag{8}$$

where, C is the diffraction correction coefficient.

The linear approach assumes that radiation and diffraction forces are linear. Therefore, the radiation and diffraction potential is negligible when the floating body dimension is considerably smaller than the wave length [43–48]. When combining Equations (4)–(8), Equation (1) can be rewritten as:

$$m\ddot{\mathbf{X}} = C\iint\limits_{S(t)} P_{dy}\mathbf{n}dS + \mathbf{F}_{\mathrm{PTO}} \tag{9}$$

The algebraic calculation of the integral in Equation (9) requires the explicit definition of the dynamic pressure P_{dy}, the infinitesimal surface element $\mathbf{n}dS$, and the limits of integration. Under the Airy's wave theory for deep water waves, the dynamic pressure at any point on the floating body in the local coordinate system can be written as:

$$P_{\mathrm{dy}} = \frac{\rho g H}{2}\frac{chk\overline{z}}{chkd}\cos(k\overline{x} - \omega t) \tag{10}$$

where, H is wave height, $\overline{x}$ is the direction of wave propagation in the local coordinate system, $\overline{z}$ is the vertical displacement of the floating body in the local coordinate system (positive upwards), and d is water depth and ω is the wave circular frequency. Wave Number k is defined by the dispersion equation $k\tanh(kd) = \omega^2/g$.

During the motion of the novel heaving point absorber on the ocean surface, the coordinates at any point on the floating body in the global coordinate system can be written as:

$$\begin{bmatrix} x \\ y \\ z \end{bmatrix} = \begin{bmatrix} X_0 \\ Y_0 \\ Z_0 \end{bmatrix} + [T]\begin{bmatrix} \overline{x} \\ \overline{y} \\ \overline{z} \end{bmatrix} \tag{11}$$

where, (X_0, Y_0, Z_0) is the coordinates of the center of gravity on the floating body in the global coordinate system and $(\overline{x}, \overline{y}, \overline{z})$ is the coordinates at any point on the floating body in the local coordinate system. $[T]$ is coordinate transformation matrix between the local coordinate system and global coordinate system.

Therefore, the dynamic pressure at any point on the floating body in the global coordinate system can be written as:

$$P_{\mathrm{dy}} = \frac{\rho g H}{2}\frac{chk(d-z)}{chkd}\cos(k(x+\overline{x}) - \omega t) \tag{12}$$

where, x is the direction of wave propagation in the global coordinate system and z is the vertical displacement of the floating body in the global coordinate system (positive upwards).

Furthermore, the force components of the wave excitation force in the vertical direction can be written as the following:

$$F_{\mathrm{ex}}^{z} = C_V\iint\limits_{S(t)} P_{dy}n_z dS \tag{13}$$

where, n_z is the projection of the normal for the wetted surface of the floating body in the vertical direction and C_V is the diffraction correction coefficient in the vertical direction.

Under the movement of the novel heaving point absorber on the ocean surface, the heave motion of the floating body does work by overcoming the damping of the water. The Underwater PTO subsystem of this absorber converts the captured wave energy of the body into electricity. In the motion model of this absorber under the Airy's wave theory, the linear damping model is used to analyze the damping force of the Underwater PTO system. Thus, the Underwater PTO damping force can be written as:

$$\mathbf{F}_{\mathrm{PTO}} = \mathrm{B}_{\mathrm{PTO}}\dot{\mathbf{X}} \tag{14}$$

where, $\mathrm{B}_{\mathrm{PTO}}$ is the Underwater PTO damping coefficient and $\dot{\mathbf{X}}$ is the heaving velocity of the floating body.

3.2. Energy Efficiency Model of the Floating Body

In the researches of the WEC, the performance of power production of the device is mainly concentrated on the energy harvesting and energy conversion. Using the capture width ratio [49,50] of the floating body as an evaluation index to measure the characteristic of energy harvesting and conversion efficiency for the novel heaving point absorber. The capture width ratio is that the ratio of the average power of the energy harvesting to the wave energy of the incident wave within the width of the body, namely the efficiency of energy harvesting in the WEC. It can be written as the following:

$$\eta = \frac{P_{WEC}}{P_0} \tag{15}$$

where, P_{WEC} is the average power of the floating body to harvests wave energy and P_0 is the wave energy of the incident wave within the width of the body.

The average power of the energy harvesting is that the instantaneous power is integrated in a period and it solves the average value. It can be written as:

$$P_{WEC} = \frac{1}{T}\int_0^T \mathbf{F}_{wave} \cdot \mathbf{V}_{WEC} dt \tag{16}$$

where, $\mathbf{F}_{\mathbf{wave}}$ is the wave force, $\mathbf{V}_{WEC}$ is the velocity of the floating body, and T is the wave period.

The average power of the energy harvesting in the vertical direction of the floating body can be written as:

$$P_{WEC}^{z} = \frac{1}{T}\int_0^T (F_{ex}^{z} + F_{PTO}^{z})Xdt = \frac{1}{T}\int_0^T ((C_V \iint\limits_{S(t)} P_{dy} n_z dS + B_{PTO}^{z} X)X)dt \tag{17}$$

where, F_{PTO}^{z} is the Underwater PTO damping force in the vertical direction and B_{PTO}^{z} is the Underwater PTO damping coefficient in the vertical direction.

The wave energy of the incident wave within the width of the floating body can be written as the following:

$$P_0 = \rho g H^2 \frac{\omega}{16k}\left[1 + \frac{2kd}{sh(2kd)}\right] \cdot B \tag{18}$$

where, B is the heading wave width of the floating body.

4. Algebraic Solution

The characteristic parameters, such as profile parameters, immersion depth, and wetted surface distribution, play an important role in the performance analysis, structural design, manufacture, and employment of the floating body. These parameters affect the motion amplitude, motion velocity, wave force, and wave energy harvesting of the floating body. Therefore, selects the profile parameters and solves the algebraic solution of the wave force is critical. Geometry shapes of floating body are two types of axisymmetric and unaxisymmetric. The axisymmetric body is a curved surface body formed by a generatrix rotating around a fixed vertical axis. The shape of the axisymmetric body is dependent on the shape of generatrix and relative position of the generatrix and the fixed axis. The axisymmetric bodies include cylinder, cone, sphere, and so on. The unaxisymmetric body is formed by a non-rotating curved surface. This body consists of rectangular, trapezoid, polyprism, and so on. In this paper, three kinds of shapes are selected for the floating body, such as rectangular, cylinder, and sphere. This section mainly solves the algebraic solution of the wave force of the above shape. Table 1 show that the algebraic solutions of the wave excitation force in the vertical direction of multi-type floating bodies. The geometric parameters of the axisymmetric body are described in cylindrical coordinates [42]. The axisymmetric bodies are described in rectangular coordinates.

Table 1. The algebraic solutions of the wave excitation force in the vertical direction of multi-type floating bodies.

The Geometry Shapes of Floating Bodies		The Coordinates of Floating Bodies	The Parameters of Profiles and Coordinates	The Wave Excitation Force in the Vertical Direction of Floating Bodies
Unaxisy-mmetric	Horizontal rectangular	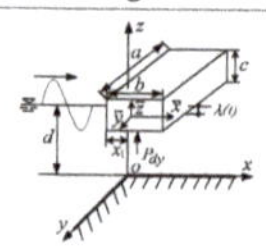	a is the length, b is the length along the wave direction, c is the height, $\lambda(t)$ is the immer-sion depth in the time of t.	$F_{ex}^{z} = C_V \frac{a\rho g H}{k} \frac{chk(d-\lambda(t))}{chkd} \cdot \cos(kx_1 - \omega t + \frac{kb}{2}) \sin \frac{kb}{2}$ (See Appendix A for calculation details.)
Axisym-metric	Vertical cylinder	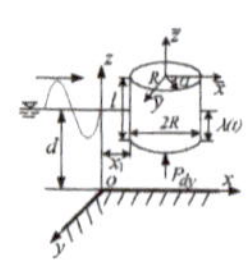	R is the radius, l is the heig-ht, $\lambda(t)$ is the immersion de-pth in the time of t, the cylin-der equation is $\overline{x}^2 + \overline{y}^2 = R^2$,the cylinder coordinate is $(r, \alpha, \overline{z})$, $\overline{z} = constant, \alpha = constant$, S is wetted surface, $S: r = R$, $0 \le \alpha \le 2\pi$, $0 \le \overline{z} \le \lambda(t)$ then, $dS = Rd\alpha d\overline{z}$.	$F_{ex}^{z} = C_V \frac{R\rho g \pi H J_0(kR)}{k} \frac{shkd - shk(d-\lambda(t))}{chkd} \cdot \cos(kx_1 + kR - \omega t)$ (See Appendix A for calculation details.)
	Sphere	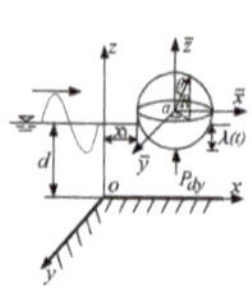	R is the radius, $\lambda(t)$ is the immersion depth in the time of t, the spherical equation is $\overline{x}^2 + \overline{y}^2 + \overline{z}^2 = R^2$, the sphere-cal coordinate is (r, θ, α),$\theta = constant$, $\alpha = constant$, S is wetted surface, $S: r = R$, $\pi - \arccos(\frac{R-\lambda(t)}{R}) \le \theta \le \pi$, $0 \le \alpha \le 2\pi$ then, $dS = R^2 \sin\theta d\theta d\alpha$	$F_{ex}^{z} = C_V \cdot \frac{\pi R^2 \rho g H}{chkd} \cdot (I_1(kR) - I_2(kR)) \cdot \cos(kx_1 + kR - \omega t)$ $I_1(kR) = \int_{\pi-\arccos(\frac{R-\lambda(t)}{R})}^{\pi} ch(kR\cos\theta) chk(d + R - \lambda(t)) \cdot J_0(kR\sin\theta) \cdot \cos\theta \sin\theta d\theta$ $I_2(kR) = \int_{\pi-\arccos(\frac{R-\lambda(t)}{R})}^{\pi} sh(kR\cos\theta) shk(d + R - \lambda(t)) \cdot J_0(kR\sin\theta) \cdot \cos\theta \sin\theta d\theta$ (See Appendix A for calculation details.)

5. Numerical Results

Based on the above analysis of the hydrodynamic performance and energy conversion characteristic for the multi-type floating bodies of the novel heaving wave energy point absorber, the wave force, heaving velocity, heaving displacement, and capture width ratio of the multi-type floating bodies were numerically simulated and comparatively analyzed.

During the numerical simulation, the parameters of the multi-type floating bodies are as follows: the length of horizontal rectangular a is 0.3 m, the length along the wave direction of horizontal rectangular b is 0.3 m, the height of horizontal rectangular c is 0.5 m, the radius of vertical cylinder R is 0.3 m, the height of vertical cylinder l is 0.5 m, the radius of sphere R is 0.3 m, the mean immersion depth is 0.15 m, the Underwater PTO damping coefficient in the vertical direction B_{PTO}^{z} is 20 KNs/m, and the mass of the novel heaving point absorber m is 10 kg. According to the model test results of previous research [51–53], the diffraction correction coefficient in the vertical direction C_V of horizontal rectangular, vertical cylinder, and sphere can be obtained, respectively.

In this study, three reference sea states are used for evaluation the energy efficiency of the novel heaving wave energy point absorber. Sea state 1: the significant wave height H_S is 0.1 m, the peak wave period T_P is 2 s; Sea state 2: the significant wave height H_S is 0.3 m, the peak wave period T_P is 3.5 s; and, Sea state 3: the significant wave height H_S is 0.5 m, the peak wave period T_P is 5 s. The first sea state covers H_S values from 0.0 to 0.1 m, the second sea state covers H_S values from 0.1 to 0.3 m, and the third sea state covers H_S values from 0.3 to 0.5 m. The combination of H_S and T_P is representative for the same wave tank, the wave climate date come from the simulation test. The model scales of the novel heaving point absorber, as follows: the Underwater PTO's diameter is 410 mm, the distance between the upper and lower impeller is 400 mm, and the overall height of the Underwater PTO is about 560 mm. The scale between the dimensions of the tested heaving point absorber and the full scale device is 1:5. Another, the significant wave height H_S is about 2.5 m in the real wave climate with the full scale device. The considered reference water depth d is 50 m. In addition, the incident wave is the linear regular wave, the angle between the direction of the linear regular wave and the direction of the novel heaving point absorber is 0°. According to the above parameters of the incident wave, the working scene of the novel heaving point absorber is the deep-water waves. Therefore, the other parameters of the incident wave are obtained, including wavelength, wave velocity, and velocity of water particles. These parameters are as follows: the wavelength is 6.2/19/39 m, the wave velocity is 3.1/5.5/7.8 m/s, and the velocity of water particles is 0.16/0.27/0.31 m/s.

The results of numerical simulation are shown in Figures 4–11. The wave force, heaving velocity, heaving displacement, and capture width ratio of the multi-type floating bodies in the second sea state is shown in Figures 4–7, respectively. Figures 8–11 show that the above parameters for comparison in the three reference sea states and the values obtained from the numerical simulation.

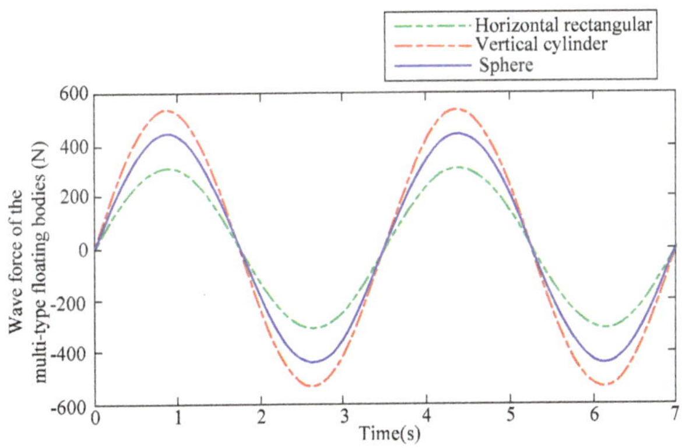

Figure 4. The wave force of the multi-type floating bodies (H = 0.3 m, T = 3.5 s).

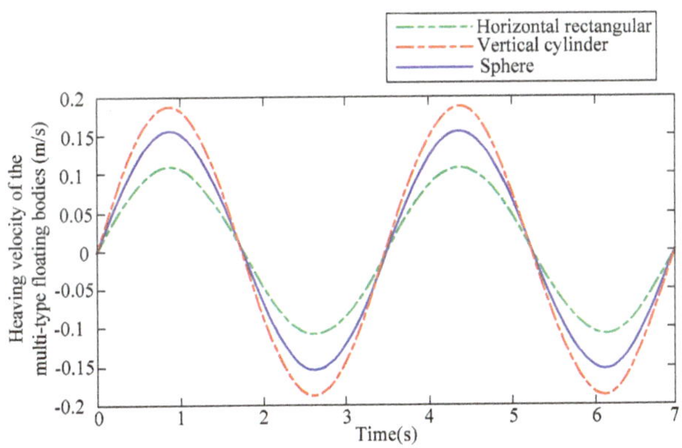

Figure 5. The heaving velocity of the multi-type floating bodies (H = 0.3 m, T = 3.5 s).

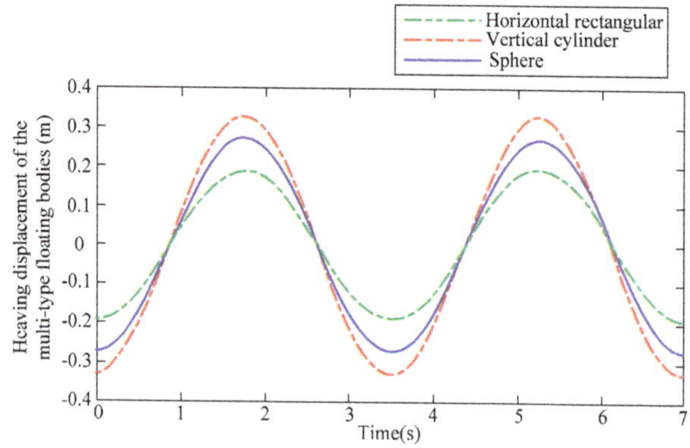

Figure 6. The heaving displacement of the multi-type floating bodies (H = 0.3 m, T = 3.5 s).

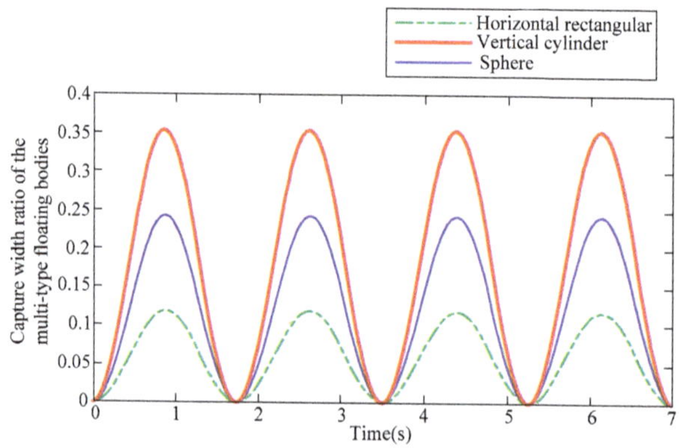

Figure 7. The capture width ratio of the multi-type floating bodies (H = 0.3 m, T = 3.5 s).

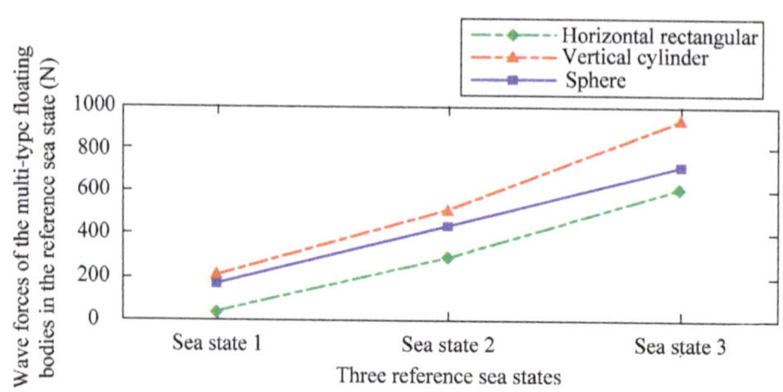

Figure 8. The wave forces of the multi-type floating bodies in the reference sea state.

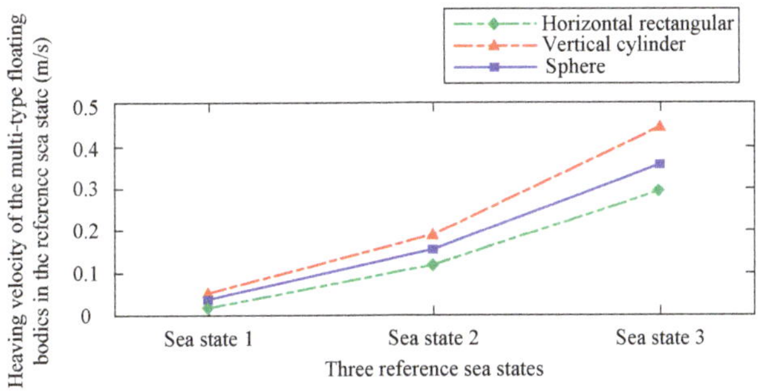

Figure 9. The heaving velocity of the multi-type floating bodies in the reference sea state.

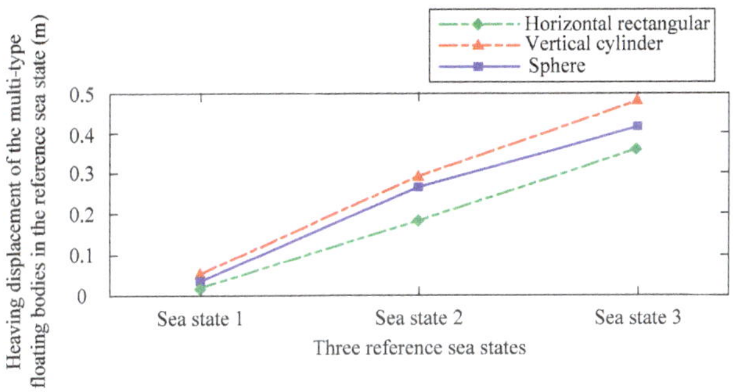

Figure 10. The heaving displacement of the multi-type floating bodies in the reference sea state.

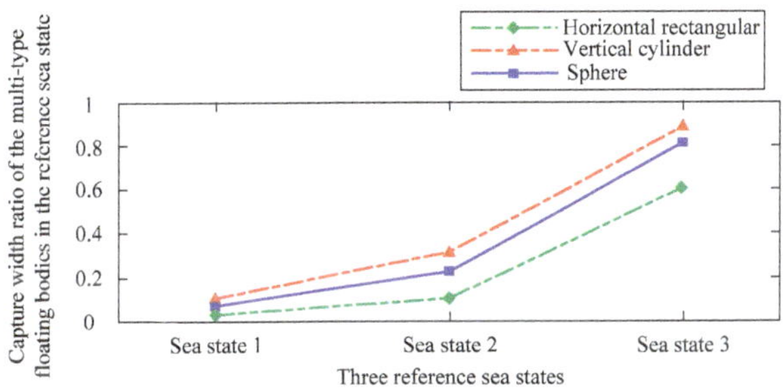

Figure 11. The capture width ratio of the multi-type floating bodies in the reference sea state.

6. Discussion

As shown in Figures 4–6, the vibration frequency curve of the wave force, heaving velocity, and heaving displacement are less of the same for the vertical cylinder buoy and the sphere buoy. However, as to the former, the curve of the wave force and the heaving displacement is steeper, the heaving velocity is faster, wave follower is better, and the generated energy is higher. Nevertheless, the wave force and heaving displacement of the horizontal rectangular floating body are smaller than those two types of floating bodies. Therefore, the generated energy and wave follower is weaker by this one less than those two types of floater bodies. In Figure 7, the capture width ratio of the vertical cylinder floating body is higher and more stable than the sphere floating body. While, the horizontal rectangular floating body is much more less than those two types of floating bodies.

In addition, as shown in Figures 8–11, the wave force, heaving velocity, heaving displacement, and capture width ratio of the floating bodying is affected by the peak wave period and the wave height at a certain sea state. The above parameters are become higher when the sea state increases. Another, the vertical cylinder floating body outperforms the other floating body in the three reference sea states. As shown in Figure 11, the curve of the capture width ratio is steeper and the value of this parameter is higher when the sea state big changes.

In summary, with the linear regular wave, the cylindrical floater vertically placed in the wave surface is the first optional shape for the novel heaving point absorber and follower is the sphere floater, which can increase the quality of power extracting and the efficient of the WEC system design. At last, the horizontal rectangular floating body is carefully selected.

7. Conclusions

This paper presents a small novel heaving point absorber of energy supply for low-power unmanned ocean devices that are based on the counter-rotating self-adaptive mechanism, with the advantages of small space device, stability, and reliable energy conversion process. For improving the efficiency of this absorber's power production, the wave force and energy efficiency are analyzed by the Froude-Krylov method and the optimal floating body is selected, the following conclusions are drawn:

(1) The structure model and working principle of the novel heaving point absorber are feasible. Based on the optimal power supply strategy, the design of the release and recovery mechanism can improve the concealment, release, and recovery rate of the unmanned ocean devices. Under the different directions of water flows, the design of the upper and lower impeller with opposite rotation directions can provide continuous relative rotational motion to the generator during rising and sinking of the Underwater PTO. The design of the locking devices and impellers' blades with circumference array can achieve the adaptively adjust the blade deflection in the heaving motion and the relative rotation of the impellers. The design of the steady blade can balance the motion of Underwater PTO and makes it not limited by the motion of heaving.

(2) The energy efficiency of the novel heaving point absorber is greatly affected by the geometric parameters, the wave force, the motion displacement, the motion velocity, and the capture width ratio of the floating body. The constitutive relation of the above parameters of the floating body is constructed. In order to calculate the above parameters, the Froude-Krylov method is used, which effectively overcomes the inaccuracy of the linear models and reduces the time consuming to simulate. The algebraic solution of wave excitation force for the axisymmetric and unaxisymmetric floating body is obtained and validated by the numerical simulation.

(3) The wave force, heaving velocity, heaving displacement, and capture width ratio of the three floating bodies are compared and analyzed by the numerical simulation. Under the same working condition, mass and mean immersion depth, the type of vertical cylindrical floater's slopes of the wave force and the steeper amplitude, faster speed of heave motion, larger amplitude of heave motion, better follow wave, larger energy produced, as well as, higher conversion efficiency. Therefore, with the linear regular wave, the cylindrical floater vertically placed on the wave surface is the first optional shape for the novel heaving point absorber and follower is the sphere floater, which can increase the quality of power extracting and the efficient of the WEC system design.

The study that was carried out in this paper focuses on the energy generation efficiency of the novel heaving point absorber for the supply power of the low-power unmanned ocean devices. Moreover, the multi-type floating bodies are optimized base on the Froude-Krylov method. Although a series of numerical analyses above has been conducted in this paper, research on the novel absorber and floating body is still not thorough enough. Therefore, a further step toward is that the verification experiments of wave tank will be done in the power production region for the novel absorber. In addition, the geometric parameters of the vertical cylinder floater will be optimized for increasing the conversion efficiency of the floating body.

Author Contributions: D.C., J.S. and Z.L. conceived the study design; D.C. and C.S. performed the numerical simulations; D.C. analyzed the data and wrote the manuscript; W.W. helped with editing the manuscript.

Funding: This research was funded by National Natural Science Foundation of China (51475465).

Conflicts of Interest: The authors declare no conflict of interest.

Abbreviations

WEC	Wave energy converter
PTO	Power take-off
B	Heading wave width of the floating body (m)
B_{PTO}	Underwater PTO damping coefficient (KNs/m)
B^z_{PTO}	Underwater PTO damping coefficient in the vertical direction (KNs/m)
C	Diffraction correction coefficient (-)
C_V	Coefficient of wave diffraction in the vertical direction (-)
D	Waterline depth of floating body (m)
d	Water depth (m)
$\mathbf{F}_D$	Diffraction force (N)
$\mathbf{F}_R$	Radiation force (N)
$\mathbf{F}_{FK_{st}}$	Static Froude-Krylov force (N)
$\mathbf{F}_{FK_{dy}}$	Dynamic Froude-Krylov force (N)
$\mathbf{F}_{PTO}$	Underwater PTO damping force (N)
$\mathbf{F}_{wave}$	Wave force (N)
F^z_{PTO}	Underwater PTO damping force in the vertical direction (N)
G	Gravity force (N)
g	Acceleration of gravity (m/s^2)
H	Wave height (m)
I	Inertia force of attached mass effect (N)
k	Wave Number (-)
m	Mass of the heaving point absorber (kg)
$\mathbf{n}$	Vector normal to the surface (-)
n_z	Projection of the normal for the wetted surface of the floating body in the vertical direction (-)
P	Pressure (Pa)
P_{st}	Hydrostatic pressure (Pa)
P_{dy}	Dynamic pressure (Pa)
P_{WEC}	Average power of the floating body to harvests wave energy (Pa)
P_0	Wave energy of the incident wave within the width of the body (Pa)
Q	Viscous damping force (N)
r	Wave amplitude (m)
S	Submerged surface (m^2)
$[T]$	Coordinate transformation matrix between local coordinate system and global coordinate system (-)
T	Wave period (s)
$\mathbf{V}_{WEC}$	Velocity of the floating body (m/s)
(X_0, Y_0, Z_0)	Coordinates of the center of gravity on the floating body in the global coordinate system
$(\bar{x}, \bar{y}, \bar{z})$	Coordinates at any point on the floating body in the local coordinate system

X	Vertical distance from the reference fluid surface to the body waterline (m)
x	Direction of wave propagation in the global coordinate system (-)
$\overline{x}$	Direction of wave propagation in the local coordinate system (-)
$\mathbf{X}$	Heaving displacement of the body from its hydrostatic equilibrium position (m)
$\dot{\mathbf{X}}$	Heaving velocity of the floating body (m/s)
$\ddot{\mathbf{X}}$	Heaving acceleration of the body (m/s^2)
Y	Vertical distance from the reference fluid surface to the fluid surface (m)
Z	Vertical distance from the reference fluid surface to any point in the fluid (m)
z	Vertical displacement of the floating body in the global coordinate system (m)
$\overline{z}$	Vertical displacement of the floating body in the local coordinate system (m)
φ	Wave angular frequency (rad/s)
ϕ	Potential flow
ϕ_{I}	Undisturbed incident flow potential
ϕ_{D}	Diffraction potential
ϕ_{R}	Radiation potential
ρ	Water density (kg/m^3)
ω	Wave circular frequency (rad/s)
η	Capture width ratio (-)

Appendix A

The calculation of wave excitation forces in the vertical direction of multi-type floating bodies.

Appendix A.1 Horizontal Rectangular Floating Body

The wave excitation forces in the vertical direction of the horizontal rectangular floating body is

$$\begin{aligned} F_{\mathrm{ex}}^{z} &= C_v \iint\limits_{s} p_z ds = C_v a \int_{x_1}^{x_1+b} \frac{\rho g H}{2} \frac{chk(d-\lambda(t))}{chkd} \cos(kx-\omega t)dx \\ &= C_v a \frac{\rho g H}{2} \frac{chk(d-\lambda(t))}{chkd} \int_{x_1}^{x_1+b} \cos(kx-\omega t)dx \\ &= C_v a \frac{\rho g H}{2k} \frac{chk(d-\lambda(t))}{chkd} [\sin(k(x_1+b)-\omega t) - \sin(kx_1-\omega t)] \\ &= C_v \frac{a\rho g H}{k} \frac{chk(d-\lambda(t))}{chkd} \cos(kx_1-\omega t+\tfrac{kb}{2}) \sin\tfrac{kb}{2} \end{aligned} \tag{A1}$$

Appendix A.2 Vertical Cylinder Floating Body

The relationship between any point $(\overline{x}, \overline{y}, \overline{z})$ on the curved surface element ds and the cylinder coordinate [54] $(r, \alpha, \overline{z})$ is

$$\begin{cases} \overline{x} = R\cos\alpha \\ \overline{y} = R\sin\alpha \\ \overline{z} = \overline{z} \end{cases} \tag{A2}$$

Then, the above equation can be written in the global coordinate system is

$$\begin{cases} x = R\cos\alpha + x_1 + R \\ y = R\sin\alpha + y_1 \\ z = z \end{cases} \tag{A3}$$

Let $\alpha = \frac{\pi}{2} + \varphi$, the dynamic pressure at any point on the cylinder surface at this time in the global coordinate system can be written as:

$$\begin{aligned} P_{\mathrm{dy}} &= \frac{\rho g H}{2} \frac{chkz}{chkd} \cos(k(R\cos\alpha + x_1 + R) - \omega t) \\ &= \frac{\rho g H}{2} \frac{chkz}{chkd} [\cos(kR\sin\varphi)\cos(kx_1 + kR - \omega t) \\ &+ \sin(kR\sin\varphi)\sin(kx_1 + kR - \omega t)] \end{aligned} \tag{A4}$$

According to the properties of the Bessel function [55] is as follows:

$$\left.\begin{aligned} \cos(kR\sin\varphi) &= J_0(kR) + 2\textstyle\sum_{m=1}^{m=\infty} J_{2m}(kR)\cos(2m\varphi) \\ \sin(kR\sin\varphi) &= 2\textstyle\sum_{m=1}^{m=\infty} J_{2m-1}(kR)\sin(2m-1)\varphi \end{aligned}\right\} \tag{A5}$$

When $m = 1$, and then:

$$\left.\begin{array}{l}\cos(kR\sin\varphi) = J_0(kR) + 2J_2(kR)\cos 2\varphi \\ \sin(kR\sin\varphi) = 2J_1(kR)\sin\varphi\end{array}\right\} \tag{A6}$$

Combining above equation, we can obtain the following equation:

$$\begin{array}{l}P_{dy} = \frac{\rho g H}{2}\frac{chkz}{chkd}[(J_0(kR) + 2J_2(kR)\cos 2\varphi)\cos(kx_1 + kR - \omega t) \\ +(2J_1(kR)\sin\varphi)\sin(kx_1 + kR - \omega t)]\end{array} \tag{A7}$$

where, $J_0(kR)$ is the first kind zero-order Bessel function, $J_1(kR)$ is the first kind one-order Bessel function, $J_2(kR)$ is the first kind two-order Bessel function. In order to convenient calculation, let $J_m(kR) = J_m$.

The wave excitation forces in the vertical direction of the vertical cylinder floating body is

$$\begin{array}{l}F_{ex}^z = C_v \iint\limits_S p_z ds = 2C_v \int_{d-\lambda(t)}^d \int_0^\pi p_z R d\alpha dz = 2C_v \int_{d-\lambda(t)}^d \int_{-\frac{\pi}{2}}^{\frac{\pi}{2}} p_z R d\varphi dz \\ = C_v \cdot \frac{R\rho g H}{chkd} \cdot \int_{d-\lambda(t)}^d chkzdz \cdot \int_{-\frac{\pi}{2}}^{\frac{\pi}{2}} \left\{\begin{array}{c}(J_0 + 2J_2\cos 2\varphi)\cos(kx_1 + kR - \omega t) \\ +(2J_1\sin\varphi)\sin(kx_1 + kR - \omega t)\end{array}\right\} d\varphi \\ = C_v \frac{R\rho g \pi H J_0(kR)}{k}\frac{shkd - shk(d-\lambda(t))}{chkd}\cos(kx_1 + kR - \omega t)\end{array} \tag{A8}$$

Appendix A.3 Sphere Floating Body

The relationship between any point $(\overline{x}, \overline{y}, \overline{z})$ on the curved surface element ds and the spherical coordinate [54] (r, θ, α) is

$$\left\{\begin{array}{l}\overline{x} = R\sin\theta\sin\alpha \\ \overline{y} = R\sin\theta\cos\alpha \\ \overline{z} = R\cos\theta\end{array}\right. \tag{A9}$$

Then, the above equation can be written in the global coordinate system is

$$\left\{\begin{array}{l}x = R\sin\theta\sin\alpha + x_1 + R \\ y = R\sin\theta\cos\alpha + y_1 + R \\ z = R\cos\theta + d + R - \lambda(t)\end{array}\right. \tag{A10}$$

Let $\alpha = \frac{\pi}{2} + \varphi$, the dynamic pressure at any point on the spherical surface at this time in the global coordinate system can be written as:

$$\begin{array}{l}P_{dy} = \frac{\rho g H}{2}\frac{chk(R\cos\theta + d + R - \lambda(t))}{chkd}\cos(k(R\sin\theta\sin\alpha + x_1 + R) - \omega t) \\ = \frac{\rho g H}{2}\frac{chk(R\cos\theta + d + R - \lambda(t))}{chkd}\left\{\begin{array}{c}\cos(kR\sin\theta\sin\varphi)\cos(kx_1 + kR - \omega t) \\ +\sin(kR\sin\theta\sin\varphi)\sin(kx_1 + kR - \omega t)\end{array}\right\}\end{array} \tag{A11}$$

According to the properties of the Bessel function [55] is as follows:

$$\left.\begin{array}{l}\cos(kR\sin\theta\sin\varphi) = J_0(kR\sin\theta) + 2\sum_{m=1}^{m=\infty} J_{2m}(kR\sin\theta)\cos(2m\varphi) \\ \sin(kR\sin\theta\sin\varphi) = 2\sum_{m=1}^{m=\infty} J_{2m-1}(kR\sin\theta)\sin(2m-1)\varphi\end{array}\right\} \tag{A12}$$

When $m = 1$, and then:

$$\left.\begin{array}{l}\cos(kR\sin\theta\sin\varphi) = J_0(kR\sin\theta) + 2J_2(kR\sin\theta)\cos 2\varphi \\ \sin(kR\sin\theta\sin\varphi) = 2J_1(kR\sin\theta)\sin\varphi\end{array}\right\} \tag{A13}$$

Combining above equation, we can obtain the following equation

$$P_{dy} = \frac{\rho g H}{2}\frac{chk(R\cos\theta + d + R - \lambda(t))}{chkd}\left\{\begin{array}{c}(J_0(kR\sin\theta) + 2J_2(kR\sin\theta)\cos 2\varphi)\cos(kx_1 + kR - \omega t) \\ +(2J_1(kR\sin\theta)\sin\varphi)\sin(kx_1 + kR - \omega t)\end{array}\right\} \tag{A14}$$

where, $J_0(kR\sin\theta)$ is the first kind zero-order Bessel function, $J_1(kR\sin\theta)$ is the first kind one-order Bessel function, $J_2(kR\sin\theta)$ is the first kind two-order Bessel function. In order to convenient calculation, let $J_m(kR\sin\theta) = J_m$.

The wave excitation forces in the vertical direction of the sphere floating body is

$$\begin{aligned} F_w^z &= 2C_v \int_{\pi-\arccos(\frac{R-\lambda(t)}{R})}^{\pi} \int_0^{\pi} p_z \cos\theta R^2 \sin\theta d\alpha d\theta \\ &= C_v \cdot \frac{R^2 \rho g H}{chkd} \cdot \int_{\pi-\arccos(\frac{R-\lambda(t)}{R})}^{\pi} \int_{-\frac{\pi}{2}}^{\frac{\pi}{2}} chk(R\cos\theta + d + R - \lambda(t)) \\ &\cdot \left\{ \begin{array}{c} (J_0 + 2J_2 \cos 2\varphi)\cos(kx_1 + kR - \omega t) \\ +2J_1 \sin\varphi \sin(kx_1 + kR - \omega t) \end{array} \right\} \cdot \cos\theta \sin\theta d\varphi d\theta \\ &= C_v \cdot \frac{\pi R^2 \rho g H}{chkd} \cdot (I_1(kR) - I_2(kR)) \cdot \cos(kx_1 + kR - \omega t) \end{aligned} \tag{A15}$$

where, $I_1(kR) = \int_{\pi-\arccos(\frac{R-\lambda(t)}{R})}^{\pi} ch(kR\cos\theta) chk(d + R - \lambda(t)) \cdot J_0(kR\sin\theta) \cdot \cos\theta \sin\theta d\theta$, $I_2(kR) = \int_{\pi-\arccos(\frac{R-\lambda(t)}{R})}^{\pi} sh(kR\cos\theta) shk(d + R - \lambda(t)) \cdot J_0(kR\sin\theta) \cdot \cos\theta \sin\theta d\theta$.

References

1. Bertaska, I.R.; Ellenrieder, K.D.V. Experimental evaluation of supervisory switching control for unmanned surface vehicles. *IEEE J. Ocean. Eng.* **2018**, *99*, 1–22. [CrossRef]
2. Mousazadeh, H.; Jafarbiglu, H.; Abdolmaleki, H.; Omrani, E.; Monhaseri, F.; Abdollahzadeh, M.R.; Mohammadi-Aghdam, A.; Kiapei, A.; Salmani-Zakaria, Y.; Makhsoos, A. Developing a navigation, guidance and obstacle avoidance algorithm for an unmanned surface vehicle (USV) by algorithms fusion. *Ocean Eng.* **2018**, *159*, 56–65. [CrossRef]
3. Zoss, B.M.; Mateo, D.; Kuan, Y.K.; Tokić, G.; Chamanbaz, M.; Goh, L.; Vallegra, F.; Bouffanais, R.; Yue, D.K. Distributed system of autonomous buoys for scalable deployment and monitoring of large waterbodies. *Auton. Robots* **2018**, *11*, 1–21. [CrossRef]
4. Banazadeh, A.; Seif, M.S.; Khodaei, M.J.; Rezaie, M. Identification of the equivalent linear dynamics and controller design for an unmanned underwater vehicle. *Ocean Eng.* **2017**, *139*, 152–168.
5. Venkatesan, R.; Sannasiraj, S.A.; Ramanamurthy, M.V.; Senthilkumar, P.; Dhinesh, G. Development and performance validation of a cylindrical buoy for deep-ocean tsunami monitoring. *IEEE J. Ocean. Eng.* **2018**, *99*, 1–9. [CrossRef]
6. Danovaro, R.; Aguzzi, J.; Fanelli, E.; Billett, D.; Gjerde, K.; Jamieson, A.; Ramirez-Llodra, E.; Smith, C.R.; Snelgrove, P.V.R.; Thomsen, L.; et al. An ecosystem-based deep-ocean strategy. *Science* **2017**, *355*, 452–454. [CrossRef] [PubMed]
7. Sun, C.; Luo, Z.; Shang, J.; Lu, Z.; Zhu, Y.; Wu, G. Design and Numerical Analysis of a Novel Counter-Rotating Self-Adaptable Wave Energy Converter Based on CFD Technology. *Energies* **2018**, *11*, 694. [CrossRef]
8. Mendez, A.; Leo, T.J.; Herreros, M.A. Current state of technology of fuel cell power systems for autonomous underwater vehicles. *Energies* **2014**, *7*, 4676–4693. [CrossRef]
9. Wang, X.; Shang, J.; Luo, Z.; Tang, L.; Zhang, X.; Li, J. Reviews of power systems and environmental energy conversion for unmanned underwater vehicles. *Renew. Sustain. Energy Rev.* **2012**, *16*, 1958–1970. [CrossRef]
10. Dhanak, M.R.; Xiros, N.I. *Handbook of Ocean Engineering*; Springer: New York, NY, USA, 2016.
11. Astariz, S.; Iglesias, G. The economics of wave energy: A review. *Renew. Sustain. Energy Rev.* **2015**, *45*, 397–408. [CrossRef]
12. Cruz, J. *Ocean Wave Energy*; Springer: Berlin, Germany, 2008; Volume 144, pp. 2451–2460.
13. Gunn, K.; Stock-Williams, C. Quantifying the global wave power resource. *Renew. Energy* **2012**, *44*, 296–304. [CrossRef]
14. Saprykina, Y.; Kuznetsov, S. Analysis of the Variability of Wave Energy Due to Climate Changes on the Example of the Black Sea. *Energies* **2018**, *11*, 2020. [CrossRef]
15. Clément, A.; McCullen, P.; Falcão, A.; Gardner, F.; Hammarlund, K.; Lemonis, G.; Lewis, T.; Nielsen, K.; Petroncini, S.; Pontes, M.-T.; et al. Wave energy in Europe: Current status and perspectives. *Renew. Sustain. Energy Rev.* **2002**, *6*, 405–431. [CrossRef]
16. Salter, S.H. Wave power. *Nature* **1974**, *249*, 720–724. [CrossRef]
17. O'Hagan, A.M.; Huertas, C.; O'Callaghan, J.; Greaves, D. Wave energy in Europe: Views on experiences and progress to date. *Int. J. Mar. Energy* **2016**, *14*, 180–197. [CrossRef]
18. Rusu, E.; Onea, F. Estimation of the wave energy conversion efficiency in the Atlantic ocean close to the European islands. *Renew. Energy* **2016**, *85*, 687–703. [CrossRef]

19. Kalogeri, C.; Galanis, G.; Spyrou, C.; Diamantis, D.; Baladima, F.; Koukoula, M.; Kallos, G. Assessing the European offshore wind and wave energy resource for combined exploitation. *Renew. Energy* **2017**, *101*, 244–264. [CrossRef]
20. Stegman, A.; Andres, A.D.; Jeffrey, H.; Johanning, L.; Bradley, S. Exploring marine energy potential in the UK using a whole systems modeling approach. *Energies* **2017**, *10*, 1251. [CrossRef]
21. Aderinto, T.; Li, H. Ocean wave energy converters: Status and challenges. *Energies* **2018**, *11*, 1250. [CrossRef]
22. Chen, B.; Bruce, T.; Greated, C.A.; Kang, H. Dynamic behavior of a wave power buoy with interior on-board linear generator. *Ocean Eng.* **2017**, *129*, 374–381. [CrossRef]
23. Weber, J.; Costello, R.; Mouwen, F.; Ringwood, J.; Thomas, G. Techno-economic WEC system optimization methodology applied to Wavebob system definition. In Proceedings of the 3th International Conference on Ocean Energy, Bilbao, Spain, 6–8 October 2010; Volume 10, p. 6.
24. Chatzigiannakou, M.A.; Dolguntseva, I.; Leijon, M. Offshore deployments of wave energy converters by Seabased industry AB. *J. Mar. Sci. Eng.* **2017**, *5*, 15. [CrossRef]
25. Fred Olsen's Lifesaver Buoy. Available online: http://boltseapower.com/company/ (accessed on 25 November 2018).
26. Moskvitch, K. News briefing: In num6ers: CETO 6 Carnegie wave energy. *Eng. Technol.* **2016**, *11*, 12–13. [CrossRef]
27. Black, J.L.; Mei, C.C.; Bray, C.G. Radiation and Scattering of Water Waves by Rigid Bodies. *J. Fluid Mech.* **1971**, *46*, 151–164. [CrossRef]
28. Mohapatra, S.C.; Guedes Soares, C. Wave forces on a floating structure over flat bottom based on Boussinesq formulation. In *Renewable Energies Offshore*; Taylor & Francis Group: London, UK, 2015; pp. 335–342.
29. Rodriguez, M.; Spinneken, J.; Swan, C. Nonlinear loading of a two-dimensional heaving box. *J. Fluids Struct.* **2016**, *60*, 80–96. [CrossRef]
30. Islam, H.; Mohapatra, S.C.; Guedes, S.C. Comparisons of CFD, experimental and analytical simulations of a heaving box-type floating structure. In *Progress in Maritime Technology and Engineering*; Taylor & Francis Group: London, UK, 2018; Volume 5, pp. 633–639.
31. Yeung, R.W. Added Mass and Damping of a Vertical Cylinder in Finite Depth Waters. *Appl. Ocean Res.* **1981**, *3*, 119–133. [CrossRef]
32. Sabuncu, T.; Calisal, S. Hydrodynamic Coefficients for Vertical Circular Cylinder at Finite Depth. *Ocean Eng.* **1981**, *8*, 25–63. [CrossRef]
33. Calisal, S.; Sabuncu, T. Hydrodynamic Coefficients for Vertical Composite Cylinders. *Ocean Eng.* **1984**, *11*, 529–542. [CrossRef]
34. Mansour, A.M.; WiUliam, A.N.; Wang, K.H. The diffraction of linear waves by a uniform vertical cylinder with cosine-type radial perturbations. Type Radial Perturbations. *Ocean Eng.* **2002**, *29*, 239–259. [CrossRef]
35. Bhatta, D.; Rahman, M. On Scattering and Radiation Problem for a Cylinder in Water of Finite Depth. *Int. J. Eng. Sci.* **2003**, *41*, 931–967. [CrossRef]
36. Kim, W.D. On the Harmonic Oscillations of a Rigid Body on a Free Surface. *J. Fluid Mech.* **1965**, *21*, 427–451. [CrossRef]
37. Bihs, H.; Kamata, A.; Lu, Z.J.; Arntsen, I.A. Simulation of Floating Bodies using a Combined Immersed Boundary with the Level Set Method in REEF3D. In Proceedings of the VII International Conference on Computational Methods in Marine Engineering, Nantes, Frances, 15–17 June 2017.
38. Koh, H.J.; Cho, I.H. Heave motion response of a circular cylinder with the dual damping plates. *Ocean Eng.* **2016**, *125*, 95–102. [CrossRef]
39. Wang, S. *The Hydrodynamic Forces and Pressure Distributions for an Oscillating Sphere in a Fluid of Finite Depth*; M.I.T. Department of Naval Architecture and Marine Engineering: Cambridge, MA, USA, 1966.
40. Finnegan, W.; Meere, M.; Goggins, J. The wave excitation forces on a truncated vertical cylinder in water of infinite depth. *J. Fluids Struct.* **2013**, *40*, 201–213. [CrossRef]
41. Ghadimi, P.; Bandari, H.P.; Rostami, A.B. Determination of the Heave and Pitch Motions of a Floating Cylinder by of the Effects of Geometric Parameters on its Dynamics in Regular Waves. *Int. J. Appl. Math. Res.* **2012**, *1*, 611–633. [CrossRef]
42. Giorgi, G.; Ringwood, J.V. Computationally efficient nonlinear froude–krylov force calculations for heaving axisymmetric wave energy point absorbers. *J. Ocean Eng. Mar. Energy* **2017**, *3*, 21–33. [CrossRef]

43. Merigaud, A.; Gilloteaux, J.C.; Ringwood, J.V. A Nonlinear Extension for Linear Boundary Element Methods in Wave Energy Device Modeling. In Proceedings of the International Conference on Ocean, Offshore and Arctic Engineering, Rio de Janeiro, Brazil, 1–6 July 2012.
44. Penalba, M.; Mérigaud, A.; Gilloteaux, J.C.; Ringwood, J.V. Influence of nonlinear Froude Krylov forces on the performance of two wave energy points absorbers. *J. Ocean Eng. Mar. Energy* **2017**, *3*, 209–220. [CrossRef]
45. Rodrigues, J.M.; Soares, C.G. Froude-krylov forces from exact pressure integrations on adaptive panel meshes in a time domain partially nonlinear model for ship motions. *Ocean Eng.* **2017**, *139*, 169–183. [CrossRef]
46. Falnes, J.; Perlin, M. Ocean waves and oscillating systems: Linear interactions including wave-energy extraction. *Appl. Mech. Rev.* **2003**, *56*, 286. [CrossRef]
47. Falnes, J. *Ocean Waves and Oscillating Systems*; Cambridge University Press: Cambridge, UK, 2002.
48. Clement, A.; Ferrant, P. Superharmonic Waves Generated by the Large Amplitude Heaving Motion of a Submerged Body. In *Nonlinear Water Waves*; Springer: Berlin/Heidelberg, Germany, 1988; pp. 423–433.
49. Yu, H.; Zhang, Y.; Chen, W. Effect of Power Take-off System on the Capture Width Ratio of a Novel Wave Energy Converter. In Proceedings of the Asian Wave & Tidal Energy Conference, Singapore, 24–28 October 2016.
50. Chen, T.X.; Wu, B.J.; Li, M. Flume experiment study on capture width ratio of a new backward bent duct buoy with a pentagon buoyancy cabin. *Ocean Eng.* **2017**, *141*, 12–17. [CrossRef]
51. Herbich, J.B. *Offshore Pipeline Design Elements*; M. Dekker: New York, NY, USA, 1981.
52. Standing, R.G. Use of potential theory in evaluating wave forces on offshore structures. In *Power from Sea Waves, Conference Institute of Mathematics and its Applications*; Academic Press: London, UK, 1980; Volume 1, pp. 175–212.
53. Garrison, C.J.; Rao, V.S. Interaction of waves with submerged objects. *J. Waterw. Harb. Coast. Eng Div.* **1971**, *97*, 259–277.
54. Korn, G.A.; Korn, T.M. *Mathematical Handbook for Scientists and Engineers*; McGraw-Hill: New York, NY, USA, 1968.
55. Watson, G.N. *Theory of Bessel Functions*; Cambridge University Press: Cambridge, UK, 1962.

© 2018 by the authors. Licensee MDPI, Basel, Switzerland. This article is an open access article distributed under the terms and conditions of the Creative Commons Attribution (CC BY) license (http://creativecommons.org/licenses/by/4.0/).

energies

MDPI

Article

Grid Integration and Power Smoothing of an Oscillating Water Column Wave Energy Converter

Gimara Rajapakse [1,*], Shantha Jayasinghe [1], Alan Fleming [1] and Michael Negnevitsky [2]

1 Australian Maritime College, University of Tasmania, Launceston, Tasmania 7250, Australia; shanthaj@utas.edu.au (S.J.); alan.fleming@utas.edu.au (A.F.)
2 Centre for Renewable Energy and Power Systems, University of Tasmania, Hobart, Tasmania 7001, Australia; michael.negnevitsky@utas.edu.au
* Correspondence: gimara.rajapakse@utas.edu.au; Tel.: +61-3-6324-9752

Received: 7 June 2018; Accepted: 13 July 2018; Published: 18 July 2018

check for updates

Abstract: This paper applies model predictive control (MPC) for the power processing of an oscillating water column (OWC) wave energy conversion (WEC) system to achieve smooth power delivery to the grid. The particular air turbine design adopted in this study produces large power pulses ranging from 0 to 1 MW in magnitude, and thus, direct connection to the grid is practically impossible, especially in weak grid conditions. Therefore, energy storage is an essential element that should be integrated into this particular WEC system in order to absorb power pulses and thereby ensure smooth delivery of power to the grid. Taking into account the repetitive nature, duration, and magnitude of the power pulses, this study has chosen "supercapacitor" as the suitable energy storage technology. The supercapacitor energy storage (SCES) is integrated into the dc-link of the back-to-back power converter of the WEC system through a bidirectional dc-dc converter. In order to achieve the desired operation of this complex power converter arrangement, a finite control set MPC strategy is proposed in this paper. Performance of the proposed energy storage system (ESS) and control strategy are evaluated through computer simulations. Simulation results show that the proposed SCES system and the control strategy are able to achieve smooth power delivery to the grid amidst power pulses coming from the generator.

Keywords: dc-dc bidirectional converter; finite control set-model predictive control (FCS-MPC); oscillating water column (OWC); supercapacitor energy storage (SCES)

1. Introduction

Oceans, which cover 70% of the Earth, offer a vast amount of renewable energy. Out of many forms, wave energy is an accessible and widely available type of ocean energy. Recent studies have shown that wave energy has the potential to make significant contributions to renewable energy targets worldwide [1–3]. Australia has been identified as a wave-energy-rich country, having the potential to extract more energy from waves than its total annual electricity demand [2]. Recognizing this potential, many companies have invested in wave energy projects in Australia [3–5]. Out of the many technologies used to harness energy from waves, oscillating water column (OWC) based wave energy conversion (WEC) technology has the advantage of having all of its moving parts above the water level, allowing relatively straightforward installation and easy access for maintenance [2,6]. These are key factors for cost reduction in any renewable energy technology and thus, with the improved performance, OWC wave energy converters will be able to compete with other renewable and fossil-fuel energies in the energy market.

The OWC concept utilizes a water column inside a partially submerged hollow concrete structure. With the incident wave, the OWC inside the chamber oscillates the air column above it and results in exhale and inhale air streams through the open end of the chamber. Energy of the air stream is then converted into electricity with the aid of a turbine coupled generator. The commonly used turbines are bidirectional which extract energy from both the inhale and exhale streams.

The Australian maritime college, in collaboration with the Wave Swell Energy Ltd., has developed a new OWC air turbine technology, which has been recognized as an efficient and simple design compared with many other OWC technologies [6,7]. This particular configuration has passive, non-return air flow valves built into its chamber, which activate during the exhale stage and equalize the pressure inside the chamber to the atmosphere, which allows the rising water column to reach its maximum height. This creates a higher differential pressure during the inhale stage resulting in high velocity air stream through the turbine. Since this particular air turbine extracts energy only during the inhale stage, the output power inherently becomes discrete pulses. These large and discrete power pulses create major operational issues, such as frequency deviations, voltage sags/swells, and instabilities if delivered to the grid without smoothing. Therefore, energy storage is an essential feature that should be incorporated into the power converter of the WEC system. Nevertheless, in contrast to wind energy or tidal energy turbines, where the rotational speed should be changed according to the wind speed or tidal flow speed, the unidirectional air-turbine used in this study does not necessarily require variable speed operation. Therefore, according to [8], the turbine speed can be regulated to remain within the optimum speed range irrespective of the air flow. This eliminates the need for complex maximum power point tracking strategies, and thus, the machine side converter controller becomes relatively less complex. However, the abovementioned large and discrete power pulses are unique to this WEC system compared to other types of OWC WEC systems, and thus, special attention has to be paid when developing associated control strategies.

As mentioned above, the promising approach to overcome the issues related to power pulses is the use of an energy storage system (ESS). The ESS absorbs power pulses to provide smooth power delivery to the grid. A comprehensive review of ESSs in renewable energy electric power systems is presented in [9]. Hamidi et al. [9] have discussed different types of batteries, supercapacitors, and hybrid energy storage system (HESS) along with their modeling techniques. They have also highlighted different utility-level power converter topologies and given six examples of ESSs in field installations. Some of the desirable qualities of ESSs are emphasized in [10], as high charge-discharge efficiency, high energy density, volume power density, high specific energy, high specific power with continuous power supply, long life, relatively cheaper value per usable kWh per cycle, and so forth. The evidence confirms that supercapacitors lead the energy storage market with high-power density, whereas the Li-ion batteries lead the industry with high energy density. Generally, one or both of these ESSs are used in renewable energy applications depending on the requirements [11,12]. This study applies a commercially available supercapacitor to smoothen the output power of the unidirectional air turbine OWC system. The supercapacitor energy storage (SCES) has high energy storage efficiency and very high cycling stability with insignificant loss of energy storage capacity [13]. A full-scale offshore OWC WEC system has been investigated in [14], in which it was confirmed that supercapacitors are capable of power smoothing satisfactorily over desirable long periods. Figure 1, illustrates a grid-connected OWC WEC system with SCES.

In [15], the authors have proposed to use a Li-ion battery energy storage (BES) system, which is directly connected to the dc-link to absorb power pulses. It has used the finite control set-model predictive control (FCS-MPC) strategy for the rectifier and inverter of the back-to-back power converter. Due to the fact that the two controllers indirectly control the battery current and the dc-link voltage is governed by the battery, direct connection is possible for batteries. If supercapacitors are used as the energy storage element, direct connection to the dc-link is not effective, as the voltage should be allowed to vary in a wide range to store more energy in the supercapacitors. Therefore, an interfacing dc-dc converter is essential for integrating supercapacitors into the power converter system. While the rectifier and inverter controllers proposed in [15] can still be used, the dc-dc converter requires an additional converter.

The use of dc-dc bidirectional converters with energy storage, such as a battery and/or supercapacitor are common in utility-level renewable energy systems, which act fast to reduce the mismatch between the supply and demand power [9]. Several types of such converters have been presented in [9], which can be used as energy storage interface systems. The half bridge (HB)-type converters are more effective than many other types as they use a lower number of passive electronics, smaller inductors, and experience lower current stresses. In [11,16,17], an ESS with a HB dc-dc bidirectional converter for a grid-connected renewable energy systems was proposed and explained converter control algorithms using proportional integral (PI) control strategy. Also, Maercos et al. [11] have provided a battery supercapacitor HESS and a literature review of 18 similar approaches with respect to their application, system, topology, rated power, comparison, sizing, control, and goals related to applications such as energy harvesting, microgrids, remote area power supply, load supply, and so forth.

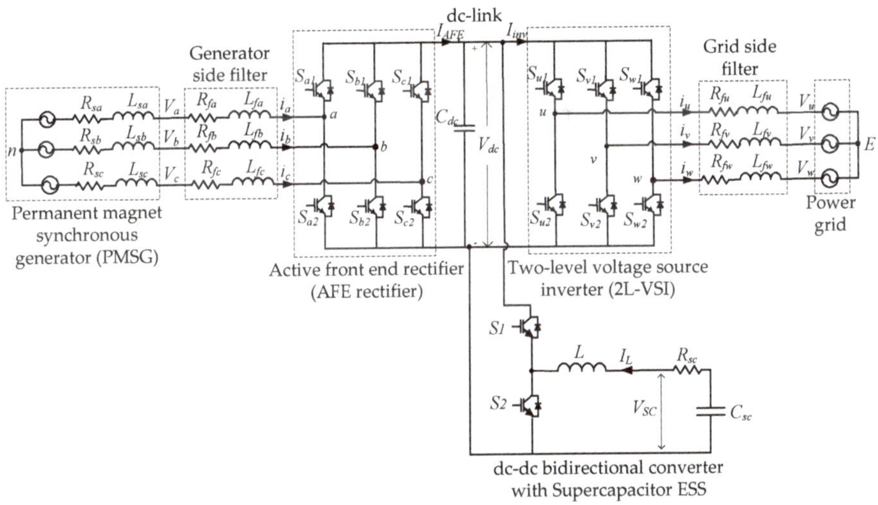

Figure 1. Schematic diagram of the grid connected oscillating water column (OWC) wave energy conversion (WEC) system with a supercapacitor energy storage system.

To gain the desirable results from any ESS, which is connected to the dc-link through an interfacing converter, one needs an effective control system. In this research, applying FCS-MPC strategy for the dc-dc bidirectional converter controller has been achieved. Figure 2 depicts the schematic diagram of the controllers that are used for the electrical power converters.

The model predictive control (MPC) approach is chosen to control the converters due to its simple, fast, accurate, and reliable nature [15,18,19]. In [20], the same control strategy has been applied to a dc-dc bidirectional converter with mode activation approach, where charge, discharge, and idle modes were designed to operate independently. In [21], a SCES coupled to the dc-link with a dc-dc bidirectional converter was proposed for the wells turbine OWC system using field-oriented control (FOC) strategy. Also, a performance comparison was done between SCES and fixed and variable speed flywheel ESSs referring to [17,22,23]. In [24], a HESS has been used for oscillating wave power generator, applying sliding mode control strategy for the whole system. Nevertheless, the feasibility of using supercapacitors for the particular WEC system considered in this study, from the control perspective, is not reported so far. Therefore, the contributions of this paper are to propose SCES for this particular WEC system, develop a FCS-MPC strategy for the entire power converter system, and verify its performance through computer simulations.

The paper is organized as follows: Section 2 presents the modeling of SCES; Section 3 presents the MPC for the electrical power converter, followed by the power management system presented in Section 4; Section 5 presents the simulation results and discusses the implications of the results; and the conclusions derived from the discussion are presented in Section 6.

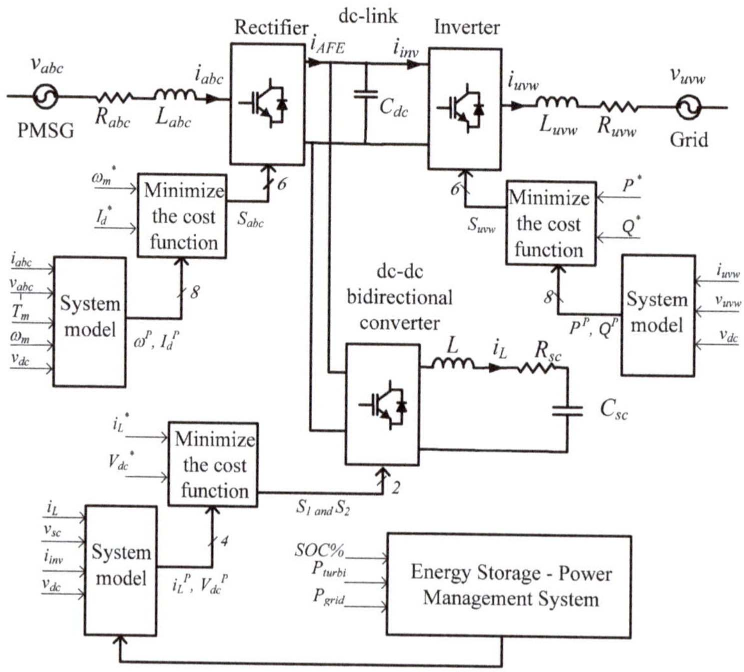

Figure 2. Schematic diagram of the proposed controllers for the grid integration of an OWC unidirectional air turbine generator with a hybrid energy storage system (HESS).

2. Supercapacitor Energy Storage

This study uses a supercapacitor bank as the energy storage element, which is connected to the dc-link through a bidirectional dc-dc converter. Supercapacitors have the advantage of long life, high-power density, attractive temperature range, and high charge-discharge efficiency [10]. The SCES is integrated into the OWC electrical power converter system to achieve smooth power delivery to the grid despite varying sea conditions. With the recent advancements of energy storage technologies, supercapacitors have become popular and commercially available for large power applications, such as WEC systems. The supercapacitor specification sheet in [25], confirms that the industrial 83 F or 165 F supercapacitor single modules with 48 Vdc can be connected in series and/or parallel configurations to gain the required voltage and capacity. Moreover, these modules provide up to 1,000,000 charge/discharge cycles. In this particular system, to attain the required voltage of the SCES (1000 V), a minimum 21 of 48 V modules are needed to connect in series, which results in the decrease of the total capacitance and increase of the internal resistance (R_{sc}) in each pole. Then, these high voltage modules can be connected in parallel to obtain the required total capacitance to gain the energy storage capacity that decreases R_{sc}. Considering R_{sc}, which represents only static losses, the stored energy and instantaneous voltage of the supercapacitor are given by the following [26]:

$$E_{sc} = \frac{1}{2} C_{sc}\, v_{sc}^2 \tag{1}$$

$$v_{sc}(t) = v_0\, e^{-\frac{t}{R_{sc}C_{sc}}} = \frac{Q_0}{C_{sc}}\, e^{-\frac{t}{R_{sc}C_{sc}}} \quad (2)$$

where E_{sc} is the supercapacitor energy, C_{sc} is the capacitance of the supercapacitor, v_{sc} is the voltage across the supercapacitor, v_0 and Q_0 are the initial voltage and charge of the supercapacitor, respectively.

The state of charge (SOC) of the supercapacitor can be estimated using Equation (3):

$$SOC_{sc} = \left(\frac{Q_0 - \int_0^t i_{sc}(t)\, dt}{Q_T} \right) \times 100 = \left(\frac{\left(C_{sc}\, v_{sc(0)}\right) - \int_0^t i_{sc}(t)\, dt}{C_{sc}\, v_{sc}} \right) \times 100 \quad (3)$$

where $i_{sc}(t)$ and Q_T are the supercapacitor current at time t and total charge, respectively. It should be noted that the method shown in Equation (3) is one of the conventional methods of estimation of SOC, and the nominal capacity decreases gradually over time due to variations in load and the internal chemical reactions. More advanced SOC estimation methods are discussed in [27,28].

The supercapacitor charge (Q_{Sc}) is equal to the product of its capacitance and voltage ($Q_{Sc} = C_{SC}\, v_{SC}$). Therefore, SOC_{sc} changes with respect to v_{SC} and vice versa. Figure 3a shows a typical supercapacitor charge/current behavior with respect to the time, and Figure 3b, shows the charge/discharge curves of a supercapacitor and Li-ion battery.

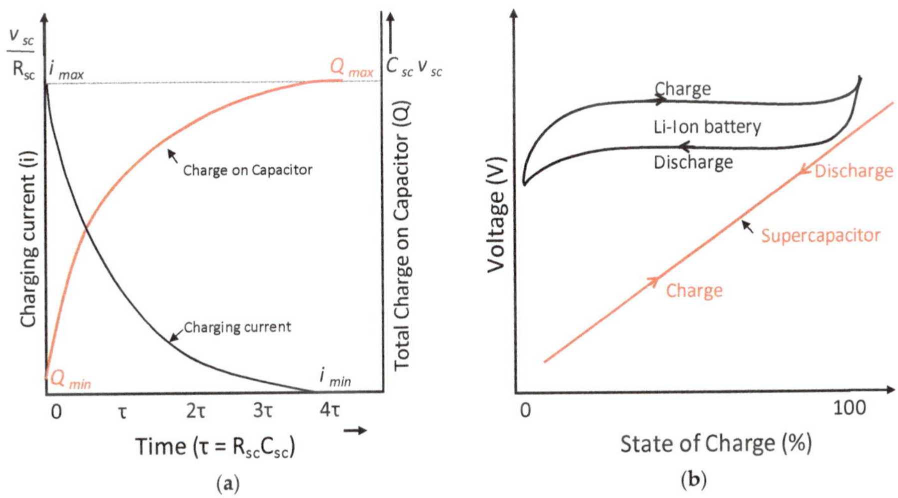

Figure 3. (**a**) Behavior of supercapacitor charge and its current with time; and (**b**) charge/discharge curves of supercapacitor and Li-ion battery.

The simple electrical model of a supercapacitor available in the MATLAB/Simulink was utilized in this study, which satisfactorily model the dynamics. More detailed models of supercapacitors can be found in [29–31].

3. Model Predictive Control-Based Converter Control Strategies

3.1. Rectifier Controller

As in [15], the input current dynamics of the rectifier were found using Kirchhoff's voltage law, and the resultant equation for the predicted current in the next sampling interval was written in the discrete time domain as follows:

$$i_s\,(k+1) = \left(1 - \frac{R_s T_s}{L_s}\right) i_s(k) + \frac{T_s}{L_s}(v_s(k) - v_{AFE}(k)) \tag{4}$$

where i_s is the rectifier input current vector, R_s is the combination of source and line filter resistance, L_s is the combination of source and line filter inductance, k is the sampling instant, T_s is the sampling time, v_s is the generator voltage vector, and v_{AFE} is the rectifier voltage vector. The generator current in the next sampling interval, $i_s(k+1)$, was predicted for each of the eight switching states using Equation (4). The current value was then converted into the dq frame currents, $i_d(k+1)$, $i_q(k+1)$, and used to predict the future angular speed of the generator, $\omega_m(k+1)$ as in [15]:

$$\omega_m(k+1) = \omega_m(k) + \frac{T_s}{J}(T_m - 1.5 n_p \Psi_{pm} i_q(k+1)) \tag{5}$$

where T_m is the mechanical torque applied on the motor, n_p is the number of pole pairs, and Ψ_{pm} is the permanent magnet synchronous generator (PMSG) flux. Then, the values $\omega_m(k+1)$ and $i_d(k+1)$ were used in the cost function, g_{rec}, given in Equation (6) to select the switching state that gives the minimum value for the cost function [15].

$$g_{rec} = |\omega^* - \omega_m(k+1)| + K.|i_d^* - i_d(k+1)| \tag{6}$$

where ω^* is the reference speed and i_d^* is the reference for the d-axis current component of the generator, which was set to zero. The arbitrary constant, K, has been used in the equation to reduce the d axis current of the generator aiming to prevent flux weakening and overheating effects on the PMSG [15].

3.2. Inverter Controller

As in [15], in the inverter controller, three-phase grid current, i_g and grid voltage, v_g were converted into the αβ stationary reference frame using the Clark's transformation, as in [32]:

$$\begin{bmatrix} f_\alpha \\ f_\beta \end{bmatrix} = \sqrt{\frac{2}{3}} \begin{bmatrix} 1 & -\frac{1}{2} & -\frac{1}{2} \\ 0 & \frac{\sqrt{3}}{2} & -\frac{\sqrt{3}}{2} \end{bmatrix} \begin{bmatrix} f_a \\ f_b \\ f_c \end{bmatrix} \tag{7}$$

where f represents current or voltage. Then, the input current dynamics of the inverter were found using Kirchhoff's voltage law. The resultant equation for the predicted current in the next sampling interval was written in the discrete time domain as follows [15]:

$$i_g(k+1) = \left(1 - \frac{R_g T_s}{L_g}\right) i_g(k) + \frac{T_s}{L_g}(v_{VSI}(k) - v_g(k)) \tag{8}$$

where v_{VSI} is the inverter voltage vector, R_g is the combination of grid and line filter resistance, and L_g is the combination of grid and line filter inductance. The grid current in the next sampling interval, $i_g(k+1)$, was predicted for each of the eight switching states using Equation (8). Then, each of the current values were converted into the dq frame currents and used to predict the power values, $P_g(k+1)$ and $Q_g(k+1)$ considering the $v_g(k+1) = v_g(k)$ as follows [15]:

$$P_g(k+1) = (v_d i_d(k+1) + v_q i_q(k+1)) \tag{9}$$

$$Q_g(k+1) = \left(v_q i_d(k+1) - v_d i_q(k+1)\right) \quad (10)$$

where i_d, i_q, v_d, and v_q are grid currents and grid voltages in a dq rotating reference frame. The values $P_g(k+1)$ and $Q_g(k+1)$ were used in the cost function, g_{inv}, given in Equation (11) to select the switching state that gives the minimum value for the cost function [15].

$$g_{inv} = \left|Q_g{}^* - Q_g(k+1)\right|^2 + \left|P_g{}^* - P_g(k+1)\right|^2 \quad (11)$$

where $P_g{}^*$ is the active power reference, which was set to 265kW, and $Q_g{}^*$ is the reactive power reference for the grid, which was set to zero.

3.3. DC-DC Converter Controller

Generally, the voltage of the ESS (battery and/or supercapacitor) is lower than the dc-link voltage and the polarity of the energy storage, and its output is set to the same with respect to the common ground. The dc-dc bidirectional converter is suited well for this type of application, because proper control of the switches regulate the dc-link power fluctuations by directing them to the energy storage [33–35]. This is a simple converter, which can be used in high-power applications where the switch rating of the converter becomes the key concern [23,36]. This converter consists of two insulated gate bipolar transistor (IGBT) switches, two anti-parallel diodes, and an inductor connected as shown in Figure 4a. This dc-dc bidirectional converter is a combination of buck and boost topologies, as shown in Figure 4b,c, respectively.

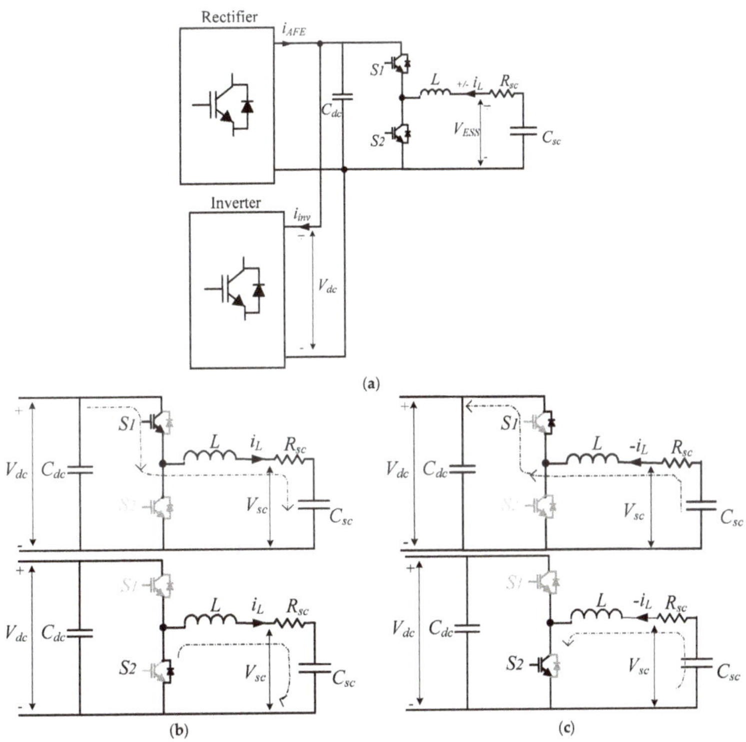

Figure 4. Half bridge (HB) dc-dc bidirectional converter of supercapacitor energy storage (SCES); (**a**) dc-link with electrical power converters; (**b**) buck mode operation of the bidirectional dc-dc converter; and (**c**) boost mode operation of the bidirectional dc-dc converter.

In this type of converter, the inductor is the main energy transfer element, which is also responsible for the output current ripple. The ESS can be either a battery or supercapacitor and is connected to the low-voltage side. The high-voltage side of the converter is connected to the dc-link. The converter is designed to operate in both buck and boost modes. This two-quadrant dc-dc converter allows inductor current flow in either direction while keeping the polarity of the dc voltage fixed during the boost and buck modes. The mode it should operate at any given time is decided by the dc-link voltage and the voltage of the energy storage. If the dc-link voltage drops below the set voltage, the ESS would supply the power. When the power generated is more than the power needed for the grid, the dc-link voltage rises above the set voltage and the converter charges the ESS absorbing the power. During the buck mode, when the $S1$ IGBT is on and $S2$ IGBT is off, the excess current from the generator charges the ESS. During the boost mode, $S1$ IGBT is off, and $S2$ IGBT is on; the ESS supplies the power to the dc-link. The opening and closing of each switch generates a pulse output, and this can be controlled to regulate the desired output voltage.

Because this converter operates in a discontinuous conduction mode to gain a smooth current/voltage output, high converter switching frequency and adequate filtering capacitance on the dc-link are required. Both the dc-link voltage and the ESS voltage could be fixed by duty cycle of IGBTs neglecting the voltage drop across IGBTs and diodes [36,37] as follows:

$$D = \frac{t_{on}}{T} = t_{on} f_{sw} \tag{12}$$

$$(1 - D) = \frac{t_{off}}{T} \tag{13}$$

where D is the duty cycle, f_{sw} is the switching frequency, and T is the switching period of the dc-dc bidirectional converter controller ($T = t_{on} + t_{off}$).

In the steady-state operation of the buck and boost modes of the dc-dc bidirectional converter with ESS, considering the average value modeling, the average inductor voltage and the net change in the inductor current are zero for periodic operation [37]. In the buck mode of operation, the derivative of the inductor current is a positive constant when the $S1$ IGBT is on as follows:

$$\frac{\Delta i_L}{\Delta t} = \frac{\Delta i_L}{DT} = \frac{v_{dc} - v_{ESS}}{L} \tag{14}$$

where v_{dc} is the dc-link voltage, v_{ESS} is the voltage across ESS, and i_L is the inductor current ($i_L = i_{ESS}$).

The derivative of i_L is a negative constant when the $S1$ IGBT is off as follows:

$$\frac{\Delta i_L}{\Delta t} = \frac{\Delta i_L}{(1 - D)T} = \frac{-v_{ESS}}{L} \tag{15}$$

Because the net change in the inductor current over one period is zero, the relationship between v_{dc} and v_{ESS} is as follows.

$$\left(\frac{v_{dc} - v_{ESS}}{L}\right) DT + \left(\frac{-v_{ESS}}{L}\right)(1 - D)T = 0 \tag{16}$$

$$v_{dc} = \frac{v_{ESS}}{D} \tag{17}$$

The voltage ratio D should be kept smaller than the unity to have efficient power conversion [37]. This is to enable the buck converter to produce an output voltage that is equal to or less than the input, while enabling the boost converter to produce an equal or greater output voltage than the input. In boost mode of operation, the derivative of i_L is a positive constant, when the $S2$ IGBT is on as:

$$\frac{\Delta i_L}{\Delta t} = \frac{\Delta i_L}{DT} = \frac{v_{ESS}}{L} \tag{18}$$

and a negative constant when the *S*2 IGBT is off as:

$$\frac{\Delta i_L}{\Delta t} = \frac{\Delta i_L}{(1-D)T} = \frac{v_{ESS} - v_{dc}}{L} \tag{19}$$

Because the net change in the inductor current over one period is zero, the relationship between v_{dc} and v_{ESS} is as follows:

$$\left(\frac{v_{ESS}}{L}\right)DT + \left(\frac{v_{ESS} - v_{dc}}{L}\right)(1-D)T = 0 \tag{20}$$

$$v_{dc} = \frac{v_{ESS}}{(1-D)} \tag{21}$$

3.3.1. Space State Model of Bidirectional DC-DC Converter

The states of the *S*1 and *S*2 switches directly affect the charging and discharging of the inductor. As shown in Figure 4, the HB dc-dc bidirectional converter operates in the buck mode only when the ESS is charging and operates in the boost mode only when it discharging [33]. The space state equations for the buck converter, as in Figure 4b, and for the boost converter, as in Figure 4c, in continuous current mode operation are discussed below, assuming that all the circuit elements are ideal.

In the buck mode, when the *S*1 IGBT is on and the *S*2 IGBT is off, the *S*2 diode is reverse = biased. Therefore, the inductor (*L*) charges. The voltage equation is as follows:

$$L\frac{di_L}{dt} = v_{ESS} - v_{dc} \tag{22}$$

When the *S*1 IGBT is off and *S*2 IGBT is off, the *S*2 diode becomes forward-biased, because *L* tries to maintain the current in the same direction and *L* starts discharging. The voltage across inductor is as follows:

$$L\frac{di_L}{dt} = v_{ESS} \tag{23}$$

Considering, the state of the *S*1 IGBT, Equations (22) and (23) give the following:

$$\frac{di_L}{dt} = \frac{1}{L}v_{ESS} - \frac{1}{L}sv_{dc} \tag{24}$$

where *s* is 1 when the *S*1 is on and *s* is 0 when the *S*1 is off.

In boost mode, when the *S*1 IGBT is off and the *S*2 IGBT is on, the *L* charges. The voltage equation is as follows:

$$L\frac{di_L}{dt} = v_{ESs} \tag{25}$$

When the *S*1 IGBT is off and the *S*2 IGBT is off, the diode *S*1 becomes forward-biased and *L* tries to maintain the current in the same direction and starts discharging as follows:

$$L\frac{di_L}{dt} = v_{ESS} - v_{dc} \tag{26}$$

Considering, the state of the *S*2 IGBT, Equations (25) and (26) give the following:

$$\frac{di_L}{dt} = \frac{1}{L}v_{ESS} + (s-1)\frac{1}{L}v_{dc} \tag{27}$$

where *s* is 1 when the *S*2 is on and *s* is 0 when the *S*2 is off.

3.3.2. Bidirectional DC-DC Converter in Discrete Time Domain

Because the sampling time is sufficiently small, and the sampling frequency is much higher than the fundamental frequency, no extrapolation is considered for this study [38]. Instead, the simple one step method was used, which selects the switching state that minimizes the error at the $(k+1)$ instant and applies it at the kth instant. Fast microcontrollers with high speed calculations, such as Tiva TM4C123G, TMS320F28377S, and so forth, allow the online implementation of these power converters.

In the buck mode ($S1$ on = L charging; $S1$ off = L discharging), using Euler's approximation method with one switching period, the predicted inductor current is as follows:

$$i_L(k+1) = \frac{T_s}{L} v_{ESS} - \frac{T_s}{L} s v_{dc} + i_L(k) \tag{28}$$

The buck mode control objective is to maintain the dc-link voltage and maintain the ESS at its upper limit by charging the SCES. During this process, the ESS SOC is designed to maintain below 80%. This constraint is included in the FCS-MPC algorithm. And the main objective of FCS-MPC is taken as to minimize the error between the inductor current reference ($i_L{}^*$) and the predicted inductor current ($i_L{}^p = i_L(k+1)$). The maximum allowable charging current of ESS, ($i_{L_{max}}$) is included as one of the constraints. The cost function, g_{buck} is as follows:

$$g_{buck} = |i_L{}^* - i_L{}^p| \tag{29}$$

The inductor current reference is as follows:

$$i_L{}^* = \frac{((P_g - P_{PMSG}) + (v_{dc_{ref}} - v_{dc}) \times 100)}{v_{ESS}} \tag{30}$$

where P_g and P_{PMSG} are grid supply power and PMSG-generated power, respectively.

In the boost mode, when $S2$ is on, the inductor will charge. When $S2$ is off, the inductor will discharge. The switch $S1$ is turned off. Using Euler's approximation method with one switching period, predicted inductor current is as follows:

$$i_L(k+1) = \frac{T_s}{L} v_{ESS} + (s-1)\frac{T_s}{L} v_{dc} + i_L(k) \tag{31}$$

The boost mode control objective is to maintain dc-link voltage at its set value by discharging the ESS. During this process, the ESS SOC should be maintained above 30%. This is also included in the FCS-MPC algorithm as a constraint, and the main objective of FCS-MPC is taken as to minimize the error between $i_L{}^*$ and $i_L{}^p$. The maximum allowable discharging current of ESS ($i_{L_{max}}$) is included as one of the constraints. The cost function, g_{boost} is as follows:

$$g_{boost} = |i_L{}^* - i_L{}^p| \tag{32}$$

Equations (12)–(32) are common for any type of ESS with a bidirectional dc-dc converter. Therefore, the proposed FCS-MPC can be used as a mutual algorithm for both a supercapacitor and Li-ion battery ESS. Figure 5 depicts the flow chart of the FCS-MPC algorithm of the dc-dc bidirectional converter that controls the charging/discharging of the EES.

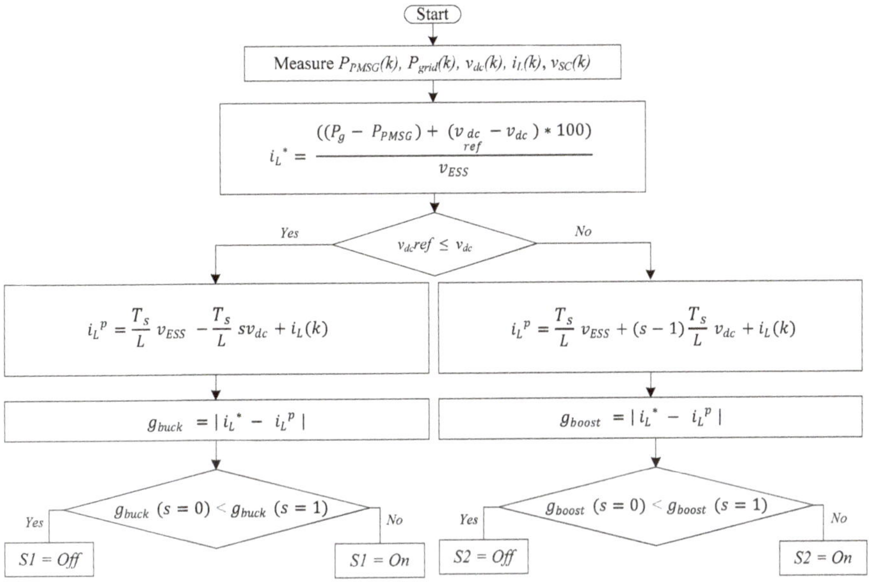

Figure 5. The flow chart of model predictive control (MPC) algorithm of a dc-dc bidirectional converter.

4. Power Management System

The OWC power conversion system, as in many other renewable energy systems, depends on the future availability of wave resources and the condition of ESS. Thus, a MPC strategy-based predictive power management system (PPMS) can be used to predict the future level of power supply to the grid based on the future availability of the renewable resources. This can be applied when determining the time period and level of power commitment to the grid. Moreover, the future SOC% of energy storage can be used to control the switching of the dc-dc bidirectional converter and protective circuit, which can be effectively utilized to coordinate control of the OWC system as in [39]. In situations where SOC% closes up the maximum limit, maintaining the dc-link voltage can be achieved using a protective circuit, such as resistors, which dissipate extra energy [39]. The PPMS has been used in [39] to manage the switching of components in stand-alone hydrogen systems with lead-acid BES and has validated PPMS over other power management systems (PMSs), such as rule-based. A similar strategy can be used to build an efficient and reliable OWC system with an ESS to minimize the energy lost and system cost with improved system dynamics. Nevertheless, wave energy forecast unit commitment estimation is beyond the scope of this paper. Instead, a simple power management strategy is used in this study by taking the average power over a 20 s period. This study aims to maintain an optimal and safe operation of OWC power convention to protect SCES from over-discharging and overcharging while maintaining the dc-link voltage in a desirable range. The PMS is used to control the dc-dc bidirectional converter to maintain the SCES SOC% between its minimum and maximum limits (30–80%) to improve its lifespan.

5. Simulation Results and Discussion

The parameters in Table A1, given in Appendix A.1, were used for the modeling and simulation of the OWC wave energy conversion system and the ESS in MATLAB/Simulink. The modeling of the air turbine, PMSG, back-to-back converter, grid, and filters can be found in [15]. The simulation results shown in Figure 6a–k were used to evaluate the FCS-MPC performance of the dc-dc bidirectional power converter along with the back-to-back converter.

The main control goal in this study was to supply a set, quality grid power output by regulating dc-bus voltage at a safe and effective voltage range while controlling the charging and discharging of the ESS within its technical limits. The other major control objective in this study, as in [15], was to maintain the OWC air turbine speed within the optimum range while extracting large discrete power pulses.

Figure 6a depicts the mechanical torque produced by the unidirectional air turbine at its design speed (650 rpm $\approx$ 68 rad/s) [6], while the turbine-generator rotor angular speed is shown in Figure 6b. As evident in this figure, the rotational speed of the air turbine is regulated to be within the optimum region throughout the simulation. This confirms that the turbine speed is kept at its control objective throughout the simulation regardless of the significant variation in extracted power (0–1 MW). Figure 6c depicts the PMSG stator current in the dq reference frame. The d-axis current is maintained closer to zero while the q-axis current varies in proportion to the torque input. This q-axis current is directly proportional to the active power drawn from the PMSG, which also varies with the extracted power as shown in Figure 6d.

Figure 6d also illustrates the grid power and SCES power. The PMSG power and SCES power waveforms confirm that the FCS-MPC of the dc-dc bidirectional converter successfully controls the power exchange between the SCES and dc-link. The grid power is maintained at its reference throughout the simulation. This is achieved by the charging and discharging the supercapacitor, compensating the generated and demand power differences. The grid voltage and grid current are shown in Figure 6e,f, respectively. The total harmonic distortion (THD) of the grid's three-phase current is less than 5% for this simulation, which allows the OWC to be integrated to the grid [40]. Figure 6g depicts the dc-link voltage, which is kept almost constant at 1200 V with more than 90% accuracy. This waveform confirms that the dc-link voltage is controlled successfully by the proposed MPC of the dc-dc bidirectional converter, adhering to the constraints. Also, according to the results, when the generated power is less than the grid power, the dc-link voltage is maintained below its set voltage by discharging the SCES, and when the generated power is more than the grid power, the dc-link voltage is maintained almost at its set level while charging the SCES. Figure 6h–j illustrates the SCES's voltage, current, and SOC, respectively. These figures confirm that the FCS-MPC and simple PMSs are capable of directing the buck and boost operation of the bidirectional converter while adhering to the ESS's technical limits, such as maximum discharge current, maximum charging current, SCES maximum voltage, cut-off voltage, minimum SOC, maximum SOC, and so forth. The dc-dc bidirectional converter, buck and boost operations, and switching pulses are shown in Figure 6k and confirm the proposed control strategy of the bidirectional dc-dc converter controlling the switching between the buck and boost modes, depending on the supply-demand power difference. The expanded figures show how the duty cycles change in the buck and boost modes.

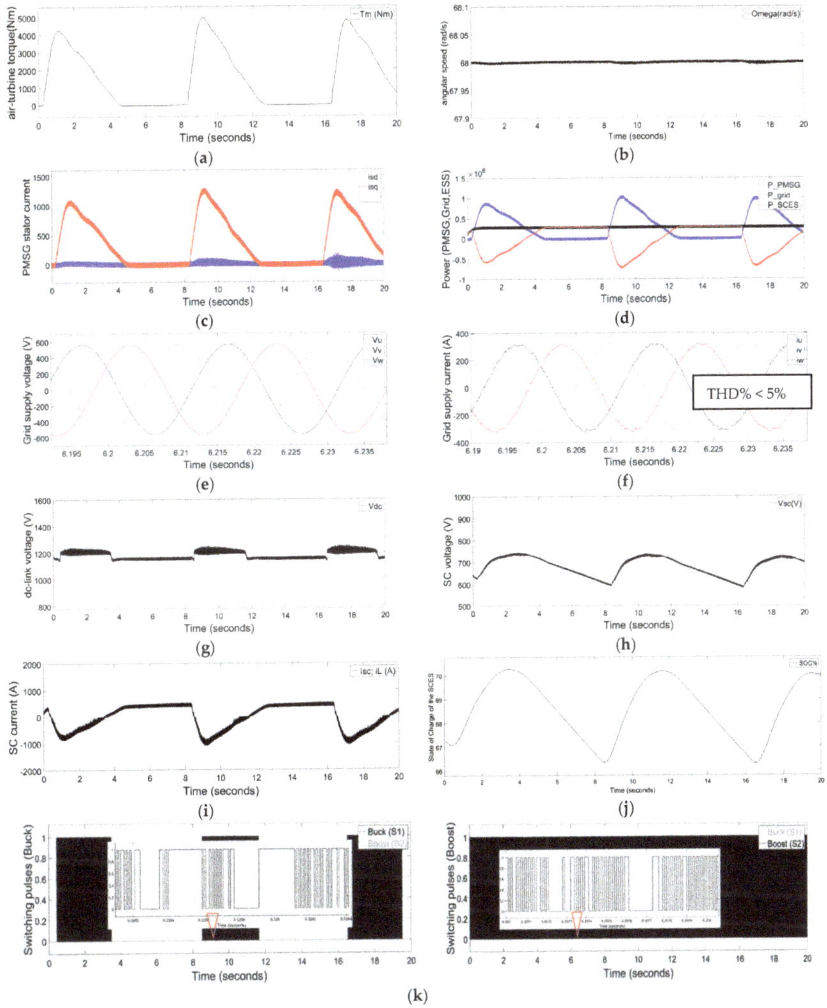

Figure 6. The simulation results of the OWC grid-integrated system with SCES; (**a**) mechanical torque profile (T_m); (**b**) generator rotor angular speed (ω_m); (**c**) PMSG stator current in *dq* frame (i_{sd} and i_{sq}); (**d**) power (PMSG, grid, SCES); (**e**) grid voltage (v_g); (**f**) grid current (i_g); (**g**) dc-link voltage (v_{dc}); (**h**) supercapacitor voltage (v_{sc}); (**i**) inductor current (i_L); (**j**) state of charge of the SCES (SOC); and (**k**) dc-dc bidirectional converter switching pulses.

6. Conclusions

This paper can be considered as the preliminary study that looks into a grid-connected OWC unidirectional air turbine generator with an SCES system. The air turbine generator is integrated into the grid using a back-to-back power converter, as in authors' previous work. This study applies and connects a commercially available, affordable-type supercapacitor to the dc-link with a dc-dc bidirectional converter. This approach confirms the smooth and quality power delivery to the grid by absorbing the large discrete power pulses generated by the PMSG. FCS-MPC is applied to the controller of the bidirectional dc-dc converter to direct charging and discharging of the SCES to maintain the dc-link voltage at the set voltage range, directly controlling the inductor current. The control algorithm used for this power converter does not consider the SCES parameters other than the SCES's voltage. Therefore, it can be used for any type of ESS with minor adjustments. This also allows the authors to carry out further research using either a Li-ion battery or supercapacitor ESS using the same converter control algorithm. Moreover, the power management strategy is addressed to maintain the SOC of the SCES at predefined technical limitations to increase the ESS's lifespan. The simulation results were analyzed to confirm the feasibility of the applied MPC strategy. The rotational speed of the turbine generator was almost constant at 68rad/sec, while the THD of the output current was below the grid code requirement of 5% during the entire simulation. These results confirm that the proposed MPC strategy for the controllers is capable of maintaining a quality power supply, adhering to the grid code irrespective of the significant intermittencies present in the extracted power from the waves under varying sea conditions.

Although these results obtained from the software simulation are very promising, further investigations and implementations are in progress for lab-scale hardware experimental validations and the development of an adaptive system against parameter variations.

Author Contributions: G.R. and S.J. conceived of and designed the simulator model and analyzed the data; G.R. wrote the paper; S.J., A.F., and M.N. critically reviewed the paper.

Acknowledgments: The authors thank Wave Swell Energy for providing the parameters to use in the simulation studies.

Conflicts of Interest: The authors declare no conflict of interest.

Appendix A

Appendix A.1. System Parameters Used in the Design

Table A1. System parameters used in the design.

PMSG	Rated Power	2 MW
	Rated rotate speed	650 rpm
	Rated Voltage	690 VAC
	Rated current	1673.5 A
	Number of pole pairs	4
	Resistance (R_s)	0.0024 Ω
	Inductances ($L_d = L_q$)	0.3552 mH (line-phase)
	Magnetic flux (Ψ_{pm})	0.666 Wb
	Inertia (J)	2 kgm^2
Grid side Filter	Inductance	1.6 mH
	Resistance	0.01 Ω
	Rated charge/discharge currents	1500 A
Grid	Voltage	690 Vrms
	Frequency	50 Hz
Generator Side Filter	Inductance	1 mH
	Resistance	0.01 Ω
DC-link	Total power	265 kW
	DC-link voltage (V_{dc})	1200 V
	DC-link current (I_{dc_max})	1000 A
Supercapacitor	Capacitance (C_{sc}) (83F; 21 in series; 4 parallel loop)	15.8 F
	Resistance (R_{sc})	52.5 mΩ
	Voltage Initial ($V_{initial}$) ($C_{sc\_nominal}$: 48 V each)	650 V
Reference Values	Active Power	265 kW
	Reactive Power	0
Other	Sample time (s) (Rec./Inv./bidirectional)	$10^{-4}/10^{-4}/7 \times 10^{-5}$
	Simulator run time	20 s

Appendix A.2. FCS-MPC for Rectifier

```
function [Sa,Sb,Sc] = fcn(vdc, is, vs, Θ, ωm, Tm)
% Sa,Sb,Sc: Switching signals for the rectifier; vdc: dc-link voltage; is: PMSG current; vS: PMSG phase voltage; Θ: angle of the reference frame (rad/s); ωm: PMSG rotational speed (rad/s); Tm: air-turbine torque.
Ts = 1e−4; % Sample time (s)
% Load parameters
Rs = 0.01; % Stator Resistance + generator side filter resistance (Ω)
Ls = 1e−3; % stator Inductance + Generator side filter inductance (H)
np = 4; %pole pairs
flux = 0.66; % PMSG flux
J = 2; % inertia
ωm_ref = 68; % 68 rad/s
K = 2e−4; % constant
id_ref = 0;
% Switching states
states = [0 0 0;1 0 0;1 1 0;0 1 0;0 1 1;0 0 1;1 0 1;1 1 1];
states2 = [0 0 0; 2 −1 −1; 1 1 −2; −1 2 −1; −2 1 1; −1 −1 2; 1 −2 1; 0 0 0];
vAFE = vdc*states2/3;
% is is derived from phase currents ia, ib and ic
%ia (k+1) = ik1a
%ib (k+1) = ik1b
%ic (k+1) = ik1c
g_opt = 1e10;
x_opt = 1;
for i = 1:8
% Current prediction at instant k+1
ik1a = (1 − Rs* Ts /Ls)*is(1) − Ts /Ls*(vAFE (i,1) − vs(1));
ik1b = (1 − Rs* Ts /Ls)*is (2) − Ts /Ls*(vAFE (i,2) − vs (2));
ik1c = (1 − Rs* Ts /Ls)*is(3) − Ts /Ls*(vAFE (i,3) − vs (3));
% abc to dq conversion
idk1 =2*(ik1a*sin(Θ) + ik1b*sin(Θ − 2*pi/3) + ik1c*sin(Θ + 2*pi/3))/3;
iqk1 =2*(ik1a*cos(Θ) + ik1b*cos(Θ − 2*pi/3) + ik1c*cos(Θ + 2*pi/3))/3;
Te = 1.5*flux* np*iqk1; % Predicted torque
ωmk1 = ωm + Ts *(Tm − Te)/J; % Predicted speed
g = abs(ωm_ref − ωmk1) + K*abs(id_ref − idk1); % Cost function
if (g < g_opt) % Minimize cost function
g_opt = g;
x_opt = i;
end
end
% Output switching states
Sa = states(x_opt,1);
Sb = states(x_opt,2);
Sc = states(x_opt,3);
```

Appendix A.3. FCS-MPC for Inverter

```
function [Su,Sv,Sw] = fcn(vdc, Pref, ig, vg)
% Su,Sv,Sw: Switching signals for the inverter; Pref: Active and reactive power reference; vdc: dc-link voltage; ig: grid current; vS: grid phase voltage.
```

```
Ts = 1e−4; % Sample time (s)
% Load parameters
Rg = 0.01; % grid side filter resistance (Ω)
Lg = 1e−3; % grid side filter inductance (H)
% Voltage vectors
v0 = 0 + 1j*0;
v1 = 2/3*vdc + 1j*0;
v2 = 1/3*vdc + 1j*sqrt(3)/3*vdc;
v3 = −1/3*vdc + 1j*sqrt(3)/3*vdc;
v4 = −2/3*vdc + 1j*0;
v5 = −1/3*vdc − 1j*sqrt(3)/3*vdc;
v6 = 1/3*vdc − 1j*sqrt(3)/3*vdc;
v7 = 0 + 1j*0;
v = [v0 v1 v2 v3 v4 v5 v6 v7];
% Switching states
states = [0 0 0; 1 0 0; 1 1 0; 0 1 0; 0 1 1; 0 0 1; 1 0 1; 1 1 1];
g_opt = 1e10;
x_opt = 1;
% Read power reference inputs at sampling instant k
Pkref = Pref(1) + 1j*Pref(2);
Pref(1) = 265e3; % grid active power reference (W)
Pref(2) = 0; % grid reactive power reference
% Read current and voltage measurements at sampling instant k
ik = (0.8165*(ig(1) − ig(2)/2 − ig(3)/2)) + 1j*(0.7071*(ig(2) − ig(3)));
vk = (0.8165*(vg(1) − vg(2)/2 − vg(3)/2)) + 1j*(0.7071*(vg(2) − vg(3)));
for i = 1:8
% Current and power prediction at instant k+1
ik1 = (1 − Rg* Ts /Lg)*ik + Ts /Lg*(v(i) − vk);
Pk1 = (real(vk)*real(ik1) + imag(vk)*imag(ik1)) + 1j*(imag(vk)*real(ik1) − real(vk)*imag(ik1));
% Cost function
g = (abs(real(Pkref − Pk1)))*(abs(real(Pkref − Pk1))) + (abs(imag(Pkref − Pk1)))*(abs(imag(Pkref − Pk1)));
% Selection of the optimal value
if (g < g_opt)
g_opt = g;
x_opt = i;
end
end
% Output switching states
Su = states(x_opt,1);
Sv = states(x_opt,2);
Sw = states(x_opt,3);
```

Appendix A.4. FCS-MPC for dc-dc Bidirectional Converter of Energy Storage

```
function [S1, S2] = fcn(vdc, vESS, iL, vdcref, PPMSG, Pg)
% S1: buck switching; S2: Boost switching; vdc: dc-link voltage; vESS: energy storage system voltage; iL: inductor current; vdcref: dc-link reference (1200V); PPMSG: air-turbine generator power; Pgrid: grid power.
Ts = 1e−5; % Sample time (s)
L = 5e−4; % Inductance (H)
vdcref = 1200; % dc-link voltage reference (V)
```

```
g_opt = 1e10;
x_opt1 = 0;
x_opt2 = 0;
%Boost converter
if (v_dc < v_dcref) % Energy deficit condition. Boost converter should be selected
i_Lref = ((P_g − P_PMSG)+(v_dcref − v_dc)*100)/v_ESS; % Inductor current reference. "100" is an arbitrary gain to speed up.
for i = 0:1
i_Lk1 = i_L + (v_ESS + ((i −1)*v_dc))* T_s /L; %Predicted inductor current
g =abs(i_Lref − i_Lk1); %Cost function for the boost converter
if (g < g_opt) %Search for the switching state which minimizes the cost function
g_opt = g;
x_opt1 = i;
x_opt2 = 0;
end
end
end
%Buck converter
if (v_dc > v_dcref) % Energy excess condition. Buck converter should be selected
i_Lref = ((P_g − P_PMSG)+(v_dcref − v_dc)*100)/v_ESS; % Inductor current reference. "100" is an arbitrary gain to speed up.
for i = 0:1
i_Lk1 = i_L - (i*v_dc − v_ESS)* T_s /L; %Inductor current prediction
g =abs(i_Lref − i_Lk1); %Cost function for the buck converter
if (g < g_opt) %Search for the switching state which minimizes the cost function
g_opt = g;
x_opt1 = 0;
x_opt2 = i;
end
end
end
% Set output signals
S1 = x_opt2;
S2 = x_opt1;
```

References

1. Aderinto, T.; Li, H. Ocean Wave Energy Converters: Status and Challenges. *Energies* **2018**, *11*. [CrossRef]
2. Hannon, M.; Griffiths, J.; Vantoch-Wood, A.; Carcas, M.; Bradley, S.; Boud, R.; Wyatt, S. *World Energy Resources; Marine Energy 2016*; World Energy Council; World Energy Resources: London, UK, 2016.
3. Carson, L.; Bradshaw, M.; Jaques, L. *Australian Energy Resource Assessment 2014*, 2nd ed.; Commonwealth of Australia (Geoscience Australia): Canberra, Australia, 2014.
4. Carson, L.; Bradshaw, M.; Jaques, L.; Che, N.; Ball, A. *Australian Energy Resource Assessment 2010*; Commonwealth of Australia (Geoscience Australia): Canberra, Australia, 2010.
5. Hughes, M.G.; Heap, A.D. National-scale wave energy resource assessment for Australia. *Renew. Energy* **2009**, *35*, 1783–1791. [CrossRef]
6. Wave Swell Energy. Sustainable Elecricity from the Ocean. Available online: http://waveswellenergy.com.au/ (accessed on 27 March 2017).
7. Australian Maritime College. Successful Ocean Wave Energy Converter Testing Leads to Commercialisation. Available online: http://www.amc.edu.au/about-amc/news-and-events/news-items/successful-tidal-turbine-testing-leads-to-commercialisation/ (accessed on 11 May 2017).

8. PerAero Turbine Design LLC. *Turbine Design Report for Wave Swell Energy Ltd.*; PerAero Turbine Design LLC: Bonita Springs, FL, USA, 2017.
9. Hamidi, S.A.; Lonel, D.M.; Nasiri, A. Modeling and Management of Batteries and Ultracapacitor for Renewable Energy Support in Electric Power Systems—An Overview. *Electr. Power Compon. Syst.* **2017**, *43*, 1434–1452. [CrossRef]
10. Rajapakse, G.; Jayasinghe, S.G.; Fleming, A.; Shahnia, F. Model Predictive Control-based Power take-off Control of an Oscillating Water Column Wave Energy Conversion System. In Proceedings of the 2017 International Conference on Substantial Energy Engineering (ICSEE 2017), Perth, Australia, 12–14 June 2017.
11. Maercos, V.M.M.; Martinez, M.A.G.; Gonzalez, F.B.; Montero, M.I.M. A Grid Connected Photovoltic Inverter with Battery-Supercapacitor Hybrid Energy Storage. *Sensors* **2017**, *17*. [CrossRef]
12. Murray, D.; Multon, A.J.B.; Ahmed, H.B. Electrical energy storage systems. In *Electrical Design for Ocean Wave and Tidal Energy Systems*; Alcorn, R., O'Sullivan, V., Eds.; The Institution of Engineering and Technology (IET): Herts, UK, 2013.
13. Kim, J.; Suharto, Y.; Daim, T.U. Evaluation of Electrical Energy Storage (EES) technologies for renewable energy: A case from the US Pacific Northwest. *J. Energy Storage* **2017**, *11*, 25–54. [CrossRef]
14. Murray, D.B.; Hayes, J.G.; Egan, M.G.; O'Sullivan, D.L. Supercapacitor Testing for Power Smoothing in a Variable Speed Offshore Wave Energy Converter. *IEEE J. Ocean. Eng.* **2011**, *37*, 301–308. [CrossRef]
15. Rajapakse, G.; Jayasinghe, S.G.; Negnevitsky, M.; Fleming, A. A Model Predictive Control-Based Power Converter System for Oscillating Water Column Wave Energy Converters. *Energies* **2017**, *10*, 1631. [CrossRef]
16. Hamidi, A.; Weber, L.; Nasiri, A. EV Chargeing Station Integrating Renewable Energy and Second Life Battery. In Proceedings of the International Conference on Renewable Energy Research and Application, Madrid, Spain, 20–23 October 2013.
17. Wu, F.; Zhang, X.; Ju, P. Application of the battery energy storage in wave energy conversion system. In Proceedings of the International Conference on Sustainable Generation and supply (SUPERGEN'09), Nanjing, China, 6–7 April 2009.
18. Rodriguez, J.; Kazmierkowski, M.P.; Espinoza, J.R.; Zanchetta, P.; Abu-Rub, H.; Young, H.A.; Rojas, C.A. State of the Art of Finite Control Set Model Predictive Control in Power Electronics. *IEEE Trans. Ind. Inform.* **2013**, *9*, 1003–1016. [CrossRef]
19. Vazquez, S.; Leon, J.I.; Franquelo, L.G.; Rodriguez, J.; Young, H.A.; Marquez, A.; Zanchetta, P. Model Predictive Control: A Review of Its Applications in Power Electronics. *IEEE Ind. Electron. Mag.* **2014**, *8*, 16–31. [CrossRef]
20. Pirooz, A.; Noroozian, R. Model Predictive Control of Classic Bidirectional DC-DC Converter for Battery Applications. In Proceedings of the 7th Power Electronics, Drive Systems & Technologies Conference (PEDSTC 2016), Iran University of Science and Technology, Tehran, Iran, 16–18 February 2016.
21. Ceballos, S.; Rea, J.; Robles, E.; Lopez, I.; Pou, J.; O'Sullivan, D.L. Control strategies for combining local energy storage with wells turbine oscillating water column devices. *Renew. Energy* **2015**, *83*, 1097–1107. [CrossRef]
22. Ceballos, S.; Rea, J.; Lopez, I.; Pou, J.; Robles, E.; O'Sullivan, D.L. Efficiency optimization in low inertia wells turbine oscillating water column devices. *IEEE Trans. Energy Convers.* **2013**, *28*, 553–564. [CrossRef]
23. Li, W.; Joos, G.; Abbey, C. A parallel bidirectional DC/DC converter topology for energy storage systems in wind applications. In Proceedings of the Conference Record of the 2007 IEEE Industry Applications Conference, New Orleans, LA, USA, 2007; pp. 171–185.
24. Zou, Y.; Cheng, K.W.E. A Vertical Flux-Switching Permanent Magnet Based Oscillating Wave Power Generator with Energy Storage. *Energies* **2017**, *10*, 887. [CrossRef]
25. Maxwell Technologies. *Maxwell_Supercapacitor_48v_DN 1009365 13*; Maxwell Technologies: San Diego, CA, USA, 2017.
26. Thounthong, P.; Sikkabut, S.; Mungporn, P.; Mobarakeh, N.; Pierfederici, S.; Davat, B. Nonlinear Control Algorithm of Supercapacitor/Li-Ion Battery Energy Storage Devices for Fuel Cell Vehicle Applications. In Proceedings of the Mecatronics-2014, Tokyo, Japan, 27–29 November 2014.
27. Hannan, M.A.; Lipu, M.S.H.; Hussain, A.; Mohamed, A. A review of lithium-ion battery state of charge estimation and management system in electric vehicle applications: Challenges and recommendations. *Renew. Sustain. Energy Rev.* **2017**, *78*, 834–854. [CrossRef]

28. Zhi, L.; Peng, Z.; Zhifu, W.; Qiang, S.; Yinan, R. State of charge estimation for Li-ion battery based on extended Kalman filter. In Proceedings of the 8th International Conference on Applied Energy—ICAE2016, Beijing, China, 8–11 October 2016; pp. 3515–3520.
29. Bernal, L.E.Z. Characterization of Double-Layer Capacitors for Power Electronics Applications. Master's Thesis, Electrical and Computer Engineering, University of Toronto, Toronto, ON, Canada, 1997.
30. Zubieta, L.; Bonert, R. Characterization of Double-Layer Capacitors for Power Electronics Applications. *IEEE Trans. Ind. Appl.* **2000**, *36*, 199–205. [CrossRef]
31. Islam, M.S.; Hossain, M.B.; Hossain, M.N.; Alam, S.B.; Chowdhury, M.E.H. Modeling of a Double-Layer Capacitor with Individual Branch Response. In Proceedings of the World Congress on Engineering and Computer Science 2010, San Francisco, CA, USA, 20–22 October 2010.
32. Bimbhra, P.S. *Generalized Theory of Electrical Machines*; Khanna Publishers: Delhi, India, 1992.
33. Vilathgamuwa, M.; Nayanasiri, D.; Gamini, S.; Hudgins, J. *Power Electronics for Photovoltaic Power Systems (Synthesis Lectures on Power Electroncs)*; Morgan & Claypool Publishers: San Rafael, CA, USA, 2015; ISBN 978-1-62705-775-2.
34. Abdullah, M.A.; Yatim, A.H.M.; Tan, C.W.; Samosir, A.S. Control of a Bidirectional Converter to Interface Ultracapacitor with Renewable Energy Sources. In Proceedings of the IEEE International Conference on Industrial Technology (ICIT), Cape Town, South Africa, 25–28 February 2013.
35. Samosir, A.S.; Yatim, A.H.M. Implementation of Dynamic Evolution Control of Bidirectional DC–DC Converter for Interfacing Ultracapacitor Energy Storage to Fuel-Cell System. *IEEE Trans. Ind. Electron.* **2010**, *57*, 3468–3473. [CrossRef]
36. Leuchter, J. Bi-Directional DC-DC Converters for Battery Buffers with Supercapacitor. In *Energy Storage in the Emerging Era of Smart Grids*; InTech: Rijeka, Croatia, 2011; p. 492. ISBN 978-953-307-269-2.
37. Hart, D.W. *Power Electronics*; McGraw-Hill: New York, NY, USA, 2011.
38. Yaramasu, V.N.R. Predictive control of Multilevel Converters for MegaWatt Wind Energy Conversion Systems. Ph.D. Thesis, Electrical and Computer Engineering, Ryerson University, Toronto, ON, Canada, 2014.
39. Brka, A.; Kothapalli, G.; Abdeli, Y.M. Predictive power management strategies for stand-alonehydrogen systems: Lab-scale validation. *Int. J. Hydrog. Energy Publ.* **2015**, *40*, 9907–9916. [CrossRef]
40. Langella, R.; Testa, A. *IEEE Recommended Practice and Requirements for Harmonic Control in Electric Power System*; University of Campania "Luigi Vanvitelli": Caserta, Italy, 2014.

© 2018 by the authors. Licensee MDPI, Basel, Switzerland. This article is an open access article distributed under the terms and conditions of the Creative Commons Attribution (CC BY) license (http://creativecommons.org/licenses/by/4.0/).

energies

MDPI

Article

Economic Feasibility of Wave Energy Farms in Portugal

Laura Castro-Santos [1], Dina Silva [2], A. Rute Bento [2], Nadia Salvação [2] and C. Guedes Soares [2,*]

1 Departamento de Enxeñaría Naval e Industrial, Universidade da Coruña, Escola Politécnica Superior, Esteiro, 15471 Ferrol, Spain; laura.castro.santos@udc.es

2 Centre for Marine Technology and Ocean Engineering (CENTEC), Instituto Superior Técnico, Universidade de Lisboa, 1049-001 Lisboa, Portugal; dina.silva@centec.tecnico.ulisboa.pt (D.S.); rute.bento@centec.tecnico.ulisboa.pt (A.R.B.); nadia.salvacao@centec.tecnico.ulisboa.pt (N.S.)

* Correspondence: c.guedes.soares@centec.tecnico.ulisboa.pt; Tel.: +351-218-417-607

Received: 22 September 2018; Accepted: 6 November 2018; Published: 14 November 2018

check for updates

Abstract: This paper develops a methodology to determine the economic feasibility of implementing offshore wave energy farms on the Portuguese continental coast. This methodology follows several phases: the geographic phase, the energy phase, the economic phase, and the restrictions phase. First, in the geographic phase, the height and the period of the waves, the bathymetry, the distance from the farm to the shore, from farm to shipyard, and from farm to port, are calculated. In the energy phase the energy produced by each wave energy converter is determined, and in the economic phase, the parameters calculated in the previous phases are used as input to find the economic parameters. Finally, in the restrictions phase, a limitation by the bathymetry will be added to the economic maps, whose value will be different depending on the floating offshore wave energy converter (WEC). In this study, three wave energy converters have been considered, Pelamis, AquaBuOY, and Wave Dragon, and several scenarios for electric tariffs have been taken into account. The results obtained indicate what the best WEC is for this study in terms of its levelized cost of energy (LCOE), internal rate of return (IRR), and net present value (NPV), and where the best area is to install wave energy farms.

Keywords: feasibility study; floating offshore wave farm; WEC; IRR; LCOE; ocean energy; marine energy

1. Introduction

The first wave power patent was from the 18th century and during the centuries, lots of types of devices have been developed [1]. Wave energy converters (WEC) are the devices that can extract the energy from ocean waves. There are many ways the WECs can be classified. However, depending on their working principle, they can be classified as oscillating water columns, oscillating bodies, and overtopping devices [2,3]. The oscillating water column device works using an air turbine (Pico [4], LIMPET, Sakata, Mutriku [5], Mighty Whale [6], Ocean Energy, SPERBOY [7], Oceanlinx [8], and REWEC3 [9]); the oscillating bodies work with a hydraulic motor, a hydraulic turbine, and a linear electrical generator (AquaBuOY [10], IPS Buoy [11], FO3 [12], PowerBuoy [12], Wavebob [13], Pelamis [14,15], PS Frog [16], SEAREV [17], AWS (Archimedes Waveswing Submerged) [18], WaveRoller [19], and Oyster [20]); and the overtopping concept works with a low-head hydraulic turbine (TAPCHAN [21], SSG (Sea Slot-cone Generator) [22], and Wave Dragon [23]). They can also be classified by considering the water depth that they were designed to operate in; fixed (less than 50 m of depth or onshore) or floating (more than 50 m of depth) or the distance to shore [24]. Other WECs are: the Wavestar, which produces electricity due to the motion of the floats that are attached by arms to the platform [25]; the SeaBeavl, which is a "taut-moored dual-body" WEC designed to improve the maintenance process [26]; and Falcão et al. analyze the hydrodynamics of IPS Buoy [27]. This paper will be focused on floating WECs, which have the advantage of operating in deep water where more wave energy can be found and in a larger range of water depths, increasing the number of locations where they can be deployed. Veigas et al. [28] studied the wave and offshore wind energy potential of a Spanish island located in the Atlantic Ocean. This study is important because it presents maps of the areas selected. Liu et al. [29] studied the energy conversion of a prototype WEC buoy in China and they considered a farm of buoys. Martinelli et al. [30] developed a method to select the design power of WECs in the first steps of the development of the product. In addition, Martinelli et al. develop an analysis based on statistic aspects of the power from OWC (Oscillating Water Column) [31]. Zanuttigh et al. [32] analyze the feasibility of the use of WECs for coastal protection and they consider the Adriatic coast. Arena et al. [33] analyze the resonant WECs. In 2011, Portugal had an experimental project for wave energy: the WaveRoller prototype, with a budget of five million euros [13]. This is the country selected to develop the case study because it is a country very involved with ocean energies, having a pilot area in its north-west coast.

The present paper will be focused on three wave energy converters: Pelamis, AquaBuOY, and Wave Dragon. These devices are not in commercial exploration and the main reason for their selection is that they represent different types of devices, different sizes, and different energy outputs. The Pelamis is an articulated structure similar to a snake and has "cylindrical sections linked by hinged joints" [2]. It has been developed in the UK and it should be installed aligned with waves [2]. It has been tested in several sizes from 1998 to 2011 [34,35]. The AquaBuOY combines the hose-pump and the IPS Buoy, being a small and modular WEC. Wave Dragon is based on the principle of wave overtopping and has "two wave reflectors focusing the incoming waves towards a doubly curved ramp, a reservoir and a set of low-head hydraulic turbines" [2,36].

Regarding the assessment, performance, and feasibility analysis: Rusu et al. [37] developed the wave energy resource for Portugal; Bozzi et al. [38] analyzed the wave energy feasibility in Italy considering three WECs: AquaBuOY, Pelamis, and Wave Dragon; Iuppa et al. [39] analyzed the case of Sicily; Guedes Soares et al. [40] considered several coastal locations to determine their efficiency; Vannucchi et al. [41] considered several Italian coasts: "Tuscany, Sardinia, Liguria et Sicily"; Dalton [42] analyzed the "non-technical barriers of wave energy in Europe"; Dalton et al. [43] considered the Pelamis in three different scenarios in Europe and the USA; and O'Connor et al. [44] analyzed the analysis of Pelamis and Wavestar in Europe.

Bozzi et al. [38] decided on the offshore location of a wave farm in Italy only by considering a particular set of points of the geography, while in the present paper all points of a particular geography are considered (in this case Portugal). Iglesias et al. [45] and Veigas et al. [46] considered several points of the geography (in this case, Galicia) but only the available energy was considered, while in the present paper the economic aspects of each point of the geography is considered (Portugal). Beels et al. [47] did not take into account the maps of the geography, and the economic calculations were very elementary, while in the present paper detailed economic aspects are calculated for all locations producing economic maps. O'Connor et al. [48] considered some economic aspects of wave energy but they did not consider the map of all the locations of the selected region, however, the present paper shows the map of all the locations. In the present method, all points of the geography are calculated, and this is very important because, for example, in a point where the wave resource is very good it may not be possible to install wave energy farms because there are restrictions. In this sense, the present method allows the addition of restrictions (in this paper only bathymetry is considered, but other restrictions can also be added).

The aim of the present paper is to develop a methodology to calculate the economic feasibility of floating offshore wave energy farms following several phases: geographic phase, energy phase, economic phase, and restrictions phase. In the geographic phase, some parameters (the height and the period of the waves [49,50], the bathymetry and the distance farm to shore, farm to shipyard, and farm to port) are calculated to be used as input values in the economic phase. The energy phase determines the energy produced by each wave energy converter (WEC). It can be calculated considering several methods, depending on the data available and the precision wanted. Afterwards, in the economic phase, the economic parameters are calculated considering the inputs provided in the previous phases. Finally, the restriction phase, imposed by the bathymetry, will be added to the economic maps, whose value will be different depending on the floating offshore WEC. The WECs considered for this study are Pelamis, AquaBuOY, and Wave Dragon, whose energy yields are calculated using several scenarios for electric tariffs. They have been selected because they represent several types of devices, although some of them are not in operation nowadays. Results indicate what the best WEC is in terms of its LCOE (levelized cost of energy), IRR (internal rate of return), and NPV (net present value) and where the best area to implement a floating offshore wave energy farm is located. The method proposed was applied to Portugal, where a high wave energy resource can be found.

2. Materials and Methods

2.1. Procedure

The method proposed is based on four phases, as shown in Figure 1: geographic phase (inputs in the figure), energy phase (power matrix and sea state location in the figure), economic phase (LCOE calculation in the figure), and restrictions phase (restriction of bathymetry in figure).

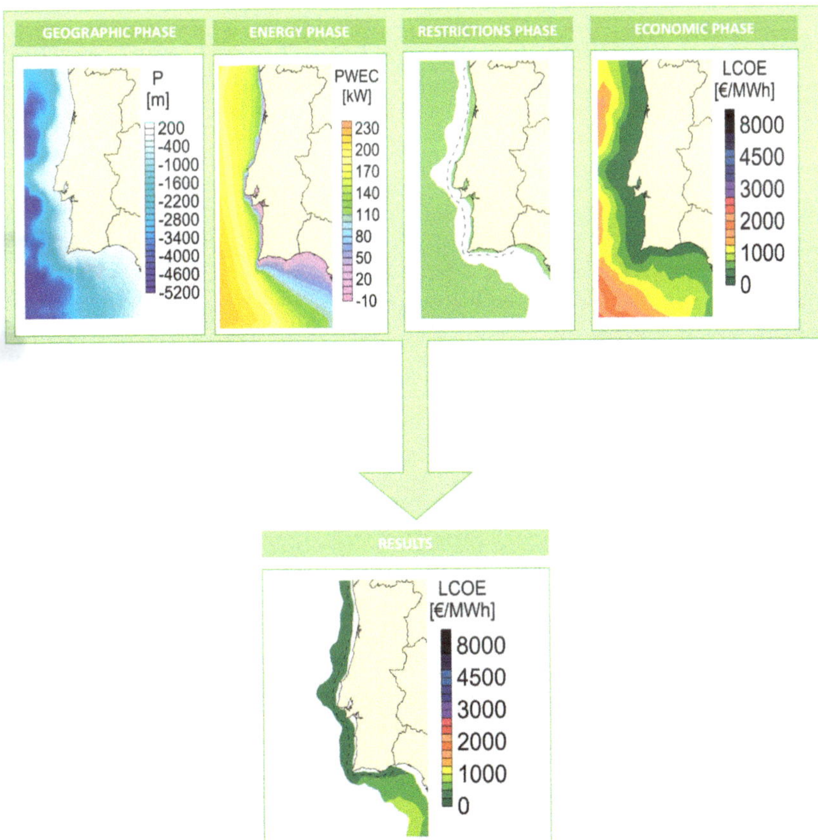

Figure 1. Method adopted [source: own]. P: depth in meters; PWEC: wave energy converter (WEC) electric power generated; LCOE: levelized cost of energy.

2.2. Geographic Phase

The geographic phase is the first stage of this methodology. The parameters that have been used as input maps to generate output economic maps were calculated in this phase. The parameters were the significant wave height and the energy period of the waves, the bathymetry, the distance from farm to shore [51], farm to shipyard, and farm to port.

The wave parameters—significant wave height (Hs) and wave period (Tm)—were obtained from a previous hindcast study [49,50] using WW III (Wave Watch III) and SWAN (Simulating Waves Nearshore) in a coupled system. In terms of the size of the grid, it is important that a floating offshore wave energy farm can be located inside the cell. The distance between offshore WECs is 87.5 times the diameter (D) of the WEC and the distance between lines of WECs is 47.5 times the diameter (D) of the WEC, as Figure 2 shows. In the WECs considered, the main dimension is the diameter of the device.

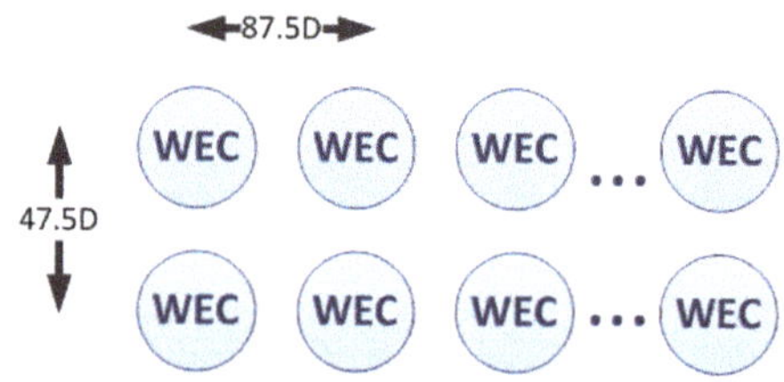

Figure 2. Distance between wave energy converters (WECs) (87.5D) and distance between lines of WECs (47.5D). [Source: own].

The restriction assessment was done using the bathymetry from GEBCO (General Bathymetric Chart of the Oceans), which was also used in the SWAN simulations of the hindcast study mentioned above.

The grid maps of the bathymetry and the parameters Hs and Tm were interpolated so as to have the size recommended for the wave energy farm implementation. The interpolation was from a resolution of $0.05° \times 0.1°$ in the case of Hs, Tm and bathymetry, and $0.5° \times 0.5°$ in the case of wind to a resolution of $0.15° \times 0.3°$. The function interp2 of MATLAB was applied to the data in a linear interpolation.

2.3. Energy Phase

In the energy phase, the energy produced by each WEC ($E_{1\text{WEC}}$) is calculated by Equation (1).

$$E_{1\text{WEC}} = NHAT \cdot P_{\text{WEC}} \cdot \eta_{availability} \cdot \eta_{transmissionlosses} \quad (1)$$

This calculation depends on the number of hours per year ($NHAT$), the WEC electric power generated (P_{WEC}), the availability ($\eta_{availability}$) and the losses due to transmission ($\eta_{transmissionlosses}$). The P_{WEC} is calculated by Equation (2).

$$P_{\text{WEC}} = \frac{1}{100} \cdot \sum_{i=1}^{n_T} \sum_{j=1}^{n_H} p_{ij} \cdot P_{ij} \quad (2)$$

The power matrix of the WEC is associated with the sea state probability of occurrence in the study location. In this equation, the n_T is the number of period and n_H is the number of wave height in the table, the p_{ij} is the probability of occurrence of the sea state corresponding to the bin defined by the line i and the column j and P_{ij} is the electric power corresponding to the same sea state or energy bin for the WEC considered [21]. Therefore, it is necessary to have the power matrix of the WEC, given by the supplier of the WEC considered, and the probability distribution matrix of the sea states for each point of the geography of the region of analysis. The devices studied here have their power matrices given in open literature.

2.4. Economic Phase

In the economic phase, two types of parameters were calculated to measure the economic viability of a wave energy farm deployment: the life-cycle cost of the wave energy farm and the economic parameters. For that, the values calculated in the geographic phase (the height of the waves, the period of the waves, the bathymetry, the distance from farm to shore, farm to shipyard, and farm to port) and energy phase (energy produced by the WEC) were used as inputs, which have different values for different points in the map (k).

The calculation of the life-cycle cost of the floating offshore wave energy farm (FOWEF) was based on the methodology developed in previous studies [22,23]. The methodology was based on the life-cycle process of floating offshore renewable energy devices, composed by six phases, attributing to each phase of the process the inherent costs, called the Life-cycle Cost System (LCS).

The phases along the life-cycle process and whose costs will be taken into account are [52]: the concept definition (C1), the design and development (C2), the manufacturing (C3), the installation (C4), the exploitation (C5), and the dismantling (C6). The Life-cycle cost system (LCS) of a floating offshore wave energy farm (FOWEF) was then calculated as Equation (3) shows:

$$LCS_{\mathrm{FOWEF}}(k) = C1(k) + C2 + C3(k) + C4(k) + C5(k) + C6(k). \tag{3}$$

Another measure of costs is the levelized cost of energy (LCOE), which takes into account the Life-cycle costs (LCS_{FOWEFt}), the energy produced by the wave energy farm (E_{t}) (without considering losses due to individual WEC efficiency or near field effects) in MWh/year and the capital cost of the project (r) [53,54]. The formula to calculate LCOE is Equation (4).

$$LCOE = \frac{\sum_{\mathrm{t}=0}^{N_{\mathrm{farm}}} \frac{LCS_{\mathrm{FOWEFt}}}{(1+r)^{\mathrm{t}}}}{\sum_{\mathrm{t}=0}^{N_{\mathrm{farm}}} \frac{E_{\mathrm{t}}}{(1+r)^{\mathrm{t}}}} \tag{4}$$

Also important is the parameter NPV (net present value, Equation (5)), which consists of the net value of the cash flows of the floating offshore wave farm, considering its discount from the beginning of the investment [54,55].

$$NPV = -G_0 + \sum_{\mathrm{t}=1}^{n} \frac{CF_{\mathrm{t}}}{(1+\mathrm{r})^{\mathrm{t}}} \tag{5}$$

The NPV depends on the cash flow ($CF_{\mathrm{t}} = R_{\mathrm{t}} - E_{\mathrm{t}}$, where E_{t} is the expenses on year and R_{t} the revenues on year), the life-cycle years (t), the initial investment (G_0), and the discount rate (r) [26].

When the NPV is equal to zero, the IRR (internal rate of return) is calculated from Equation (6) [25,27].

$$-G_0 + \sum_{\mathrm{t}=1}^{n} \frac{CF_{\mathrm{t}}}{(1+\mathrm{IRR})^{\mathrm{t}}} = 0 \tag{6}$$

The WACC (weighted average cost of capital) has been developed based on Equation (7).

$$WACC = \frac{MV_e \cdot R_e + MV_d \cdot R_d \cdot (1-T)}{MV_e + MV_d} \tag{7}$$

It is dependent on the variables of total equity (MV_e) and its costs (R_e), total debt (MV_d) and its costs (R_d), and the tax shield ($(1-T)$).The floating offshore wave energy farm (FOWEF) studied is economically feasible if the NPV > 0, the IRR > WACC, and LCOE presents low values.

2.5. Restrictions Phase

Once all the economic maps have been calculated, it is important to restrict the area where the wave energy farm is to be installed. This is due to the fact that there could be a good region in economic terms (IRR, NPV, and LCOE), but with limitations of usage (seismic fault lines, environmental protected areas, offshore electric cable lines, navigation areas, etc.). In this paper, the bathymetry is the only restriction taken into account.

The bathymetry restriction will be from $(D_c + 20)$, D_c being the maximum draft of all the floating offshore WECs considered, to 1000 m of depth, which is considered an adequate value to install this type of wave farm. The type of floating offshore wave platform also restricts the economic maps.

Finally, territorial waters (22.2 km) will also be shown in the maps because there is no law about the offshore space for offshore waves at the moment.

2.6. Case Study

The case study is the offshore area of Portugal, as shown in Figure 3, characterized by deep waters, as shown in Figure 4, and good wave energy resources.

Figure 3. Portugal and Spain.

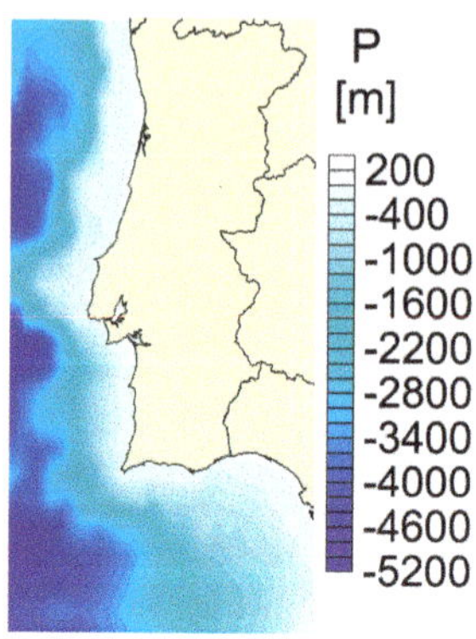

Figure 4. Bathymetry of Portugal, with P being the depth in meters [56].

It is important to know the life-cycle of the project, which is considered 20 years, and the size of the grid, which is considered 16 km × 33 km.

Several floating offshore wave energy platforms have been taken into consideration: Pelamis, AquaBuOY, and Wave Dragon, as shown in Figure 5. The characteristics of the wave farm will depend on the type of WEC considered, as Table 1 shows. The final configuration (number of WECs per line, number of lines, total number of WECs in the farm) was assembled considering the total power of the farm, which in all the cases is close to 110 MW.

Table 1. Characteristics of the wave energy farms depending on the WEC taken into consideration.

	Concept	Pelamis	AquaBuOY	Wave Dragon	Units
NAF	Number of WECs per line	7	21	3	WECs/line
NL	Number of lines	21	21	9	lines
N_{WEC}	Total number of WECs in the farm	147	441	27	-
D	Main dimension	4	6	54	m
P_{WEC}	Power of each WEC	0.75	0.25	4	MW
P_{FOWEF}	Total power of the farm	110.25	110.25	108.00	MW
		110,250	110,250	108,000	kW

(a) (b) (c)

Figure 5. Types of WECs considered: Pelamis (**a**) [57], AquaBuOY (**b**) [58], and Wave Dragon (**c**) [23].

Due to the fact that Portugal does not have a specific electric tariff for wave energy, several tariffs were considered and are presented in Table 2.

Table 2. Electric tariffs considered for a floating offshore wave energy farm.

Scenario	Electric Tariff (€/MWh)
Scenario 1	300
Scenario 2	400
Scenario 3	600

The restriction considered for bathymetry was 50 m, based on adding +20 m to the maximum draft of all the WECs considered, as shown in Figure 6.

The energy produced by a particular WEC is dependent on its power matrix, as shown in Tables 3–5, and on the number of occurrences of each sea state at the point considered, as shown in Table 6.

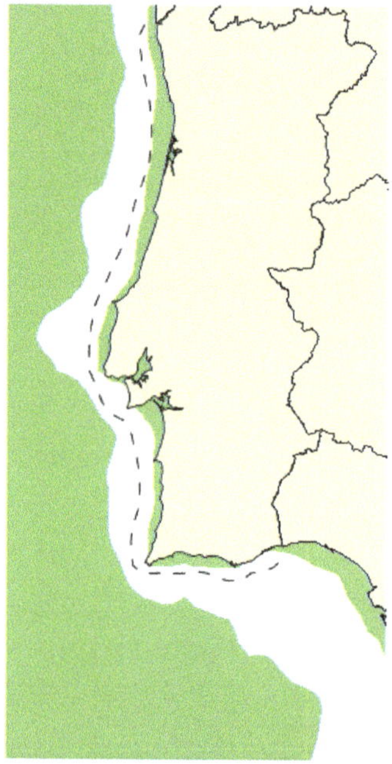

Figure 6. Bathymetry restriction. The white area represents the region selected considering the bathymetry restriction. The dashed line represents the territorial waters. [Source: own].

Table 3. Power matrix of the Pelamis [59]. *Hs*: wave height; *Te*: wave energy period.

***Te* (s)**	**Power Matrix (in kW)**																
***Hs* (m)**	**5**	**5.5**	**6**	**6.5**	**7**	**7.5**	**8**	**8.5**	**9**	**9.5**	**10**	**10.5**	**11**	**11.5**	**12**	**12.5**	**13**
0.5	0	0	0	0	0	0	0	0	0	0	0	0	0	0	0	0	0
1	0	22	29	34	37	38	38	37	35	32	29	26	23	21	0	0	0
1.5	32	50	65	76	83	86	86	83	78	72	65	59	53	47	42	37	33
2	57	88	115	136	148	153	152	147	138	127	116	104	93	83	74	66	59
2.5	89	138	180	212	231	238	238	230	216	199	181	163	146	130	116	103	92
3	129	198	260	305	332	340	332	315	292	266	240	219	210	188	167	149	132
3.5	0	270	354	415	438	440	424	404	377	362	326	292	260	230	215	202	180
4	0	0	462	502	540	546	530	499	475	429	384	366	339	301	267	237	213
4.5	0	0	544	635	642	648	628	590	562	528	473	432	382	356	338	300	266
5	0	0	0	739	726	731	707	687	670	607	557	521	472	417	369	348	328
5.5	0	0	0	750	750	750	750	750	737	667	658	586	530	496	446	395	355
6	0	0	0	0	750	750	750	750	750	750	711	633	619	558	512	470	415
6.5	0	0	0	0	750	750	750	750	750	750	750	743	658	621	579	512	481
7	0	0	0	0	0	750	750	750	750	750	750	750	750	676	613	584	525
7.5	0	0	0	0	0	0	750	750	750	750	750	750	750	750	686	622	593
8	0	0	0	0	0	0	0	750	750	750	750	750	750	750	750	690	625

Table 4. Power matrix of the AquaBuOY [59]. *Hs*: wave height; *Tp*: wave peak period.

Tp (s)	Power Matrix (in kW)												
***Hs* (m)**	**5**	**6**	**7**	**8**	**9**	**10**	**11**	**12**	**13**	**14**	**15**	**16**	**17**
1	0	0	8	11	12	11	10	8	7	0	0	0	0
1.5	0	13	17	25	27	26	23	19	15	12	12	12	7
2	0	24	30	44	49	47	41	34	28	23	23	23	12
2.5	0	37	47	69	77	73	64	54	43	36	36	36	19
3	0	54	68	99	111	106	92	77	63	51	51	51	27
3.5	0	0	93	135	152	144	126	105	86	70	70	70	38
4	0	0	0	122	176	198	188	164	137	112	91	91	49
4.5	0	0	0	223	250	239	208	173	142	115	115	115	62
5	0	0	0	250	250	250	250	214	175	142	142	142	77
5.5	0	0	0	250	250	250	250	250	211	172	172	172	92

Table 5. Power matrix of the Wave Dragon [59]. *Hs*: wave height; *Tp*: wave peak period.

Tp (s)	Power Matrix (in kW)												
***Hs* (m)**	**5**	**6**	**7**	**8**	**9**	**10**	**11**	**12**	**13**	**14**	**15**	**16**	**17**
1	160	250	360	360	360	360	360	360	320	280	250	220	180
2	640	700	840	900	1190	1190	1190	1190	1070	950	830	710	590
3	0	1450	1610	1750	2000	2620	2620	2620	2360	2100	1840	1570	1310
4	0	0	2840	3220	3710	4200	5320	5320	4430	3930	3440	2950	2460
5	0	0	0	4610	5320	6020	7000	7000	6790	6090	5250	3950	3300
6	0	0	0	0	6720	7000	7000	7000	7000	7000	6860	5110	4200
7	0	0	0	0	0	7000	7000	7000	7000	7000	7000	6650	5740

Table 6. Example of the number of occurrences for different sea states (in % from the total) for a point of Portugal. *Hs*: wave height; *Tp*: wave peak period.

Tp (s) *Hs* (m)	5	6	7	8	9	10	11	12	13	14	15	16	17
1	0.63	0.41	0.93	1.26	0.70	0.38	0.11	0.06	0	0.03	0.02	0	0
2	2.43	5.93	3.38	5.67	6.77	8.50	7.23	2.68	0.06	0.77	0.22	0.06	0.02
3	0.01	1.27	6.20	2.07	1.10	2.79	7.73	6.05	0.07	2.19	0.41	0.22	0.04
4	0	0	0.44	2.98	0.58	0.38	1.55	3.66	0.07	2.28	0.39	0.24	0.03
5	0	0	0	0.53	1.10	0.30	0.25	1.11	0.03	1.57	0.30	0.24	0.01
6	0	0	0	0.02	0.23	0.54	0.09	0.13	0.01	0.68	0.47	0.19	0.05
7	0	0	0	0	0.02	0.16	0.20	0.03	0.01	0.09	0.11	0.22	0.04

The electric power generated by each WEC, considering the power matrix of each WEC and the number of occurrences for different sea states for each point of the geography, are presented in Figure 7.

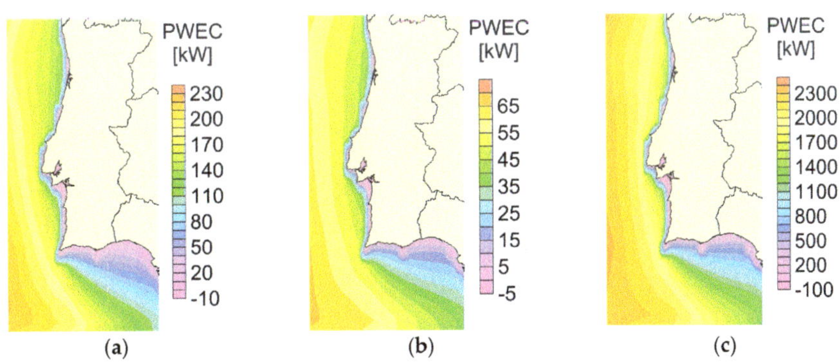

Figure 7. Power of Pelamis (**a**), AquaBuOY (**b**), and Wave Dragon (**c**). [Source: own].

3. Results

The results obtained for each WEC in terms of LCOE, but without the bathymetry restriction, are displayed in Figure 8a–c. The Wave Dragon, as shown in Figure 8c, was the one that presents the best value for this parameter, with 316.90 €/MWh, followed by 735.94 €/MWh for the Pelamis, as shown in Figure 8a, and 2967.85 €/MWh for the AquaBuOY, as shown in Figure 8b.

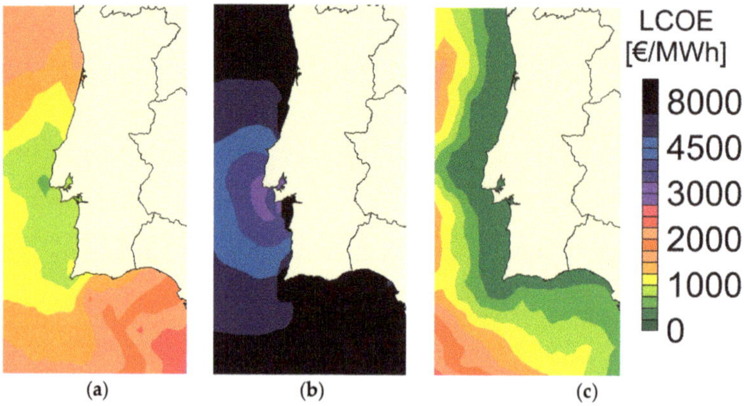

Figure 8. Results for LCOE without bathymetry restrictions for Pelamis (**a**), AquaBuOY (**b**), and Wave Dragon (**c**) platform. [Source: own].

However, it is important to notice that the installation of the WEC depends on the bathymetry, so the previous maps shown in Figure 9 cannot be used in their totality for the choice of the best deployment place. In this context, a restriction for bathymetry has been considered, as shown in Figure 9a–c. However, in this study, this will not affect the results for the best LCOE.

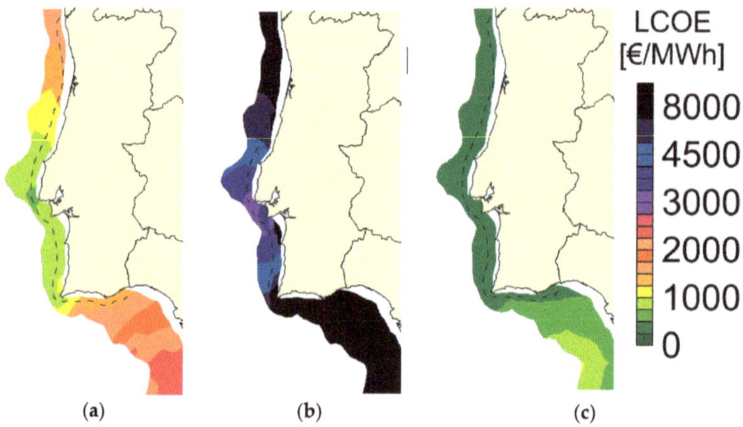

Figure 9. Results for LCOE with bathymetry restrictions for Pelamis (**a**), AquaBuOY (**b**), and Wave Dragon (**c**). [Source: own].

Considering Scenario 1 for the electric tariff with 300 €/MWh, the implementation of a wave farm is not economically feasible whatever the WEC used. The values of IRR for all the WECs considered are lower than the WACC value, and the values of NPV are less than zero, the opposite of the feasibility conditions. For IRR, the values are around 4.74% for Wave Dragon, 11.94% for Pelamis, and −50.94% for the AquaBuOY. In terms of NPV, the Wave Dragon had −67.56 M€, the Pelamis −561.52 M€, and the AquaBuOY 2732.93 M€, as shown in Table 7.

Table 7. Results for the 300 €/MWh tariff. IRR: internal rate of return; NPV: net present value.

Type of WEC	IRR	NPV
Wave Dragon	4.74%	−67.56 M€
Pelamis	11.94%	−561.52 M€
AquaBuOY	−50.94%	2732.93 M€

Scenario 2, which takes into account a 400 €/MWh electric tariff, has better results than the previous one. The WACC, which is dependent of the location (the total equity (MV_e) and the total debt (MV_d) depends of the life-cycle costs of each location), has values from 6% to 7%, which is lower than the IRR value of Wave Dragon, as shown in Figure 10c, with 9.35%. For Pelamis, as shown in Figure 10a, this parameter assumes the value of −5.67% and for AquaBuOY, as shown in Figure 10b, a value of −36.29%. For the NPV, the Wave Dragon, as shown in Figure 11c, has a value of 246.80 M€ (NPV > 0), the Pelamis, as shown in Figure 11a, a value of −429.67 M€, and AquaBuOY, as shown in Figure 11b, a value of 2637.39 M€. Looking at the results, the Wave Dragon is the WEC that encompasses all the requirements for a viable wave farm project.

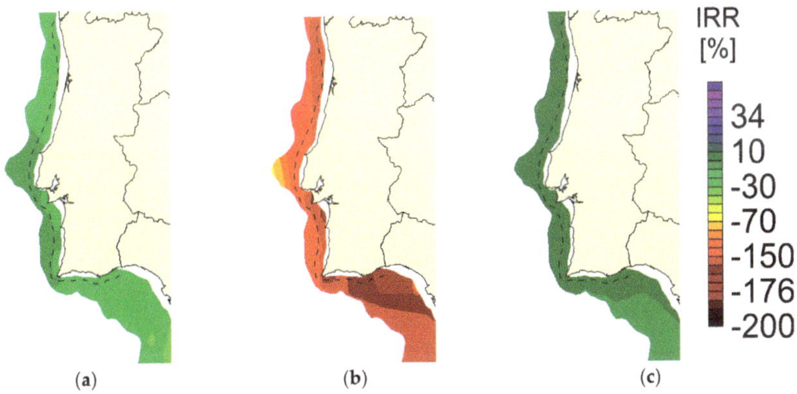

Figure 10. Results for IRR considering a tariff of 400 €/MWh and the bathymetry restriction for Pelamis (**a**), AquaBuOY (**b**), and Wave Dragon (**c**). [Source: own].

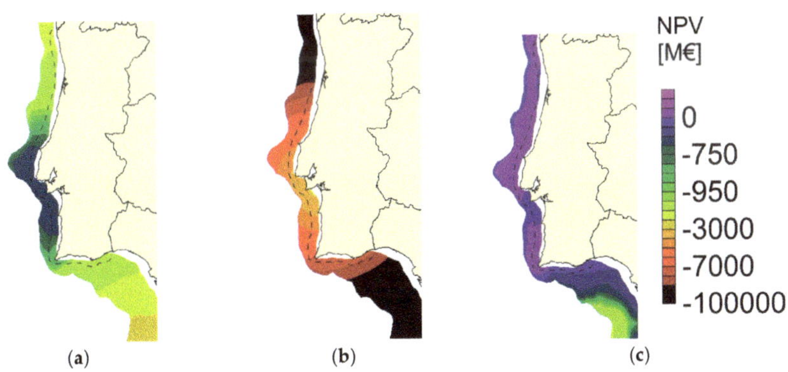

Figure 11. Results for NPV considering a tariff of 400 €/MWh and the bathymetry restriction for Pelamis (**a**), AquaBuOY (**b**), and Wave Dragon (**c**). [Source: own].

The last scenario, Scenario 3, which takes into consideration a 600 €/MWh electric tariff, gives the best results of all the three scenarios. For IRR, the Wave Dragon, as shown in Figure 12c, presents the best value with 17.25%, followed by Pelamis, as shown in Figure 12a, with 2.49%, and AquaBuOY, as shown in Figure 12b, with −22.97%.

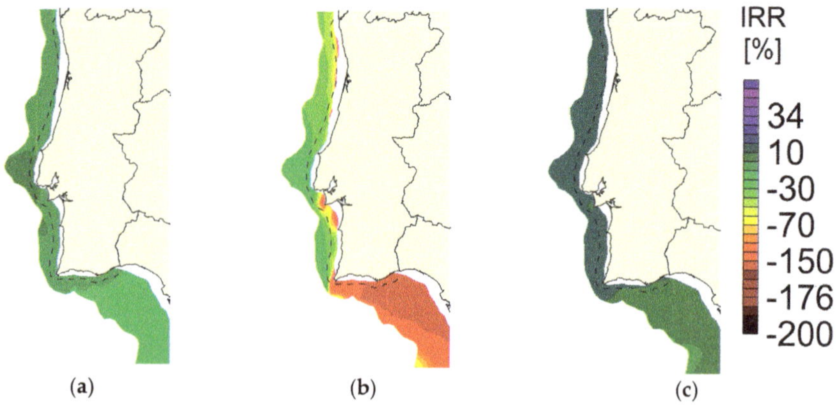

Figure 12. Results for IRR considering a tariff of 600 €/MWh and the bathymetry restriction for Pelamis (**a**), AquaBuOY (**b**), and Wave Dragon (**c**). [Source: own].

In terms of NPV, the best value for Scenario 3 is 881.24 M€ for the Wave Dragon, as shown in Figure 13c, followed by −168.38 M€ for the Pelamis, as shown in Figure 13a, and −244.73 M€ for the AquaBuOY, as shown in Figure 13b. Therefore, with this electric tariff, a wave farm project with Wave Dragon WECs would be economically feasible.

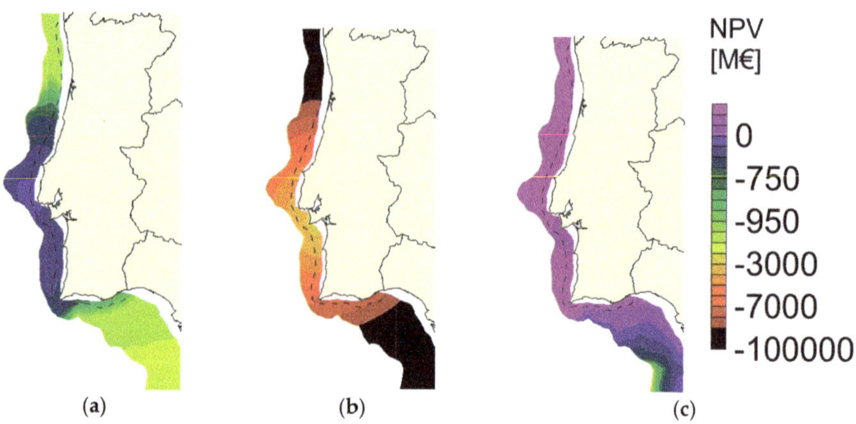

Figure 13. Results for NPV considering a tariff of 600 €/MWh and the bathymetry restriction for Pelamis (**a**), AquaBuOY (**b**), and Wave Dragon (**c**). [Source: own].

With respect to the best area for the wave farm installation, the area located close to Lisbon, in the center of Portugal, seems to be the better choice, as can be seen in all the economic maps with bathymetry restrictions. In the future, this would be a good location to install this type of offshore wave energy technology, when the reduction of costs, due to the commercial phase, guarantees the economic feasibility of the project.

4. Conclusions

The goal of this paper was to develop a method to calculate the economic feasibility of floating offshore wave energy farms. Therefore, their internal rate of return (IRR), net present value (NPV), and levelized cost of energy (LCOE) have been calculated. The method proposed has four phases: geographic phase, energy phase, economic phase, and bathymetry restriction phase. The aim of the geographic phase is to calculate the input values that will be used in the economic phase: the significant height of the waves, the period of the waves, the bathymetry, the distance from farm to shore, farm to shipyard, and farm to port. The second phase is the energy phase, which determines the energy produced by each WEC. This can be calculated by several methods, depending on the data available and the precision required. The next phase is the economic phase, where the economic parameters are calculated using the parameter results of the previous phases as inputs. Finally, there is the bathymetry restriction phase, where the restriction by bathymetry will be added to the economic maps, whose value will be different depending on the floating offshore WEC.

The case study was the Portuguese continental coast, which has a good wave energy resource. Different WECs were used for the evaluation: the Pelamis, AquaBuOY, and Wave Dragon, as well as different scenarios for electric tariffs.

The results of LCOE, IRR, and NPV indicate what the best WEC to use in a wave farm is and what the best location is to install it. The Wave Dragon has the best LCOE, with 316.90 €/MWh, followed by Pelamis, with 735.94 €/MWh, and AquaBuOY, with 2967.85 €/MWh. This is due to the fact that Wave Dragon generates more energy for the location selected, as its power matrix shows. This is a result of being a larger device and benefiting from the effect of the scale of the energy produced, which then makes it cheaper. According to the IRR and the NPV results, only Scenario 2 (400 €/MWh) and 3 (600 €/MWh) for the electric tariff are economically feasible when using a Wave Dragon platform.

The layout considered is similar for all the WECs, considering that each device is separated from the other, taking into account the main dimensions of the platform. In the future, it can be improved considering the different energy production systems that each WEC has.

This study presents a novelty compared to other studies because it takes into account the economic aspects of wave energy, not only their technical aspects. It is very important to know the areas where the farms can be installed in economic terms, and whose use can help to analyze maritime planning of the countries in the future.

Portugal has good wave energy potential. In the future, areas close to Lisbon would be good locations to install this type of offshore technology, when the reduction of costs, due to the commercial phase, guarantees the economic feasibility of the project.

Author Contributions: All authors have been involved in either calculations or writing, as follows: Introduction, L.C.-S., D.S. and C.G.S.; Methodology, L.C.-S.; Case study, L.C.-S., D.S., A.R.B., N.S. and C.G.S.; Results, L.C.-S.; Discussion, L.C.-S., D.S., A.R.B., N.S. and C.G.S.

Funding: This research received no external funding.

Conflicts of Interest: The authors declare no conflict of interest.

References

1. Lindroth, S.; Leijon, M. Offshore wave power measurements—A review. *Renew. Sustain. Energy Rev.* **2011**, *15*, 4274–4285. [CrossRef]
2. Falcão, A.F.O. Wave energy utilization: A review of the technologies. *Renew. Sustain. Energy Rev.* **2010**, *14*, 899–918. [CrossRef]
3. Guedes Soares, C.; Bhattacharjee, J.; Tello, M.; Pietra, L. *Review and Classification of Wave Energy Converters*; Guedes Soares, C., Garbatov, Y., Sutulo, S., Santos, T.A., Eds.; Taylor & Francis Group: Oxfordshire, UK, 2012.
4. OWC Pico Plant Web Page OWC Pico Plant. Available online: http://www.pico-owc.net/cms.php?page=542&wnsid=dbb177dd9668f08318207830330904df (accessed on 5 September 2018).

5. Caballero Santos, C. Estudio de Plantas de Producción de Energías Renovables con Aprovechamiento de la Energía del Mar. Bachelor's Thesis, Universidad Carlos III de Madrid, Madrid, Spain, 2011.
6. Clément, A.; McCullen, P.; Falcão, A.F.O. Wave energy in Europe: Current status and perspectives. *Renew. Sustain. Energy Rev.* **2002**, *6*, 405–431. [CrossRef]
7. Sperboy Sperboy Webpage. Available online: http://www.sperboy.com (accessed on 1 September 2018).
8. Ocean Linx Ocean Linx Wavepage. Available online: http://www.oceanlinx.com/ (accessed on 5 September 2018).
9. Arena, F.; Romolo, A.; Malara, G.; Fiamma, V.; Laface, V. The First Full Operative U-OWC Plants in the Port of Civitavecchia. In *ASME 2017 36th International Conference on Ocean Offshore & Arctic Engineering, Volume 10: Ocean Renewable Energy*; ASME: New York, NY, USA, 2017; p. V010T09A022.
10. Weinstein, A.; Fredrikson, G.; Jane, M.; Group, P.; Denmark, K.N.R. *AquaBuOY—The Offshore Wave Energy Converter Numerical Modeling and Optimization*; IEEE: San Diego, USA, 2003; pp. 1854–1859.
11. Baddour, E. Energy from Waves and Tidal Currents. Available online: http://s3.amazonaws.com/zanran_storage/tecnet.pte.enel.it/ContentPages/106131895.pdf (accessed on 18 September 2018).
12. Magagna, D.; Uihlein, A. Ocean energy development in Europe: Current status and future perspectives. *Int. J. Mar. Energy* **2015**, *11*, 84–104. [CrossRef]
13. Wavebob Wavebob Webpage. Available online: http://www.wavebob.com (accessed on 1 September 2018).
14. Carcas, M.C. The OPD Pelamis WEC: Current status and onward programme. *Int. J. Ambient. Energy* **2003**, *24*, 21–28. [CrossRef]
15. Rusu, E.; Guedes Soares, C. Coastal impact induced by a Pelamis wave farm operating in the Portuguese nearshore. *Renew. Energy* **2013**, *58*, 34–49. [CrossRef]
16. McCabe, A.P.; Bradshaw, A.; Meadowcroft, J.A.C.; Aggidis, G. Developments in the design of the PS Frog Mk 5 wave energy converter. *Renew. Energy* **2006**, *31*, 141–151. [CrossRef]
17. Ruellan, M.; Ben Ahmed, H.; Multon, B.; Josset, C.; Babarit, A.; Clement, A. Design Methodology for a SEAREV Wave Energy Converter. *IEE Trans. Energy Convers.* **2010**, *25*, 760–767. [CrossRef]
18. Awsocean Awsocean Webpage. Available online: http://www.awsocean.com/ (accessed on 1 September 2018).
19. AW-Energy Web Page AW-Energy. Available online: http://aw-energy.com/about-waveroller/wave-farms (accessed on 1 September 2018).
20. Babarit, A.; Hals, J.; Muliawan, M.J.; Kurniawan, A.; Moan, T.; Krokstad, J. Numerical benchmarking study of a selection of wave energy converters. *Renew. Energy* **2012**, *41*, 44–63. [CrossRef]
21. Pérez-Collazo, C.; Greaves, D.; Iglesias, G. A review of combined wave and offshore wind energy. *Renew. Sustain. Energy Rev.* **2015**, *42*, 141–153. [CrossRef]
22. Veigas, M.; López, M.; Iglesias, G. Assessing the optimal location for a shoreline wave energy converter. *Appl. Energy* **2014**, *132*, 404–411. [CrossRef]
23. Page, W.D.W. Wave Dragon Technology. Available online: http://www.wavedragon.net (accessed on 7 September 2016).
24. Ram, K.; Narayan, S.; Ahmed, M.R.; Nakavulevu, P.; Lee, Y.H. In situ near-shore wave resource assessment in the Fiji Islands. *Energy Sustain. Dev.* **2014**, *23*, 170–178. [CrossRef]
25. Kramer, M.M.; Marquis, L.; Frigaard, P. Performance Evaluation of the Wavestar Prototype. In Proceedings of the 9th EWTEC 2011: Proceedings of the 9th European Wave and Tidal Conference, Southampton, UK, 5–9 September 2011.
26. Elwood, D.; Yim, S.C.; Prudell, J.; Stillinger, C.; von Jouanne, A.; Brekken, T.; Brown, A.; Paasch, R. Design, construction, and ocean testing of a taut-moored dual-body wave energy converter with a linear generator power take-off. *Renew. Energy* **2010**, *35*, 348–354. [CrossRef]
27. Falcão, A.F.O.; Cândido, J.J.; Justino, P.A.P.; Henriques, J.C.C. Hydrodynamics of the IPS buoy wave energy converter including the effect of non-uniform acceleration tube cross section. *Renew. Energy* **2012**, *41*, 105–114. [CrossRef]
28. Veigas, M.; Carballo, R.; Iglesias, G. Wave and offshore wind energy on an island. *Energy Sustain. Dev.* **2014**, *22*, 57–65. [CrossRef]
29. Liu, Z.; Qu, N.; Han, Z.; Zhang, J.; Zhang, S.; Li, M.; Shi, H. Study on energy conversion and storage system for a prototype buoys-array wave energy converter. *Energy Sustain. Dev.* **2016**, *34*, 100–110. [CrossRef]
30. Martinelli, L.; Zanuttigh, B.; Kofoed, J.P. Selection of design power of wave energy converters based on wave basin experiments. *Renew. Energy* **2011**, *36*, 3124–3132. [CrossRef]

31. Martinelli, L.; Zanuttigh, B.; Kofoed, J.P. Statistical analysis of power production from OWC type wave energy converters. In Proceedings of the 8th European Wave and Tidal Energy Conference, Uppsala, Sweden, 7–10 September 2009; pp. 295–303.
32. Zanuttigh, B.; Martinelli, L.; Castagnetti, M.; Ruol, P.; Kofoed, J.P.; Frigaard, P. Integration of wave energy converters into coastal protection schemes. In Proceedings of the 3rd International Conference on Ocean Energy, Bilbao, Spain, 6–8 October 2010; pp. 1–7.
33. Arena, F.; Barbaro, G.; Fiamma, V.; Laface, V.; Malara, G.; Romolo, A.; Strati, F.M. Resonant Wave Energy Converters: Concept development. *Energia Ambiente e Innovazione* **2015**, *61*, 52–57.
34. Pizer, D.J.; Retzler, C.; Henderson, R.M.; Cowieson, F.L.; Shaw, M.G.; Dickens, B.; Hart, R. Pelamis WEC—Recent advances in the numerical and experimental modelling programme. In Proceedings of the 6th European Wave and Tidal Energy Conference, Glasgow, UK, 29 August–2 September 2005.
35. Yemm, R.; Pizer, D.; Retzler, C.; Henderson, R. Pelamis: Experience from concept to connection. *Philos. Trans. A. Math. Phys. Eng. Sci.* **2012**, *370*, 365–380. [CrossRef] [PubMed]
36. Kofoed, J.P.; Frigaard, P.; Friis-Madsen, E.; Sørensen, H.C. Prototype testing of the wave energy converter wave dragon. *Renew. Energy* **2006**, *31*, 181–189. [CrossRef]
37. Rusu, E.; Guedes Soares, C. Numerical Modelling to Estimate the Spatial Distribution of the Wave Energy in the Portuguese Nearshore. *Renew. Energy* **2009**, *34*, 1501–1516. [CrossRef]
38. Bozzi, S.; Archetti, R.; Passoni, G. Wave electricity production in Italian offshore: A preliminary investigation. *Renew. Energy* **2014**, *62*, 407–416. [CrossRef]
39. Iuppa, C.; Cavallaro, L.; Foti, E.; Vicinanza, D. Potential wave energy production by different wave energy converters around Sicily. *J. Renew. Sustain. Energy* **2015**, *7*, 061701. [CrossRef]
40. Guedes Soares, C.; Bento, A.R.; Goncalves, M.; Silva, D.; Martinho, P. Numerical evaluation of the wave energy resource along the Atlantic European coast. *Comput. Geosci.* **2014**, *71*, 37–49. [CrossRef]
41. Vannucchi, V.; Cappietti, L. Wave energy assessment and performance estimation of state of the artwave energy converters in italian hotspots. *Sustainability* **2016**, *8*, 1300. [CrossRef]
42. Dalton, G.J. Non-technical barriers to wave energy in Europe. In Proceedings of the 8th European Wave and Tidal Energy Conference, Uppsala, Sweden, 7–10 September 2009; pp. 1–28.
43. Dalton, G.J.; Alcorn, R.; Lewis, T. Case study feasibility analysis of the Pelamis wave energy convertor in Ireland, Portugal and North America. *Renew. Energy* **2010**, *35*, 443–455. [CrossRef]
44. O'Connor, M.; Lewis, T.; Dalton, G. Techno-economic performance of the Pelamis P1 and Wavestar at different ratings and various locations in Europe. *Renew. Energy* **2013**, *50*, 889–900. [CrossRef]
45. Iglesias, G.; López, M.; Carballo, R.; Castro, A.; Fraguela, J.A.; Frigaard, P. Wave energy potential in Galicia (NW Spain). *Renew. Energy* **2009**, *34*, 2323–2333. [CrossRef]
46. Veigas, M.; López, M.; Romillo, P.; Carballo, R.; Castro, A.; Iglesias, G. A proposed wave farm on the Galician coast. *Energy Convers. Manag.* **2015**, *99*, 102–111. [CrossRef]
47. Beels, C.; Troch, P.; Kofoed, J.P.; Frigaard, P.; Vindahl Kringelum, J.; Carsten Kromann, P.; Heyman Donovan, M.; De Rouck, J.; De Backer, G. A methodology for production and cost assessment of a farm of wave energy converters. *Renew. Energy* **2011**, *36*, 3402–3416. [CrossRef]
48. O'Connor, M.; Lewis, T.; Dalton, G. Operational expenditure costs for wave energy projects and impacts on financial returns. *Renew. Energy* **2013**, *50*, 1119–1131. [CrossRef]
49. Silva, D.; Bento, A.R.; Martinho, P.; Guedes Soares, C. High Resolution local wave energy modelling in the Iberian Peninsula. *Energy* **2015**, *91*, 1099–1112. [CrossRef]
50. Silva, D.; Martinho, P.; Guedes Soares, C. Wave energy distribution along the Portuguese continental coast based on a thirty three years hindcast. *Renew. Energy* **2018**, *127*, 1064–1075. [CrossRef]
51. Bento, A.R.; Rusu, E.; Martinho, P.; Guedes Soares, C. Assessment of the changes induced by a wave farm in the nearshore wave conditions. *Comput. Geosci.* **2014**, *71*, 50–61. [CrossRef]
52. Castro-Santos, L.; Filgueira-Vizoso, A.; Carral-Couce, L.; Formoso, J.Á.F. Economic feasibility of floating offshore wind farms. *Energy* **2016**, *112*, 868–882. [CrossRef]
53. Castro-Santos, L.; Diaz-Casas, V. Economic influence of location in floating offshore wind farms. *Ocean Eng.* **2015**, *107*, 13–22. [CrossRef]
54. Short, W.; Packey, D.; Holt, T. *A Manual for the Economic Evaluation of Energy Efficiency and Renewable Energy Technologies*; National Renewable Energy Laboratory (NREL): Golden, CO, USA, 1995.

55. Hertenstein, J.H.; Mckinnon, S.M. Solving the puzzle of the cash flow statement. *Bus. Horiz.* **1997**, *40*, 69–76. [CrossRef]
56. Castro-Santos, L.; Prado, G.; Costa, P.; Estanqueiro, A. Methodology to design an economic and strategic offshore wind energy Roadmap in Portugal. In Proceedings of the 10th Deep Sea Wind R&D Conference, Trondheim, Norway, 24–25 January 2013; pp. 168–176.
57. EMEC Pelamis Wave Power. Available online: http://www.emec.org.uk/about-us/wave-clients/pelamis-wave-power/ (accessed on 7 September 2016).
58. Finavera Finavera Webpage. Available online: http://www.finavera.com/ (accessed on 1 September 2018).
59. Silva, D.; Rusu, E.; Guedes Soares, C. Evaluation of Various Technologies for Wave Energy Conversion in the Portuguese Nearshore. *Energies* **2013**, *6*, 1344–1364. [CrossRef]

© 2018 by the authors. Licensee MDPI, Basel, Switzerland. This article is an open access article distributed under the terms and conditions of the Creative Commons Attribution (CC BY) license (http://creativecommons.org/licenses/by/4.0/).

energies

MDPI

Article

Hydrodynamic Analysis of a Marine Current Energy Converter for Profiling Floats

Shuang Wu [1,2], Yanjun Liu [1,2,*] and Qi An [1,2]

1 Institute of Marine Science and Technology, Shandong University, Qingdao 266237, Shandong, China; wushuang2017@mail.sdu.edu.cn (S.W.); 201716023@mail.sdu.edu.cn (Q.A.)

2 Key Laboratory of High Efficiency and Clean Mechanical Manufacture, Ministry of Education, Jinan 250061, Shandong, China

* Correspondence: lyj111@sdu.edu.cn; Tel.: +86-531-8836-9277

Received: 12 July 2018; Accepted: 22 August 2018; Published: 24 August 2018

check for updates

Abstract: With the continuous improvement of people's interest in ocean exploration, research on deep-water profiling floats has received more and more attention. Energy supply is the key factor that restricts the working hours of deep-water floats. For this consideration, a marine current energy converter for deep-water profiling floats is proposed in this paper. A spiral involute blade is designed so that energy can be captured in two directions. Specifically, in the shallow sea area, the energy of the radial current is captured, and in the deep-sea area, the axial relative flow energy of the floats' autonomous up and down motions is captured. This captured energy is then converted into electrical energy to charge the battery and extend the working time of the floats. The novel spiral involute blade has unique hydrodynamic characteristics. The turbine's self-starting performance and its capacity coefficient are the main research topics studied using the computational fluid dynamics technique. Through numerical analysis and simulation, the self-starting response range and energy capture were obtained. This paper verifies the feasibility of this innovative idea using a theory analysis and provides the basis for future prototype testing and further applied research.

Keywords: marine current energy; spiral involute blade; hydrodynamic analysis; numerical simulation

1. Introduction

Since entering the 21st century, many countries have shifted their strategic focus to the oceans. United States proposed the establishment of Integrated Ocean Observing System (IOOS) [1,2], the European Union proposed the "Leader SHIP 2020" initiative [3], the United Kingdom proposed the "Oceans 2025" plan [4], and China implemented the "Strengthening Marine Powers" strategy [5], and so on. The implementation and realization of these marine strategies cannot be separated from the three-dimensional observation of oceans with marine observation equipment. As of 2016, there were more than 20 types of profiling floats in normal operation, mainly designed by the United States and France [6], and their maximum profile depth was mainly concentrated at about 2000 m.

However, the vast ocean far from the mainland has a depth of more than 2000 m, and the average depth of the global ocean reaches 3850 m. In the North Atlantic, the temperature and salinity relationship at 2000 m changes rapidly, and it is very difficult to use a delay mode correction for the data obtained from the profiling floats [7]. It can be seen that it is necessary to extend the observation depth of the automatic profiling floats to obtain deeper observational data. At present, the international deep-sea automatic profiling floats mainly include Deep Sounding Oceanographic Lagrangian Observer (Deep-SOLO), developed by the Scripps Institution of Oceanography, and Autonomous Profiling Explorer Deep (APEX-Deep), developed by Teledyne Webb [8,9], with a maximum dive depth of 6000 m. In 2017, our team successfully developed the 4000-m deep-sea Argo profiling float and conducted a sea trial in the Mariana Trench (as shown in Figure 1).

(a) (b)

Figure 1. (a) The lifting process; (b) the deployment of our profiling float.

However, as the operational depth of floats increases, the required energy increases significantly, and due to the corresponding increase in the environmental pressure, the wall thickness of the float shell must be increased in order to meet the pressure requirements, resulting in an increase in mass and volume. As a result, the energy consumption of the device is increased. Traditional battery systems are large and have limited capacity. Lithium batteries also have the potential risk of contaminating the marine environment. Therefore, we have adopted high-efficiency hydraulic pump buoyancy drive technology and proposed an innovative concept for integrated ocean current energy generation. We have also added a marine energy conversion device to our existing float. The energy of the ocean current is converted into electricity to charge the battery and the working time of the buoy is extended.

The specific design concept uses a two-way energy-absorption turbine on the top of the float (as shown in Figure 2). In different water depths and sea areas, the horizontal movement of the current and the vertical movement of the float in the vertical direction can drive the rotation of the blade and convert the kinetic energy into electrical energy. In the shallow sea areas where ocean current energy density is high, the rotation of the blades is mainly driven by the horizontal currents; in the relatively calm deep-sea areas, the motion of fluids caused by the float's autonomous bobbing and submerging movements drives the blades to rotate. To the best of our knowledge, the use of wave energy to power floats has a definite application basis, the main working principle is that the buoy of the wave energy converter vibrates up and down under the action of waves, which is then stored and converted into electrical energy sufficient for marine lighting [10–12]. These ocean energy converters are usually not less than two meters in diameter. We have not yet seen relevant research results on the application of ocean energy in deep-sea profiling floats, and its volume is relatively small, the characteristic diameter of the cylindrical profiling float mentioned in Figure 1 does not exceed 0.4 m. Moreover, the two-way energy capture of the turbine is also a major innovation in this paper.

This paper will mainly introduce the design and analysis of the motion and the force of the marine energy converter blades used for profiling floats and will describe a numerical simulation of the hydrodynamics to determine the feasibility and applicable conditions of the innovative application.

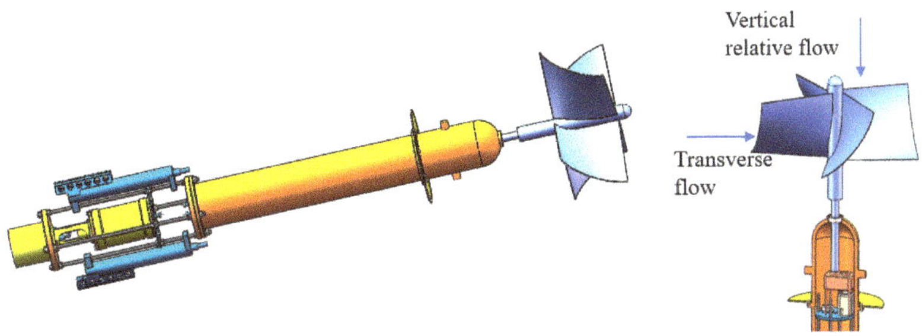

Figure 2. (**a**) 3D drawing of profiling float; and (**b**) Schematic diagram of the internal profile of the turbine connecting mechanism.

2. Blade Design and Analysis

As the energy capture device, the turbine blades are the most important components that need to be studied and designed. They are also the focus of this paper. Presently, there are mainly two types of turbines used in ocean energy utilization technology, classified as the horizontal axis type and the vertical axis type, according to the relative position of the flow direction and the rotation axis [13–15]. The blades needed in this paper can achieve energy capture in both the horizontal and vertical directions, that is, self-starting rotation and energy conversion can be realized in both the horizontal and vertical directions, so can the existing two forms of turbines be combined? To answer this question, a spiral involute blade has been designed.

2.1. Blade Design

The Indian Institute of Technology studied the helical Savonius rotor for the purpose of alleviating the problem of uneven blade torque [16]. The tidal current rotor has good self-starting characteristics and the maximum energy efficiency is about 0.18. Gorlov proposed a new spiral blade [17,18] and the Gorlov Helical Turbine Company introduced the GHT vertical axis spiral tidal turbine. The spiral blade has a specified torsion angle, good self-starting performance, and smooth operation. Ocean Renewable Power (Portland, ME, USA) has developed spiral vane turbines that adapt to different environments, such as RivGen, TidGen, and OCGen [19]. Some of these spiral vane turbines have been field-tested. All of the above previous research work applied to a vertical axis tidal current energy conversion device [20]. In our design, taking into consideration the particularity of the turbine's bidirectional flow drive, the vertical shaft spiral involute blade was designed. The schematic diagram of the design scheme is shown in Figure 3. According to the overall size of the connected buoy, the diameter D_2 of the rotor shaft is designed to be 60 mm, the radial outer diameter D_1 is 300 mm, th axial arc length is 310 mm, the rotation angle is 30 degrees, and the involute base circle radius is $r_0 = 0.233D_1$. The involute equation is:

$$\begin{cases} X = r_0 \cdot (t_0 \cdot sin(t_0) + \cos(t_0)) \\ Y = r_0 \cdot (sin(t_0) - t_0 \cdot \cos(t_0)) \end{cases} \quad (1)$$

where t_0 is the extent of the flare angle of the involute.

2.2. Movement and Stress Analysis

Taking the blade cross section as the analysis object, according to the radial flow and the axial relative flow, the force and motion analysis coordinate systems of the blade are established.

The axial relative flow conditions are shown in Figure 4a. The center point of the axis of rotation is the origin of the coordinates, and the global coordinate system OXY is established. The blade rotates counterclockwise around the axis of the turbine rotation, where the stable rotation speed is ω_1, the radius of rotation is R_1, and the positive direction of the X-axis coincides with the incoming flow direction. The local coordinate system oxy is established in the direction of the radial direction of the blade section, as shown in Figure 4a. The origin o of the local coordinate system can be determined by the radius of the blade and the angle of the blade rotation, which can be expressed as:

$$(X, Y)_o = (R_1 \cos\theta, R_1 \sin\theta) \tag{2}$$

where θ reflects the position of the entire blade in rotation. At any time t, there is: $\theta = \omega_1 t$. Supposing the distance between any point on the blade p to o is r, then the included angle between p and the x-axis is φ and the angle between the negative direction of the x-axis and the positive direction of the X-axis is β, which is called the elevation angle. The position of any point on the blade is:

$$(X, Y)_p = (R_1 \cos\theta - 2r\cos(\beta + \varphi), \quad R_1 \sin\theta - 2r\sin(\beta + \varphi)) \tag{3}$$

The combined velocity of the blades relative to the flow is V_{R1}. The flow velocity V and the rotation speed of the blades around the turbine axis satisfy the following expression:

$$\vec{V_{R1}} = \vec{V} + \omega_1 \vec{R_1} \tag{4}$$

The angle between the combined speed V_{R1} and the x-axis is the angle of attack α. Therefore:

$$V_{R1} = \sqrt{V^2 + (\omega_1 R_1)^2 + 2\omega_1 R_1 V \sin\theta} \tag{5}$$

$$\alpha = \arctan(\omega_1 R_1 \sin(\beta - \theta) - V\cos\beta / \omega_1 R_1 \sin(\beta - \theta) - V\cos\beta) \tag{6}$$

Similar to the previous analysis process, for radial flow conditions shown in Figure 4b, the section perpendicular to the axis of rotation is taken as the analysis object in order to establish the coordinate system O′X′Y′, and the center point of the rotation axis is taken as the coordinate origin. Then the coordinate position of any point is:

$$(X', Y')_p = (R_2 \cos\theta' - r_2 \cos(\theta' - \varphi'), \quad R_2 \sin\theta' - r_2 \sin(\theta' - \varphi')) \tag{7}$$

and the combined velocity is:

$$V_{R2} = \sqrt{V^2 + (\omega_2 R_2 \cos\theta')^2 + 2V\omega_2 R_2 \sin\theta'} \tag{8}$$

Generally, the direction of the drag force D in relation to the blade is the same as the direction of the combined velocity V_{Ri}. The lift force L is perpendicular to the drag force D, and the drag and lift are:

$$D = C_D \cdot \frac{1}{2} \rho {V_{Ri}}^2 S \tag{9}$$

$$L = C_L \cdot \frac{1}{2} \rho {V_{Ri}}^2 S \tag{10}$$

where C_D and C_L are the drag coefficient and the lift coefficient, respectively, and the values of the two are related to the position of the material of the blade and the shape and position of the blade. ρ is the density of the fluid and S is the inflow area of the blade. The vector expression of the resultant force of a single blade is:

$$\int \Delta p(V, \omega, \theta) dS = D + L \tag{11}$$

where $\Delta p(V,\omega,\theta)$ is the pressure difference between the upstream and the back surface of the blade and the differential pressure value varies with the flow velocity V, the turbine rotation speed ω, and the position angle θ. The average torque of the blade in one cycle is:

$$M_F = \frac{1}{2\pi}\int_0^{2\pi} T(\theta)d\theta \tag{12}$$

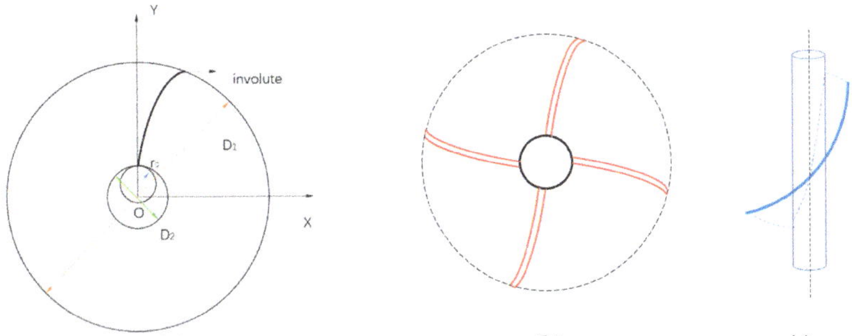

Figure 3. Blade design schematics for (**a**) Involute blade size determination; (**b**) Blade axial top view schematic; and (**c**) Radial blade schematic.

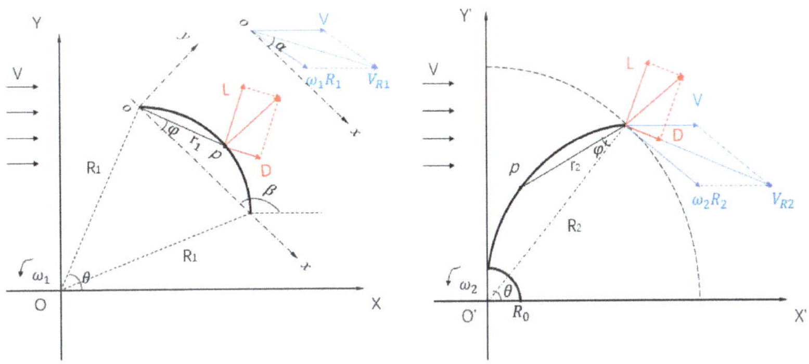

Figure 4. Schematic diagrams of the blade section motion and the force analysis: (**a**) is driven by axial flow and (**b**) is driven by radial flow.

3. Hydrodynamic Performance Analysis

3.1. Hydrodynamic Equation

At a specific flow velocity, in addition to the hydrodynamic torque M_F, the turbine is connected to the load and the movement of the turbine can be controlled through the change of the load; this is called the load moment M_A. The turbine is also subjected to the system's frictional resistance moment M_f. According to the D'Alembert principle and Newton's second law of motion, the hydrodynamic equation of motion is:

$$M_F - M_A - M_f = J\frac{d\omega}{dt} \tag{13}$$

where J is the rotational inertia of the turbine and the effect of the frictional resistance torque is ignored in the numerical simulation. The load torque is generated by the motor and acts on the rotating shaft in the form of damping torque. According to the empirical value, the load torque is linearly related to the square of the rotational speed. The proportional coefficient of the two is represented by the load factor b.

Since the load torque and rotational angular speed of the turbine are not constant, the output power is also a function of time:

$$P = \sum_{t} M_t \cdot \omega \tag{14}$$

The energy utilization rate, C_p, is used to measure the turbine's efficiency in extracting energy from currents:

$$C_p = \frac{P}{\frac{1}{2}\rho v V^3 S} \tag{15}$$

3.2. Numerical Models Establisment

The motion model equation was calculated and solved using the three-dimensional incompressible unsteady Reynolds-Averaged Navier-Stokes equation and the standard k-ε turbulence model [21]. The Reynolds average momentum equation can be expressed as follows:

$$\begin{cases} \frac{\partial \overline{u_i}}{\partial x_i} = 0 \\ \rho\left(\frac{\partial \overline{u_i}}{\partial t} + \overline{u_k}\frac{\partial \overline{u_i}}{\partial x_k}\right) = -\frac{\partial \overline{p}}{\partial x_i} + \frac{\partial}{\partial x_j}\left(u\frac{\partial \overline{u_i}}{\partial x_j}\right) + \frac{\partial R_{ij}}{\partial x_j} \end{cases} \tag{16}$$

where $i, j = 1, 2, 3$, denotes the three spatial coordinates of xyz. The Reynolds stress tensor is defined as:

$$R_{ij} = u_T\left(\frac{\partial \overline{u_i}}{\partial x_j} + \frac{\partial \overline{u_j}}{\partial x_i}\right) - \frac{2}{3}\rho k \delta_{ij} \tag{17}$$

Several authors have proven that the standard k-ε turbulence model is suitable for solving problems involving rotating blades [22,23]. Therefore, we used this model in the present investigation.

A turbine driven by a flow current can be defined with two types of cases. In a shallow sea area, the turbine is subjected to the impact of the horizontal current, while the buoy maintains a constant velocity of 1.0 m/s. In a nearly calm deep-sea area, the turbine only moves vertically along the buoy.

A hexahedral structure grid was used to divide the grid. After the grid quality check and the grid convergence verification, the total number of final elements was 4,866,240. The grid partitioning is shown in Figure 5. In order to capture the translation and rotation response simultaneously and facilitate the mesh update, the computational domain is divided into two zones: an accompanying moving zone and a dynamic mesh zone. The accompanying moving zone around the structure is a cylindrical surface, which can follow the translational and rotational motion of the system. The rest of the computational domain is defined as a dynamic mesh zone, whose grids deform and perform any needed adaptation during the calculation. Data is transferred between the overlapping interfaces of the two zones. A User-Defined Function (UDF) was used to control the movement of the buoy and the pressure changes as the depth decreased. The specific functions are shown in Appendix A. The left and the top sections of the fluid domain are set as the velocity inlet. The left side flow velocity simulation range is set to 0.1 m/s~2.0 m/s and the upper flow velocity is kept constant at 1.0 m/s according to the designed operating speed of the profiling float. The right and the bottom sections of the fluid domain are set as pressure outlets. In order not to affect the full development of the flow field behind the turbine and reduce the flow field boundary effect, the computation domain used for simulation is a cuboid region with size of 30 $D_1 \times$ 30 $D_1 \times$ 80 D_1. Table 1 shows the computational information. The Semi-Implicit Method for Pressure Linked Equations (SIMPLE) algorithm was used for pressure-velocity coupling. All of the transport equations were discretized using the finite-volume method with a second-order upwind scheme for all transport equations [24].

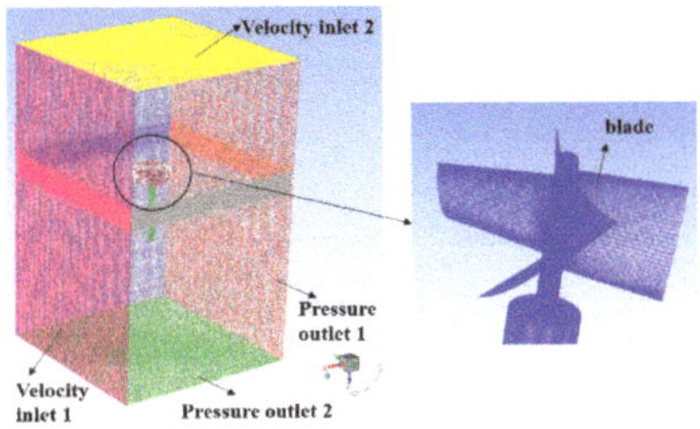

Figure 5. Meshing and boundary conditions.

Table 1. Analysis conditions.

Description	Analysis Condition
Working fluid	Water (1025 kg/m^3)
Left side velocity inlet	0.1~2.0 m/s
Upper flow velocity inlet	1.0 m/s
Pressure outlets	Relative pressure = 0 Pa

3.3. Method Validation

The numerical method used in this work has been validated against the numerical and experimental results of the study on the performance of tidal current turbines in our earlier studies [25]. In the numerical studies, tidal current turbines designed by our team compared with the experimental results observed by Pascal and Bahaj [26,27], and the free-to-rotate tidal turbine was compared with the experimental results reported in their results. The comparison of the results under the same variable parameter settings is shown in Figure 6. The numerical results coincided well with the experimental data, and the numerical method was verified to be capable of solving such a complicated issue regarding rotation and translation simultaneously.

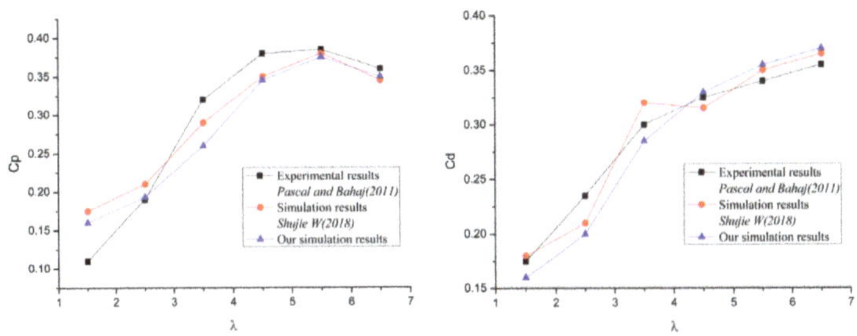

(a) (b)

Figure 6. (**a**) Comparison of the Cp vs. λ response of the experimental results of [26,27] and the simulation results of [25]; (**b**) comparison of the Cd vs. λ response of the experimental results of [26,27] and the simulation results of [25].

4. Numerical Simulation Results Analysis

4.1. Self-Starting Analysis

The initial state of the turbine is static. With the impact of the incoming flow, when the torque of the water flow is greater than the load torque, the turbine can start to rotate. This is also the point at which the turbine suffers the largest resistance moment. Once the turbine is started, the resistance moments that are experienced during operation will become smaller than they were initially.

Figure 7 shows the time history curve of the turbine rotation angular velocity for different radial flow velocities. It can be seen that an increase of the incoming flow velocity can effectively increase the starting torque of the turbine, shorten the time of the starting phase, improve the self-starting capability, and stabilize the turbine at a higher rotational speed. When the radial flow velocity is V = 0.8 m/s, the steady rotation speed can reach 6.5 rad/s. When the flow velocity decreases to V = 0.45 m/s, the start-up time is longer, about 8 s, the impeller rotation cannot reach the periodic stable state, and the rotation changes irregularly. However, when the flow velocity is less than 0.3 m/s, the turbine is only disturbed by the water flow and has no self-starting characteristics.

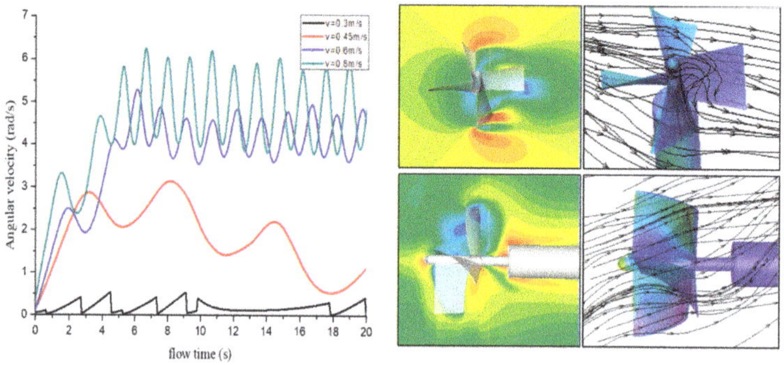

(**a**) (**b**) (**c**)

Figure 7. (**a**)Time-domain curves of angular velocity at different flow velocities; (**b**) Velocity profile of the cross section of the flow field around the turbine at a certain time; (**c**) A streamline diagram of the velocity field around the turbine at a certain time.

Figure 8 shows the time history curve of the turbine rotation angular velocity for different load factors at flow velocity of 0.1 m/s. Different load damping coefficients mean different load moments. As can be seen from the Figure 8, the smaller the load damping coefficient, the earlier the turbine starts to stabilize. In addition, the larger the load damping coefficient, the smaller the angular velocity at which the turbine can reach the steady phase. When b = 0 N·ms^2/rad^2, it means that there is the ideal state without load, and the angular velocity is also maximum at this time; when b = 0 N·ms^2/rad^2 to b = 1 N·ms^2/rad^2, the angular velocity decreases obviously. And then with the increase of b, the angular velocity value decreases slowly. It can be seen that the load plays an important role in limiting the speed of the turbine.

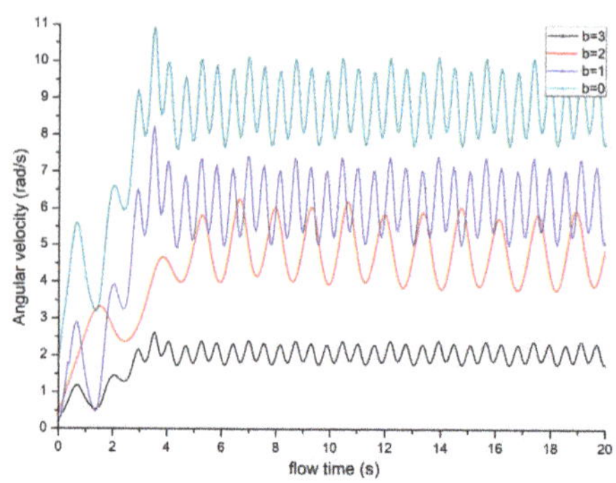

Figure 8. Time-domain curves of angular velocity at different load factors.

In addition, considering the profiling float in the actual working process, the wave and the current impact will cause the body to tilt, so that there is an angle between the turbine shaft and the direction of the flow, which will affect the self-starting performance of the turbine. Figure 9 shows the time history curve of the impeller rotation angular velocity corresponding to a flow velocity V = 0.8 m/s and body motion inclination angles β of 5°, 10°, 15°, and 20°, where the black curve represents the result of no inclination as a comparison. Due to the presence of the angle of inclination, the effective speed at which the turbine rotates is reduced, and the self-starting of the turbine at the same flow velocity is more difficult. The angular velocity variation of each rotation cycle of the impeller is greater, and the variation of the angular velocity in one cycle reaches about 5 rad/s. There are two angular velocity peaks in one rotation cycle. The calculated results show that when the range of the inclination angle does not exceed 20°, the initial self-starting velocity of the turbine is 0.43 m/s.

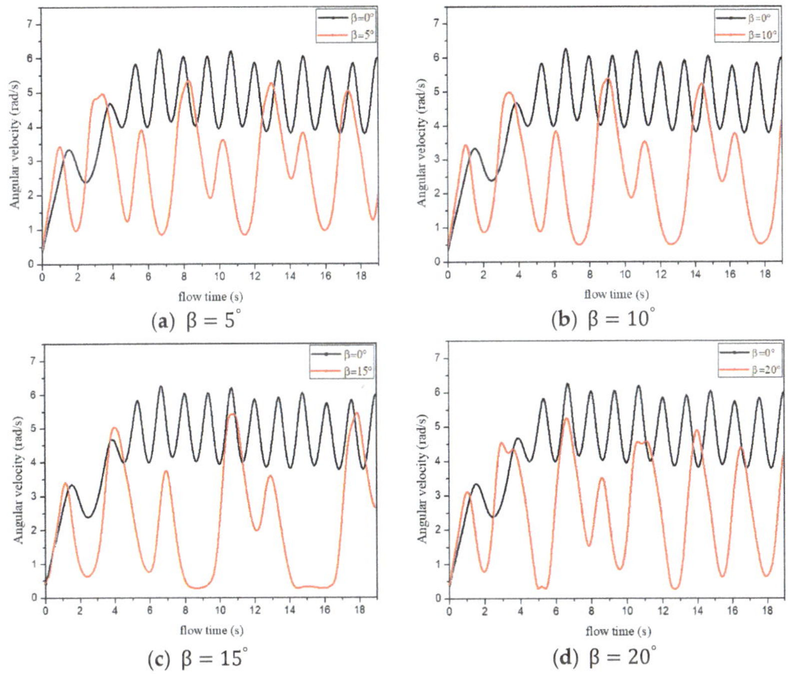

Figure 9. Time history curve of the turbine rotation angular velocity at different inclination angles.

According to the above analysis, the higher the flow velocity, the smaller the load factor, the smaller the buoy inclination angle, the better the self-starting performance as well as the higher rotary angular velocity of the turbine. However, the load fluctuation is greater. If the turbine's own moment of inertia is large and there is no effective braking capacity, the start-stop process will cause great damage to the structural strength of the turbine. In actual work, the braking performance of the turbine needs to be considered, and the excessive load will not be conducive to the rotation of the turbine, so the load and the speed need to find the best fit to ensure the relative stability of the self-starting and rotating of the device.

4.2. Energy Utilization Analysis

The curves of the lift L and the resistance D of the turbine can be obtained by numerical calculation. According to the calculation formula of the second part, the resulting curve of the total hydrodynamic torque and the load moment can be obtained. Figure 10 shows the results of flow velocity V = 0.6 m/s. Since the dynamic characteristics of the system are periodic, the integral of inertia of the right side of the motion balance Equation (13) should be zero within one revolution,

$$\int_{0^\circ}^{360^\circ} J\frac{d\omega}{dt}d\theta = 0 \tag{18}$$

At this point, the average hydrodynamic torque acting on the blade during one revolution of the turbine should be equal to the average load moment at the output $\overline{M_F} = \overline{M_A}$.

It can be seen from Figure 10 that after the turbine enters a steady state, both the hydrodynamic torque and the load moment pulsate up and down around 0.08 Nm. The pulsation frequencies of the two are essentially the same. For the severity of pulsation, the hydrodynamic torque is significantly higher than the load torque. The pulsation amplitude of the hydrodynamic torque can reach 0.17 Nm or more, while the load torque is only about 0.11 Nm.

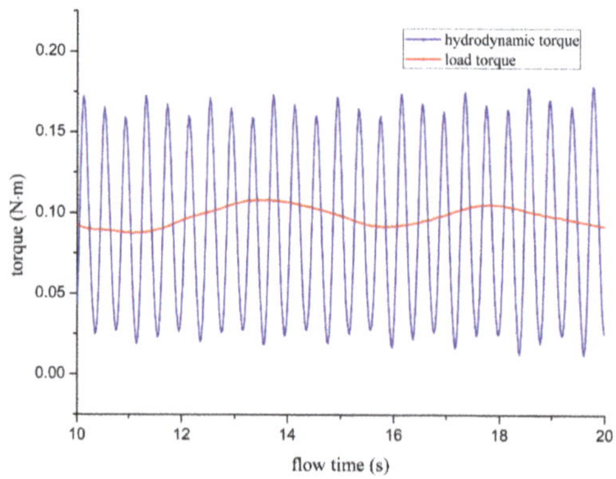

Figure 10. Time history curve of the hydrodynamic moment and the load moment.

The velocity ratio $\lambda = \omega R/V$ and the energy utilization coefficient C_p corresponding to the different velocity ratios are calculated as shown in Figure 11. It can be seen that as the velocity ratio increases, the energy utilization rate first increases and then decreases. Near the speed ratio λ = 2, C_p reaches the maximum value of 0.208.

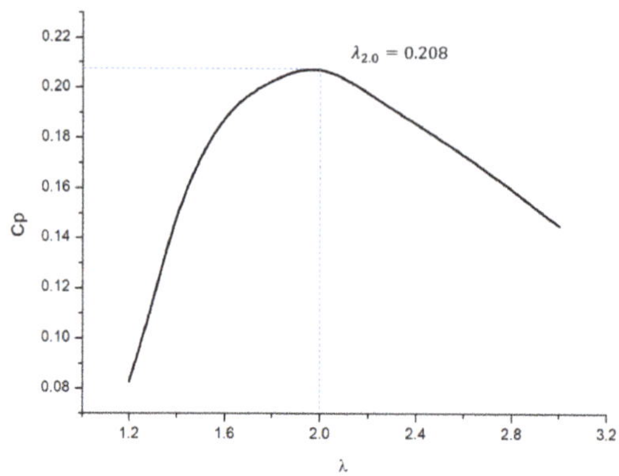

Figure 11. The energy utilization coefficient's variation with the change in the velocity ratio.

The relationships between the energy utilization coefficient and velocity ratio when the incoming flow inclination angles are 10° and 20° are as shown in Figure 12. The results show that the energy utilization decreases with the increase of the inclination angle, and the peak inflection point is advanced. The maximum value is 0.169 when the inclination angle is 10° and the corresponding velocity ratio is about 1.8. The maximum value is 0.148 when the inclination angle is 20° and the corresponding velocity ratio is about 1.7.

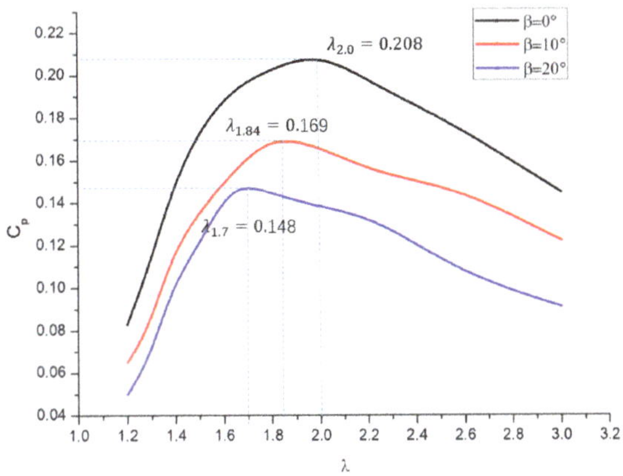

Figure 12. The relationship between the energy utilization coefficient and the velocity ratio at different inclination angles.

Considering that the wind turbine already has more mature research results, Figure 13 compares the results of this paper with the empirical curve of the four-blade fan energy utilization coefficient [28,29]. It can be seen from the figure that the numerical results are very close to the empirical curve of the fan. In the future, the energy utilization effect will be further optimized by changing the number and the parameters of the blades.

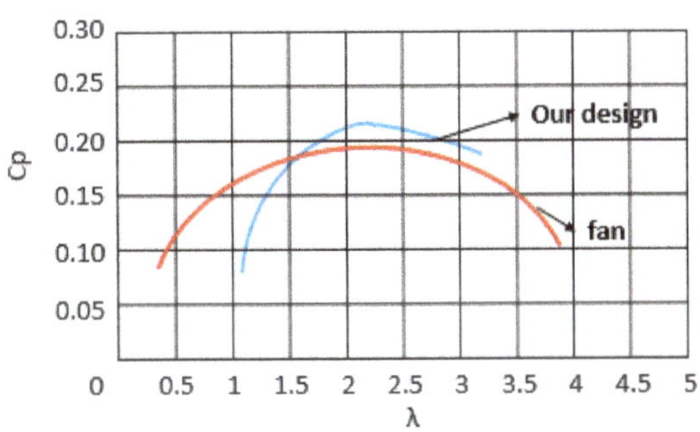

Figure 13. Comparison of the fan energy utilization coefficient curve with the same number of blades.

5. Conclusions

A marine current energy converter for deep-sea profiling floats is proposed in this paper, and the helical involute blade is designed to harness energy from the radial current and the axial relative flow. The blades are analyzed using computational fluid dynamics technique to determine its motion and hydrodynamic performance. This paper not only provides a specific theoretical research basis for the application of ocean energy resources on profiling floats, but also expands the new ideas of ocean energy utilization technologies for different scales and different application environments. The main findings of the present work are summarized as follows.

Through the analysis of fluid structure interaction, the minimum self-starting velocity of the turbine is about 0.3 m/s, the minimum flow velocity during steady operation is 0.6 m/s, and the maximum energy utilization coefficient is 0.208. When the profiling float is inclined by the current, the minimum self-starting flow velocity is 0.43 m/s within 20°, and the energy utilization coefficient is about 0.15. According to China's existing ocean energy resource data [30], the self-stating requirements of the turbine can be met, thus ensuring the feasibility of this innovative application. By comparing with the empirical curve of the energy utilization coefficient of the fan impeller, the numerical results are also considerable.

In future research, an overall experimental test will be conducted in order to further optimize the structural parameters and ensure the reliability and adaptability of the device, so as to fully prepare for the final engineering application.

Author Contributions: Conceptualization, S.W. and Y.L.; Data curation, S.W. and Q.A.; Investigation, S.W. and Y.L.; Software, S.W. and Q.A.; Writing—original draft, S.W.; Writing—review & editing, Y.L.

Funding: This research was supported by the Natural Science Foundation of Shandong Province (ZR2016WH02), the National Nature Science Foundation of China (U1706230) and the National Key R&D Program of China (2016YFE0205700).

Acknowledgments: The authors are grateful for the financial support provided by the Natural Science Foundation of Shandong Province.

Conflicts of Interest: The authors declare no conflict of interest.

Nomenclature

Parameter	Symbol
Turbine diameter (mm)	D_1
Rotor shaft diameter (mm)	D_2
Involute base circle radius (mm)	r_0
Rotational speed (rad/s)	ω_i ($i = 1, 2$)
Rotational radius (mm)	R_i ($i = 1, 2$)
Position angle (°)	θ, θ'
Elevation angle (°)	φ, β, φ'
Combined velocity (m/s)	V_{Ri} ($i = 1, 2$)
Flow velocity (m/s)	V
Attack angle (°)	α
Drag/lift force (N)	D, L
Drag/lift coefficient (N)	C_D, C_L
Fluid density (kg/m^3)	ρ
Inflow area (m^2)	S
Hydrodynamic torque (N·m)	M_F
Load moment (N·m)	M_A
Resistance moment (N·m)	M_f
Rotational inertia (kg·m^2)	J
Load factor (N·ms^2/rad^2)	b
Output power (W)	P
Energy utilization rate	C_p
Reynolds stress tensor	R_{ij}

Appendix A

User defined function:

```
#include "udf.h"
#include "dynamesh_tools.h"
static real omega_x=0.0;

DEFINE_CG_MOTION (or,dt,vel,omega,time,dtime)
{int s_id=7;
real zx=1448.6e-6;
real domega_x;
#if! RP_HOST
face_t f;
cell_t c;
real f_glob [ND_ND], m_glob[ND_ND],x_cg[ND_ND],dv_x,dv_y,dv_z,domega_y,domega_z;
real total_m=0.0;
Domain *domain;
Thread *tf1;
int i;
domain=Get_Domain (1);
tf1=Lookup_Thread(domain,7);
for (i=0; i<ND_ND;i++)
x_cg[i]=DT_CG (dt)[i];
Message0("%fn", x_cg [0]);
Compute_Force_And_Moment (domain, tf1, x_cg, f_glob, m_glob,0);
total_m=m_glob [0];
domega_x=dtime*(total_m/zx);
omega_x+=domega_x;
omega [0] =omega_x;
vel [0] =-1.0;
#endif
node_to_host_real(vel,ND_ND);
node_to_host_real(omega,ND_ND);}
DEFINE_PROFILE(p_h,thread,position)
{real x[ND_ND];
real h;
face_t f;
begin_f_loop(f,thread)
{F_CENTROID(x,f,thread);
h=3.365-x [0];
F_PROFILE(f,thread,position)=998.1*9.81*h; }
end_f_loop(f,thread) }
DEFINE_CG_MOTION (ve,dt,vel,omega,time,dtime)
{vel [0] = -1.0;}
```

References

1. State Key Laboratory of Marine Geology, Tongji University. *Seafloor Observation System: An International Review*; MGLab: Shanghai, China, 2006.
2. IOOS U.S. *Integrated Ocean Observing System: A Blueprint for Full Capability;* Version 1.0.; IOOS Office: Maryland, MD, USA, 2010.
3. Available online: https://www.marineinsight.com/shipping-news/leadership-2020 (accessed on 2 January 2017).
4. Available online: https://www.westernchannelobservatory.org.uk/ (accessed on 1 January 2017).
5. Available online: http://theory.people.com.cn/n1/2017/1129/c40531-29674382.html (accessed on 29 November 2017).
6. Shaolei, L. The position of the Beidou profile buoy in the global Argo real-time ocean observing network is at stake. *Argo Newslett.* **2016**, *42*, 16–18.
7. Renqing, L.; Jianping, X. Argo: A successful decade. *Basic Sci. China* **2009**, *4*, 15–21.
8. Teledyne Webb Research. Variable Bouyancy Profiling Float. U.S. Patent 8,875,645 B1, 4 November 2014.
9. Petzrick, E.; Truman, J.; Fargher, H. Profiling From 6000 m with the APEX-Deep Float. *Sea Technol.* **2014**, *55*, 27.
10. Chiba, S.; Waki, M.; Kornbluh, R.; Pelrine, R. Innovative Power Generators for Energy Harvesting Using Electroactive Polymer Artificial Muscles. *SPIE Proc.* **2008**, *6927*. [CrossRef]

11. Saupe, F.; Gilloteaux, J.C.; Bozonnet, P.; Creff, Y.; Tona, P. Latching Control Strategies for a Heaving Buoy Wave Energy Generator in a Random Sea. *IFAC Proc. Vol.* **2014**, *47*, 7710–7716. [CrossRef]
12. Bastien, S.P.; Sepe, R.B.; Grilli, A.R.; Grilli, S.T.; Spaulding, M.L. Ocean wave energy harvesting buoy for sensors. In Proceedings of the 2009 IEEE Energy Conversion Congress and Exposition, San Jose, CA, USA, 20–24 September 2009; pp. 3718–3725.
13. Bai, Y.; Du, M.; Zhou, Q.; Jie, M.; He, W. Proceeding of Tidal Current Energy Conversion System. *Ocean Dev. Manag.* **2016**, *33*, 57–63.
14. Han, S.; Park, J.; Lee, K.; Park, W.; Yi, J. Evaluation of vertical axis turbine characteristics for tidal current power plant based on in situ experiment. *Ocean Eng.* **2013**, *65*, 83–89. [CrossRef]
15. Liu, H.; Ma, S.; Li, W.; Gu, H.; Lin, Y.; Sun, X. A review on the development of tidal current energy in China. *Renew. Sustain. Energy Rev.* **2011**, *15*, 1141–1146. [CrossRef]
16. Kamoji, M.A.; Kedare, S.B.; Prabhu, S.V. Performance tests on helical Savonius rotors. *Renew. Energy* **2009**, *34*, 521–529. [CrossRef]
17. Gorban, A.N.; Asme, A.M.; Gorlov, A.M.; Silantyev, V.M. Limits of the turbine efficiency for free fluid flow. *J. Energy Resour. Technol.* **2001**, *123*, 311–317. [CrossRef]
18. Gorlov, A.M. Helical turbines for the Gulf Stream: Conceptual approach to design of a large-scale floating power farm. *Mar. Technol.* **1998**, *35*, 175–182.
19. Ocean Renewable Power Company [EB/OL]. 2014. Available online: http://orpc.co/default.aspx (accessed on 1 January 2018).
20. Zangiabadi, E.; Edmunds, M.; Fairley, I.; Togneri, M.; Williams, A.; Masters, I.; Croft, N. Computational Fluid Dynamics and Visualisation of Coastal Flows in Tidal Channels Supporting Ocean Energy Development. *Energies* **2015**, *8*, 5997–6012. [CrossRef]
21. Shih, T.; Liou, W.; Shabbir, A.; Yang, Z.; Zhu, J. A new k-ε eddy viscosity model for high Reynolds number turbulent flows. *Comput. Fluids* **1995**, *3*, 227–238. [CrossRef]
22. Liu, W.; Xiao, Q.; Cheng, F. A bio-inspired study on tidal energy extraction with flexible flapping wings. *Bioinspir. Biomim.* **2013**, *8*, 036011. [CrossRef] [PubMed]
23. Le, T.Q.; Lee, K.S.; Park, J.S.; Ko, J.H. Flow-driven rotor simulation of vertical axis tidal turbines: A comparison of helical and straight blades. *Int. J. Nav. Archit. Ocean Eng.* **2014**, *6*, 257–268. [CrossRef]
24. Liu, Z.; Qu, H.; Shi, H. Numerical Study on Self-Starting Performance of Darrieus Vertical Axis Turbine for Tidal Stream Energy Conversion. *Energies* **2016**, *9*, 789. [CrossRef]
25. Shujie, W.; Xiaoli, Y.; Peng, Y.; Junzhe, T.; Xiancai, S. Numerical simulation and experimental research on performance of horizontal axis tidal turbine. *Acta Energ. Sol. Sin.* **2018**, *39*, 1203–1209.
26. Galloway, P.W.; Myers, L.E.; Bahaj, A.S. Experimental and numerical results of rotor power and thrust of a tidal turbine perating at yaw and in waves. In Proceedings of the World Renewable Energy Congress 2011, Linkoping, Sweden, 8–13 May 2011.
27. Galloway, P.W.; Myers, L.E.; Bahaj, A.S. Quantifying wave and yaw effects on a scale tidal stream turbine. *Renew. Energy* **2014**, *63*, 297–307. [CrossRef]
28. Available online: http://www.360doc.com/content/14/0107/17/12109864_343364720.shtml (accessed on 7 January 2014).
29. Zhu, J.Y.; Huang, H.L.; Shen, H. Self-starting aerodynamics analysis of vertical axis wind turbine. *Adv. Mech. Eng.* **2015**, *7*, 1–12. [CrossRef]
30. Chu, Q.; Zhang, X.; Sun, J. Numerical simulation of the tidal current in the East China Sea. *Mar. Forecasts* **2015**, *32*, 51–60.

© 2018 by the authors. Licensee MDPI, Basel, Switzerland. This article is an open access article distributed under the terms and conditions of the Creative Commons Attribution (CC BY) license (http://creativecommons.org/licenses/by/4.0/).

energies

MDPI

Article

The Effect of Control Strategy on Tidal Stream Turbine Performance in Laboratory and Field Experiments

Carwyn Frost [1,*], Ian Benson [1], Penny Jeffcoate [2], Björn Elsäßer [3] and Trevor Whittaker [1]

1 School of Natural and Built Environment, Queens University Belfast, David Keir Building, Queen's University, Belfast BT7 1NN, UK; ian.benson@qub.ac.uk (I.B.); t.whittaker@qub.ac.uk (T.W.)
2 Sustainable Marine Energy, Edinburgh EH6 6QW, UK; penny.jeffcoate@sustainablemarine.com
3 Danish Hydraulics Institute (DHI), Ports and Offshore Technology, Agern Allé 5, 2970 Hørsholm, Danmark; bje@dhigroup.com
* Correspondence: c.frost@qub.ac.uk; Tel.: +44-289-097-4012

Received: 9 May 2018; Accepted: 5 June 2018; Published: 12 June 2018

check for updates

Abstract: The first aim of the research presented here is to examine the effect of turbine control by comparing a passive open-loop control strategy with a constant rotational speed proportional–integral–derivative (PID) feedback loop control applied to the same experimental turbine. The second aim is to evaluate the effect of unsteady inflow on turbine performance by comparing results from a towing-tank, in the absence of turbulence, with results from the identical machine in a tidal test site. The results will also inform the reader of: (i) the challenges of testing tidal turbines in unsteady tidal flow conditions in comparison to the controlled laboratory environment; (ii) calibration of acoustic Doppler flow measurement instruments; (iii) characterising the inflow to a turbine and identifying the uncertainties from unsteady inflow conditions by adaptation of the International Electrotechnical Commission technical specification (IEC TS): 62600-200. The research shows that maintaining a constant rotational speed with a control strategy yields a 13.7% higher peak power performance curve in the unsteady flow environment, in comparison to an open-loop control strategy. The research also shows an 8.0% higher peak power performance in the lab compared to the field, demonstrating the effect of unsteady flow conditions on power performance. The research highlights the importance of a tidal turbines control strategy when designing experiments.

Keywords: tidal energy; experimental testing; acoustic Doppler profiler; Strangford Lough

1. Introduction

The economic feasibility of offshore wind energy has reached unprecedentedly low strike prices in the recent Contracts for Difference (CfD) auction in the UK at £57.50/MWh in 2022/23, dropping 43% since 2012 [1]. The offshore renewable energy sector must drive to deliver other significant technologies that can achieve competitive levelised cost of electricity (LCOE). To deliver further cost reductions in an economically sustainable manner, developers must improve their technical understanding of a technology and gain experience of its performance in the real marine environment.

Even though numerical models and various design codes and methods for rotational power generators such as wind turbines and hydropower turbines are well established and in general yield excellent results, these models still require validation through experimental testing [2,3]. It is noted from this model scale testing that Reynolds independence is a key consideration for non-dimensional comparison.

To date there is limited published research on the experimental testing of prototype devices in the marine environment. However, Gaurier et al. [4] published a study comparing the test results from the same turbine undertaken in different lab facilities. They conclude that turbulence characteristics in lab facilities need to be described more adequately to better assess performance results. Furthermore, Mycek et al. studied the effect of turbulence intensity on a model turbine in a circulating flume. The range of turbulence intensities used (3% and 15%) may be typical of what is found at a tidal test site and the results show that this change in intensity has a near negligible effect on the time averaged performance of the turbine [5].

To support the industry in understanding the challenges of tidal turbine performance assessment a series of studies have been undertaken by Queen's University Belfast at a significant scale in tank, lake and tidal environments. The experimental campaigns are described in Table 1 and related publications are cited.

Table 1. Tidal Turbine Testing (TTT) project history.

Project	Date	Experiment Description	Publications
TTT	2013–2014	◦ Tandem pushing tests in Montgomery Lake ◦ Moored tandem tests in Strangford Narrows	[6–8]
TTT 2	2014–2015	◦ MaRINET (Marine Renewables Infrastructure Network) transnational access to Consiglio Nazionale delle Ricerche-Istituto nazionale per studi ed esperienze di architettura navale (CNR-INSEAN) for turbine tests ◦ MaRINET transnational access to Strangford Narrows for SCHOTTEL Hydro STG 50 turbine (SCHOTTEL, Spay, Germany) tests	[9–12]
TTT 3	2015–2017	◦ Moored single turbine testing in Strangford Narrows ◦ Towing tank testing in CNR-INSEAN with Wave and Yaw Angles	[13]

From this previous research and literature review it has been reported that experimental testing in an unsteady marine environment results in power performance disparities when compared with steady flow testing. Jeffcoate et al. [7] reports a 24% reduction in C_P between steady and unsteady flows. Meanwhile, Starzmann et al. [11] report comparable power performance between steady and unsteady flows (only a 5% change in C_P); however, the thrust performance was higher in the unsteady conditions relative to the steady conditions. In addition to this previous work, applicable studies have been conducted by other groups at a similar scale. The work of Forbush et al. [14] reports an 8% increase in performance with turbulent flow; however, the cross-flow tidal turbine used in this research sets it apart as a unique result, which may be specific to the rotor type. Blackmore et al. [15] performed repeat experiments in a circulating flume, the inflow characteristics were varied using a grid. The inflow velocity was maintained, while the Turbulence Intensity (TI) and length scale varied. The findings showed a 10% reduction in C_P with an increase in TI, conversely an increase in length scale resulted in an increase in C_P.

In all field experiment cases, the cause of power disparity between lab and field is partly attributed to uncertainty in the experiments. Further consideration of the source of uncertainty has identified two critical contributing factors:

i. A crucial factor in the accuracy of performance measurements arises from the control strategy imposed on the tidal turbine. The capability of the control system to keep the rotor operating near its optimum Tip Speed Ratio (TSR), when driven by an unsteady and non-uniform inflow velocity, and avoiding 'stall', is a key element in a successful commercial turbine. Similarly, the control strategy adopted during an experiment can increase the error bounds in derived performance indicators, such as C_P. Control strategies used in previous tidal turbine performance testing vary. For devices at low Technology Readiness Level (TRL) scales [16], such as those used in laboratory scale experiments, an open-loop control may suffice [9]. However, more commonly, a closed-loop, proportional–integral–derivative (PID) feedback control strategy is found [17,18] in higher TRL devices. It is recognised that in order to develop the industry, advanced control strategies require development and testing in highly parameterised conditions. Efforts are being made in this field in both research and industry. The use of overspeed, pitch or stall control strategies with peak power tracking, or surface mapping algorithms for condition monitoring purposes are under development [19,20]. Furthermore, future options may include the possibility of feed-forward algorithms such as those being trialled in the wind sector [21].

ii. The second source of increased uncertainty comes from the increased variability of the inflow velocity. This is of particularly significance since the power density scales with the cube of the inflow velocity. Furthermore, both genuine variability and sampling errors are compounded in real velocimetry data. Separating and quantifying their effects requires care in calibration of instruments and data analysis. In order to promote consistent best practice in the power performance testing of Tidal Energy Converters (TECs), Johnstone et al. published best practices for the wave and tidal sector [22] which specifies the requirements for clear uncertainty analysis. Further to this the IEC (Geneva, Switzerland) published a Technical Specification IEC/TS 62600-200 [23]. The specification provides the methodology for determining an average value for velocity at a site, enabling the time average performance of a turbine to be captured and reported to a common standard. The IEC specification has been used in other research projects and across the industry [10,13,24] and will be used to guide the data analysis in this paper.

In summary, the literature review has highlighted experiments where turbine testing in the lab has been compared to the field. The control strategy and inflow measurements are significant parameters that must reflect the device TRL and testing environment. In order to inform the sector of the significance of these parameters, the TTT turbine will be deployed in the lab and field, the derived performance results will be compared for two control strategies, and in a steady and unsteady inflow environment. The lab experiments will be used as an opportunity to calibrate and derive uncertainty metrics for the inflow, this will then be applied to the field experiments and performances compared.

2. Experimental Setup

Under the TTT 3 project, both the laboratory and field tests were conducted using the same instrument, maintaining continuity between experiments. This was achieved by using the same turbine, velocity instruments and data acquisition system. The laboratory experiments were undertaken in the controlled steady environment of the towing tank facility CNR-INSEAN, Italy. The field test experiments were performed in the uncontrolled, unsteady environment of the tidal test site in Strangford Narrows, Northern Ireland (Lat.: 54.381801°, Long.: −5.556743°).

The 1.5 m diameter turbine is pictured during operations in Strangford Lough in 2016 (Figure 1); the device dimensions can be found in the supplementary appendix of previous work [6] and remain consistent in this work. The rotor blades are an Eppler E387 airfoil geometry, each blade is 0.575 m long and features a pitch distribution similar to industrial wind turbine blades, with a root pitch angle of 32° which recedes to 13° at the tip of the blade. As this is not a commercial blade and is not designed to be scaled the significance of Reynolds independency is limited.

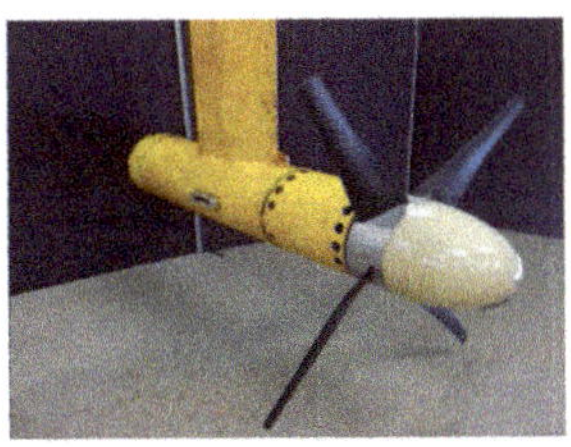

Figure 1. Tidal Turbine Testing (TTT) device and rotor.

2.1. Flow Instrumentation

Acoustic Doppler profilers (ADPs) are often used for tidal flow resource assessment, and are the given method for characterising inflow conditions by the IEC [23]. The Nortek 2 MHz 'Aquadopp' used in the experiments was the primary instrument for determining inflow velocity in these experiments.

The experimental setup shown in this section relates to all work as part of the TTT 3 project. Please refer to referenced papers for further details on previous experimental setups [6,9]. The supporting gantry also hosts a Nortek 'Vector' Acoustic Doppler Velocimetry (ADV) with the sample volume located at the apex of swept area at 16 Hz, this provided insight into the inflow turbulence. The deployment parameters for both instruments are given below. In accordance with the IEC standards for power performance assessment, the inflow measurements were taken at an upstream distance of between 2–5 equivalent diameters from the plane of rotation [23].

The geometrical setup can be seen in Figure 2 and is detailed in Table 2, where the parameter D2 has two values: one for the towing tank and one for Strangford Lough tests, respectively. This difference was due to the location of suitable mounting points in the towing carriage.

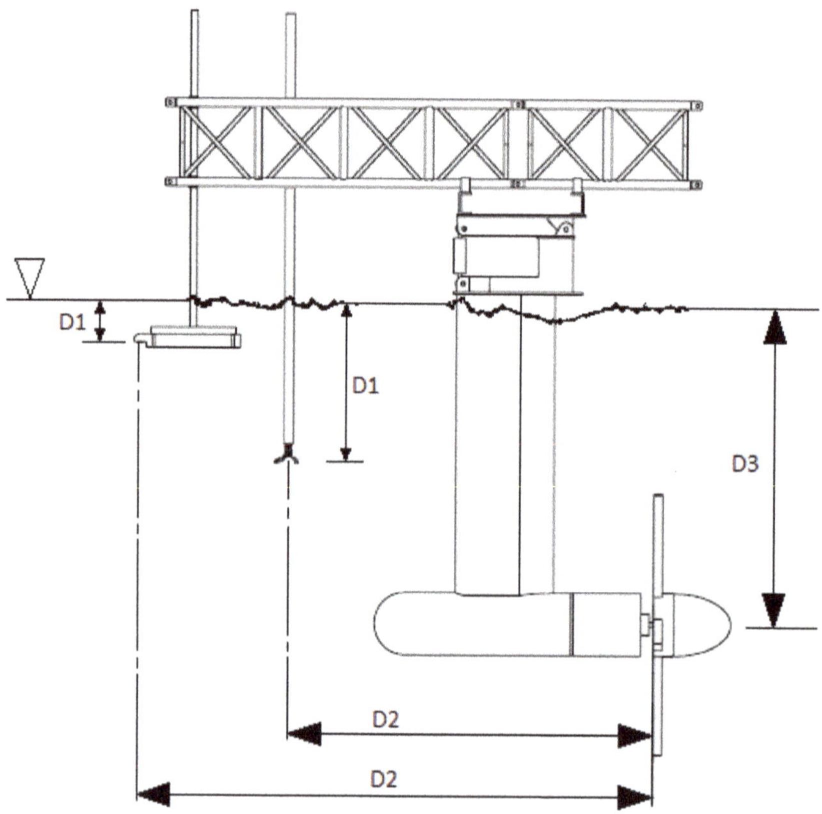

Figure 2. Turbine, Acoustic Doppler profiler (ADP) and Acoustic Doppler Velocimetry (ADV) deployment Configuration.

Table 2. Turbine, Acoustic Doppler profiler (ADP) and Acoustic Doppler Velocimetry (ADV) deployment parameters.

Parameter	ADP		ADV	
	Strangford	CNR-INSEAN	Strangford	CNR-INSEAN
Distance D1 (m)	0.30 m	0.30 m	0.86 m	0.86 m
Distance D2 (m)	2.99 m	3.80 m	3.10 m	3.7 m
Distance D3 (m)	1.75 m	1.75 m	1.75 m	1.75 m
Power	high	high	high	high
Transmit length	N/A	N/A	8 mm	8 mm
Number of cells	20	20	N/A	N/A
Cell Size (m)	0.25 m	0.25 m	N/A	N/A
Blanking Distance (m)	0.25 m	0.25 m	N/A	N/A
Co-ordinate System	Beam	Beam	Beam	Beam
Sample Frequency (Hz)	1 Hz	1 Hz	16 Hz	16 Hz
Sample Period (s)	120–600 s	90–140 s	120–600 s	90–140 s

2.2. Flow Instrumentation Validation

Many ADP instruments have previously been independently calibrated and validated to verify the quality of factory settings. Shih et al. [25] demonstrated the close agreement between calibrations for two different ADP suppliers, RDI (Poway, CA, USA) & SonTek (San Diego, CA, USA) in a towing tank facility in 2000. More recently, Oberg et al. [26] performed similar experiments with the most recent instrument firmware, again using RDI and SonTek instruments. Little work has been found which independently calibrates the Nortek Aquadopp ADP used in this experiment; however, Elsäβer et al. showed discrepancies in time averaged velocities of up to 0.19 ms^{-1} between two instruments collocated in the field [27]. In order to ensure instrument accuracy, the ADP (2 MHz Aquadopp) was calibrated during the towing tests under conditions as close as were possible to the field. Seeding of the towing tank was required to improve the Signal to Noise Ratio (SNR).

To validate this ADP calibration against another independent instrument while working in the field, the Nortek Vector ADV was likewise calibrated during the tank work. 16 Hz Vector data was averaged over 1-s bins to match the ADP (2 MHz Aquadopp) data bins. All the calibration results are given in Table 3 and show the deviation of the instruments from their factory calibration. Each calibration equations were derived for U, V and W from the towing tests for the Aquadopp (ADP) and Vector (ADV) instruments. The linear equations demonstrate the precision and bias drift since the previous calibration (in mm/s) as yielded by the Nortek transform matrices for each instrument. Note the signs are chosen to make the output components have the same axis convention for both instruments. U is the main direction of inflow, along the turbine axis.

Table 3. Deviation of Nortek instruments from factory calibration.

Nortek Transform Matrix Output	ADP—Aquadopp	ADV—Vector
x	$U = -1.0124x + 4.97$	$U = -1.0042x + 6.4$
y	$V = -1.0124y + 0$	$V = +1.0042y + 0$
z	$W = -1.0124z + 0$	$W = -1.0042z + 0$
Bias in U (1 σ)	−0.03%	+0.05%
Precision in U (1 σ)	±0.6%	±0.9%

In summary, the calibration performed in the towing tank agrees closely with the factory calibration, showing less than a percentage difference in bias and precision. The same individual instruments, deployment parameters and locations were used during both the laboratory and fieldwork testing. Further validation of the results has been undertaken using the fieldwork data, further detail can be found in Appendix A. This provides confidence in instrument accuracy and draws attention to the high spatial variability in tidal flows, as discussed by Elsäßer et al. [27].

2.3. Control Strategy

The control of wind turbines is an established and mature area of technology. The control strategy involves multiple inputs and outputs [28] and more than one control loop, often extending to variables intended to optimise the performance of an array and its grid interface.

In contrast, the original TTT 2 experimental design was an open-loop load control, as shown in Figure 3. Control was provided by a binary array of resistors (5 values, each the double of the previous: 10.25 ohms, 20.5 ohms etc.). Thus, 32 load values could be placed on the DC-bus by switching combinations in and out. Electromechanical contactors select the load on command from the central Data Acquisition System (DAQ), based on a National Instruments Compact Rio (cRio) running Labview and further custom-made interface electronics and cabling.

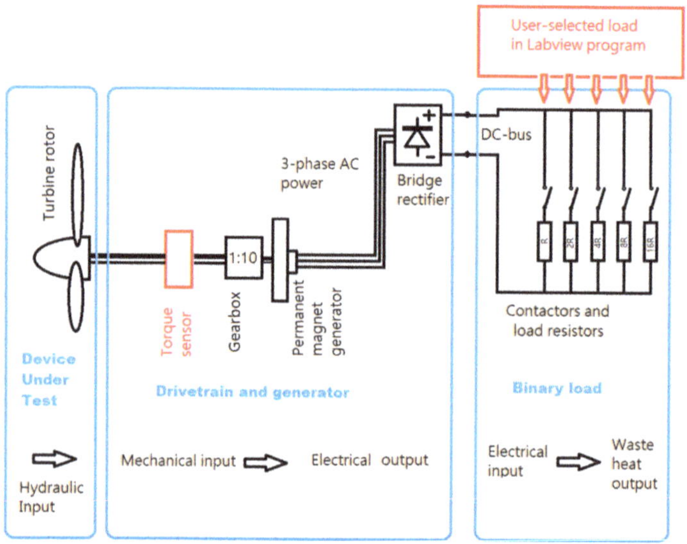

Figure 3. TTT 2 open-loop control and binary load.

In developing the control system, a single control loop was included with only one process variable as the input (the rotor shaft speed) and one output (the electrical demand placed on the alternator). A review of Proportional–Integral–Derivative (PID) control theory, developed in the mid 20th century, is presented by Bennet [29]. In order to incorporate a closed-loop PID load control, a linear regulator was used to place an electronic load on the turbine at the DC output terminals. The upgraded PID control loop is shown in Figure 4. An absolute rotary encoder was included to provide higher resolution of shaft speed for the PID loop.

The PID control system was run as a Labview virtual instrument in the DAQ programme, which was written specifically for the project. Most of the programme's other features were identical for both TTT 2 (open-loop) and TTT 3 (PID; see Table 1 for project details). To control the turbine for the TTT 3 system, the user selects a chosen shaft speed (set point) and the system then attempts to maintain that speed independently of flow features in the field, or of towing speed in the laboratory. The PID parameters used were identical in both the field and towing tank; all the results under PID control were collected with the parameters set as in Table 4.

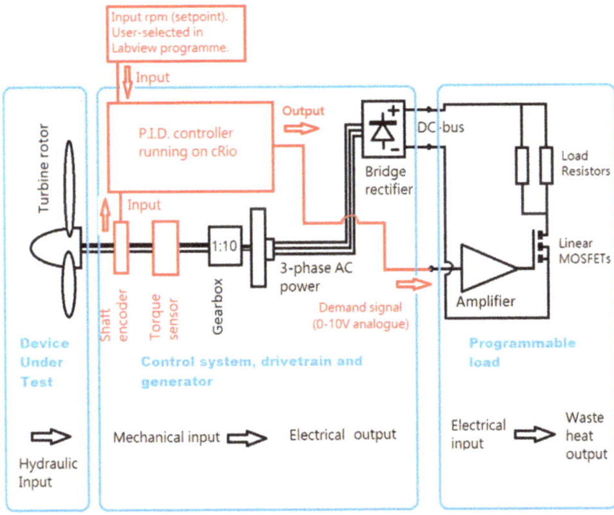

Figure 4. TTT 3 closed-loop Proportional–Integral–Derivative (PID) control and programmable load.

Table 4. Proportional-Integral-Differential parameters.

Symbol	Description	Value
Kc	Proportional constant	2.2 (no dimensions)
Ti	Integration time constant	0.003 (min)
Td	Differential time constant	0.001 (min)

The proportional constant, Kc or gain was either positive or negative, depending on the rotational direction of the driveshaft. This also applied to the set RPM (Rotations per Minute) value in the control system. The tuning of the loop parameters was done in the field with the Zeigler and Nichols 'ultimate cycling' method [30], followed by some minor optimisation by trial and error. It is worth noting that the ultimate cycling method gave adequate robustness and precision of control for our purposes and was only slightly adjusted with subsequent trial and error.

3. Non-Dimensional Performance Characteristics

As previously noted, the DAQ features a National Instrument Compact Rio, the cRio synchronises the various instrument and control data streams into a common format, timestamps them and outputs them as a data file for post-processing.

To compare and analyse the performance of the turbine between the various experimental campaigns, the non-dimensional performance characteristics will be derived. These are shown in Equations (1) and (2) and provide the Tip Speed Ratio (TSR) and Coefficient of Power (C_P).

$$\mathrm{TSR}_i = \frac{\overline{\omega}_i \times r}{\overline{U}_i} \tag{1}$$

$$C_{P,i} = \frac{\overline{P}_i}{\frac{1}{2} \times \rho \times A \times \overline{U}_i^{\,3}} \tag{2}$$

where the rotational speed of the turbine ω was in rad/s, the turbine radius, $r = 0.75$ m, water density, $\rho = 1025\ \mathrm{kg/m^3}$ and the turbine area, $A = 1.767\ \mathrm{m^2}$. The mean power, $\overline{P}$ was derived from the product of the rotational speed and mechanical torque measurements averaged over the period of the test window. Both non-dimensional performance indicators require a value for the inflow velocity to the turbine ($\overline{U}_i$). To derive the mean current velocity, the method of bins, employed by the IEC 62600:200 [23], was used. This method has been outlined in previous work; however, the method has since been developed by the authors to account for bias introduced by the Doppler noise and thus determine the associated uncertainty in the inflow velocity. Further detail on the method and its development can be found in Appendix B and previous research [13].

4. Results

The following section is presented in two subsections, time series results and derived performance (time averaged) results. In each subsection, the comparison between field and lab results will be made and a comparison of the two turbine control strategies (open loop and PID control).

4.1. Time Series Results

Figure 5 is a time series plot comparing the experimental data in the field (Strangford Lough) and the lab (CNR-INSEAN); the two experimental setups feature the PID controller and were conducted under the TTT 3 experiments in 2016/17. Turbine rotational speed, fluid inflow velocity and turbine output power are plotted as time series. Figure 5a shows the rotational velocity from the PID controller was set to 50 RPM for the field and lab. The stability in both sets of results are good with standard deviations of $2\sigma_{\mathrm{RPM}} = 1.57$ and 0.45 respectively for the field and lab. Subplot (b) shows the relative velocity of the water passing the turbine. One of the challenges of ADP and ADV deployments in towing tanks is maintaining a sufficient level of seeding in the water to suit the acoustic reflection and achieve strong SNR, this is not an issue in the field. For the lab velocity, the carriage encoder was used to derive the velocity and subsequent performance characteristics. The ADP results from the field show high flow variation ($2\sigma_{\mathrm{Vel}} = 0.18$) and this instability is carried through to the mechanical power of the turbine (see Figure 5c, while the steady conditions in the lab result in a steady power output.

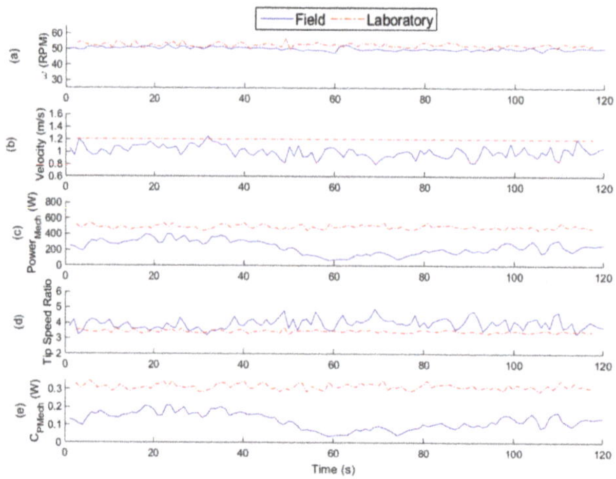

Figure 5. Time series plot of (**a**) turbine rotational speed; (**b**) inflow velocity; and (**c**) mechanical power. (**d**) Tip Speed Ratio and (**e**) Coefficient of Mechanical Power using PID controller in the lab and field.

The derived non-dimensional performance characteristics are shown in subplots (d) and (e). For the field results it is clear the TSR fluctuates with the inverse of the inflow velocity as expected. More significantly, the slight difference in TSR results in a significant difference in the C_P. The lab results have a stable TSR and C_P with $2\sigma_{\mathrm{TSR}} = 0.16$ and $2\sigma_{\mathrm{CP}} = 0.03$, while the fluctuating field results have lower mean value and higher standard deviation ($2\sigma_{\mathrm{TSR}} = 0.70$ and $2\sigma_{\mathrm{CP}} = 0.07$).

Figure 6 shows the time series results from the open loop controller and PID controller from Strangford Lough field campaigns in TTT and TTT 3. Time series were selected which had the same time-averaged velocity and the time series (and thus average) extends to circa four minutes in both cases. The TTT deployment in 2013 used an open loop control strategy setting a constant demand torque to the alternator, while the TTT 3 deployment in 2016 used the PID (closed loop) control strategy with a set RPM, fluctuating the demand torque to maintain the constant RPM. Note the PID Control case exhibits the same results as shown in Figure 5. The open-loop control RPM time series (Figure 6a) has a higher standard deviation ($2\sigma_{RPM} = 6.37$) than the equivalent closed-loop control time series ($2\sigma_{RPM} = 1.49$). However, maintaining a constant RPM does not result in a constant power output due to the fluctuations in the inflow velocity as shown in Figure 6b. The inflow velocity for both experiments is derived from the IEC method of bins using the ADP data, as described previously. The mechanical power time series, Figure 6c, shows little difference in signal fluctuation between the two control methods, as confirmed by the standard deviations ($2\sigma_{Power} = 110.58$ and 109.88 respectively for open and PID control). The cause of the fluctuations has now been isolated to the inflow velocity fluctuations, as opposed to the type of control strategy, and can be considered independent of fluctuations in RPM. The further significance of these two control strategies can be seen in the derived performance characteristics. Figure 6d,e show the TSR and C_P it is clear that while the PID experiences a more stable TSR value ($2\sigma_{TSR} = 1.02$ and 0.70 Open and PID control respectively) the C_P for both control strategies have similar fluctuations ($2\sigma_{CP} = 0.06$ and 0.07 Open and PID control respectively).

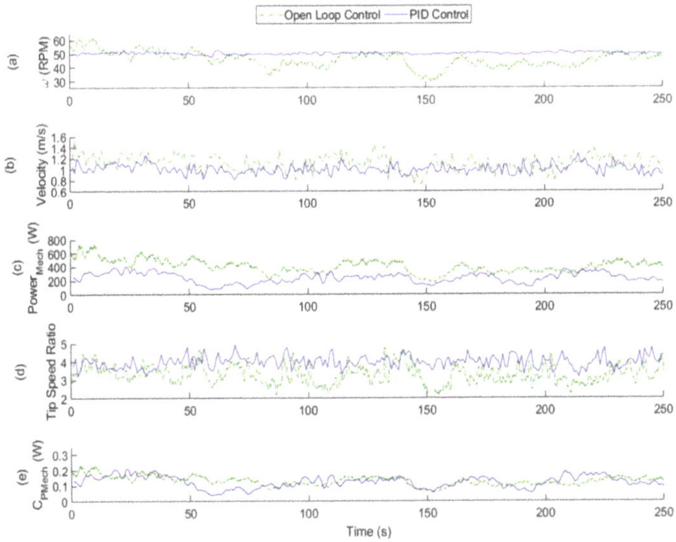

Figure 6. Strangford Lough time series plot of (**a**) turbine rotational speed; (**b**) inflow velocity; and (**c**) mechanical power. (**d**) Tip Speed Ratio (**e**) Coefficient of Mechanical Power.

4.2. Derived Performance Results

The power performance characteristics of the Eppler rotor using the same experimental setup and PID control settings in CNR-INSEAN and Strangford Lough are shown in Figure 7. The varying inflow experienced at Strangford Lough has been plotted in the figure using 0.1 ms^{-1} velocity bins associated with each of the time averaged reading (this is in accordance with the IEC standard [23]). Two trends are noted in to the uncertainty bounds; firstly, with increasing TSR, the uncertainty increases. This indicates the uncertainty from rotational speed, ω and inflow velocity $\overline{U}_i$ are the dominating sources. Secondly, the uncertainty bounds for low velocity bins ($U_{0.8}$–$U_{1.0}$) are comparably greater than higher velocity bins ($U_{1.1}$–$U_{1.4}$) at the same or similar TSR values. This may be due to the Signal to Noise Ratio (SNR) of the instrument being poor at these lower velocities, increasing uncertainty. The two curves show very close agreement at low TSR; however, as the power curves reach their peak, there is an 8.0% difference in C_P and a 4.5% difference in the TSR for peak C_P. The separation between the curves is maintained as the turbine approaches freewheeling. The uncertainty bounds in the results overlap, but there is also a number of data points well below the performance curve. This deficit in power performance is due to the unsteady inflow parameters and this outcome agrees with Starzmann et al. [11] and is also comparable with results from the flume experiments by Blackmore et al. [15]. This shows that turbine power performance is adversely affected by unsteady inflow. Further consideration of the inflow turbulence metrics and their significance on hydrodynamic performance must be considered.

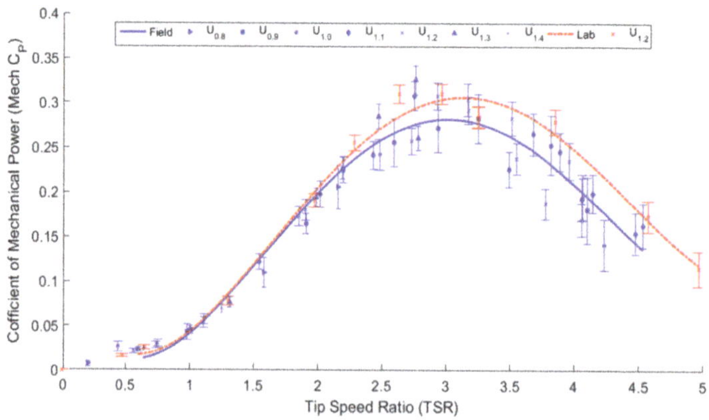

Figure 7. Mechanical Power performance for Eppler Rotor in the lab and field using PID Controller.

In its nature, a tidal test site will have a significantly varying inflow, this can be seen by the data points corresponding to the average velocity over which they were taken. Table 5 further shows this variation in velocities in accordance with the IEC standards for power performance of a tidal turbine [23]. Each data set in the velocity bins comes from a time averaged result of at least 2 min.

Table 5. Tabulated results for Strangford Lough.

Velocity Bin (ms^{-1})	Mean Current Velocity, ($\overline{U}_i$)	Mean Power Output, ($\overline{P}_i$)	Mean SD of Power Output, ($\overline{SD}_P$)	Number of Data Sets, n
0.65–0.75	0.720	37.972	7.088	3
0.75–0.85	0.802	100.929	45.999	6
0.85–0.95	0.922	113.202	59.554	4
0.95–1.05	1.012	193.498	94.951	13
1.05–1.15	1.100	215.158	91.195	16
1.15–1.25	1.183	264.867	78.834	8
1.25–1.35	1.254	83.518	15.557	2

In Figure 8, the average mechanical power performance of the PID control system can be seen to have a higher performance curve than the equivalent open loop control system. At peak performance, the PID control turbine is 13.7% more efficient than the open-loop control turbine.

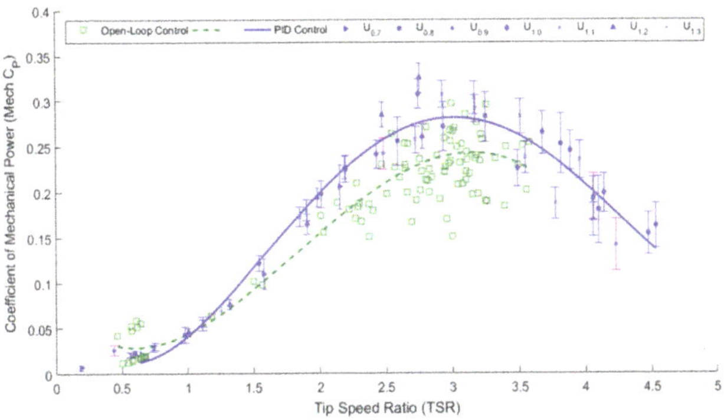

Figure 8. Mechanical Power Performance of turbine in Strangford Lough using two control strategies.

In addition to the increase in performance, the inclusion of the closed-loop control system improves the distribution of the data points across the C_P-TSR curve. The open-loop control results have greater scatter, resulting in a comparatively poorer fit ($R^2 = 0.84$ and 0.96 for Open and PID Control respectively, Degrees of Freedom = 4). Furthermore, on the right-hand side of the curve, the closed-loop control system has reached a higher TSR, providing a fuller picture of the Eppler rotor performance curve, this is due to the control system having a higher variability in resistance, thus getting closer to freewheeling.

5. Discussion

The comparison of performance in turbine response during experiments in the lab and field show that the performance characteristics do change in the presence of unsteady inflow conditions. While there is a performance drop in C_P between the lines of best fit of the steady lab conditions and the unsteady field conditions, the uncertainty bounds of the data points overlap suggesting the difference is with the region of uncertainty. In order to improve these uncertainty bounds, it is suggested that increasing the number of samples/sample rate or increasing the number of bins in the projected area of the rotor is required. These options are limited when using an ADP due to their correlation with the Doppler noise source. This highlights the importance of instrument selection and setup for deriving the inflow performance.

The method used to account for uncertainties shows robustness in its application in the field and lab. The use of ADPs to measure inflow in the field is appropriate in high flow environments, when characterising full-scale turbine performance, in accordance with the IEC standard. However, awareness of the ADPs' contribution to uncertainty in the derived performance characteristics is important. Due to the difficulty in maintaining sufficient seeding material in towing tanks, the appropriateness of ADPs remains a challenge; however, as shown in steady flow conditions, the carriage velocity is sufficiently accurate. Alternatively, Particle Image Velocimetry (PIV) measurements would be the favoured method in towing tanks.

The comparison of the PID controller with a set RPM and open-loop controller in unsteady inflow conditions (Strangford Lough) showed a 13.7% difference in peak C_P. The reasons for this distinction in performance are shown in the time series results. Figure 6d shows that restricting the RPM with the PID controller reduces the fluctuation in the TSR over the period of the experiment. As the open-loop control experiences greater fluctuations in TSR, during experiments near peak performance the turbine will be operating at suboptimal TSR in the C_P-TSR curve. This accounts for the reduction in scatter and higher performance at optimal TSR. However, the PID control with set RPM still results in a fluctuating TSR. In a commercial system where characterising the performance curve isn't the objective, a more appropriate control strategy would use a set TSR, or peak power tracking. This may require knowledge of both the turbine rotational speed and the inflow velocity. This is a consideration for further work.

6. Conclusions

The experimental work presented in this paper successfully completed two experimental campaigns in CNR-INSEAN, Italy and Strangford Lough, Northern Ireland. The continuity of experimental equipment between the two sites provided the opportunity to investigate ADP and ADV calibration agreement with factory settings. It was found that both instruments closely agreed with the factory transformation matrices.

The research has furthered the development of the IEC bin method [23] by the inclusion of a Doppler noise bias correction factor and a method for calculating uncertainties in the derived performance characteristics, with particular attention to uncertainties associated with inflow characterisation.

The work has shown that the effect of unsteady inflow on the derived turbine power performance has a slight detrimental impact of circa 8%. This is in line with the findings of Starzmann et al. [11] and Blackmore et al. [15] and can be considered within the experimental accuracy. To clarify this further will require narrowing the confidence intervals and possible methods of doing this have been suggested.

The two control types, representative of typical strategies in experimental and prototype devices, have shown that a PID feedback controller with set RPM helps achieve a distinctly higher performance curve than the open loop control system, when applied to the turbine operating in unsteady flow conditions such as Strangford Lough. This is critical to consider in experimental design. The fluctuating TSR, as a result of unsteady flow and constrained RPM, remains a weakness in the system if in the presence of large inflow fluctuations. For this reason, it is recognised as a suitable control strategy for characterising a turbine in unsteady flow conditions. Continued development of the PID control is an area of further work, using peak power tracking algorithms and a pre-defined power curve. The inclusion of these strategies will bring the device further in-line with existing industry strategies and make experimental output more relatable to the sector.

Author Contributions: Conceptualization, C.F., I.B., P.J., B.E. and T.W.; Data curation, C.F., I.B. and P.J.; Formal analysis, C.F. and I.B.; Funding acquisition, P.J. and B.E.; Investigation, C.F. and I.B.; Methodology, C.F., I.B., P.J. and B.E.; Project administration, C.F., B.E. and T.W.; Software, C.F. and I.B.; Supervision, B.E. and T.W.; Validation, I.B.; Visualization, C.F. and I.B.; Writing–original draft, C.F.; Writing–review & editing, I.B., P.J., B.E. and T.W.

Funding: This research was funded by Invest Northern Ireland, through the Centre for Advanced Sustainable Energy (CASE).

Acknowledgments: The authors would like to thank Francesco Salvatore, Luigi Fabbri and all the other CNR-INSEAN staff who supported the work. The authors would also like to thank Cuan Boake of Applied Renewables Research Ltd., Ralf Starzmann of Schottel Hydro and Graeme Mackie of Oceanflow Energy.

Conflicts of Interest: The authors declare no conflict of interest. The funding sponsors had no role in the design of the study; in the collection, analyses, or interpretation of data; in the writing of the manuscript, and in the decision to publish the results.

Nomenclature

Parameter	Symbol
Area of turbine (m^2)	A
Velocity Bin Identifier	i
Time Instant Identifier	j
Depth Profile Bin Identifier	k
Number of samples	n
Density (kg/m^3)	ρ
Extracted Power	P
Extracted Torque	Q
Turbine Radius (m)	r
Sample Identifier	s
Extracted Thrust	T
Mean Velocity (m/s)	$\overline{U}$
Rotational Speed (rad/s)	ω

Standard Deviation σ
Two Standard Deviation (95% confidence interval) 2σ

Appendix A

ADP Towing Tank Calibration Results

To test the calibration equations shown in Section 2.2, the disagreement between Vector and Aquadopp velocity measurements in Strangford Lough were examined over 30 runs. These measurements were completely independently of all those used for the calibration. The inflow velocity in the Strangford was not necessarily well aligned with the turbine axis during all tests, so the magnitudes and the yaw angles for the inflow velocities were analysed. Poorer agreement between the two instruments was found at yaw angles of 15 degrees or more.

In Table A1, the variable S expresses the disagreement between concurrent Vector and Aquadopp velocity measurements, as a percentage. The values predicted using the calibration data alone, as well as those obtained from the validation data, are tabulated.

Table A1. Precision and bias of ADP and ADV.

Variable S (Percentage Disagreement between Vector and Aquadopp)	Prediction from Calibration	Whole Validation Dataset	Validation for Yaw within ±15 Degrees
Bias = Mean(S)	−0.08%	+1.03%	+0.27%
Precision = Standard Deviation(S)	±1.1%	±2.53%	±2.02%

The Aquadopp is evidently a very good unbiased estimator of mean inflow velocity for yaw angles under 15 degrees.

The validation data utilises effectively 1.7 Aquadopp depth cells, where 5.7 are used over the full swept area. This means that the precision in the inflow velocity expressed as a standard deviation is expected to be very close to 1% when averaged over the swept area.

Appendix B

B.1. Deriving Single Measurement for Inflow Velocity

The IEC 62600:200 technical specification sets out the approach for determining the power performance of a device using ADP inflow data. The 'method of bins' enables the velocities in the shear profile across the projected area of the rotor to be represented as an area weighted and power averaged single point measurement.

Figure A1 illustrates this method, the projected area of the turbine is slices into a series of sections representing the depth bins of the ADP. Meanwhile, the tubular sections represent each sample, or time. For the set-up in the TTT 3 project, each depth bin is 0.25 m deep and each time period is 1 s. Each section has an area A_K and a velocity measurement $U_{i,j,k,n}$ (see nomenclature for definitions).

The IEC approach to obtaining a single value for the flow across the swept area of the turbine during a given test case follows this sequence of equations [23]. Firstly, the power-weighted and area average for the projected area of the turbine for each period is calculated in Equation (A1), and illustrated by the thicker lines in Figure A1.

$$\hat{U}_{i,j,n} = \left[\frac{1}{A} \times \sum_{k=1}^{s} U_{i,j,k,n}^{3} \times A_k\right]^{\frac{1}{3}} \tag{A1}$$

The equation uses the instantaneous velocity measurement in each of profiler depth bins ($U_{i,j,k,n}$), which was cubed and weighted by the area of the depth bin (A_k). The sum of these is then divided by the total swept area (A) and cube rooted. This provides the power-weighted current velocity ($\hat{U}_{i,j,n}$) for each period.

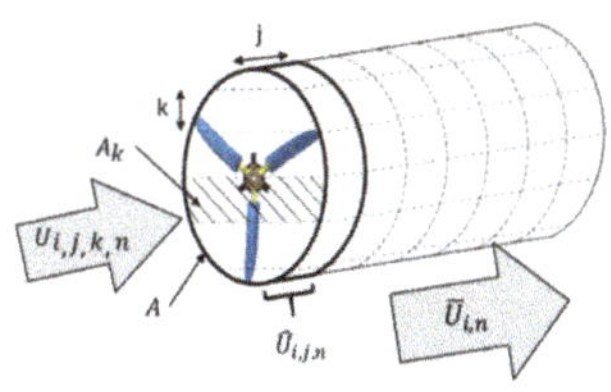

Figure A1. Power-weighted current velocity calculation illustration.

It is at this point that the adaptation of the IEC 'method of bins' is introduced. All ADP measurements are subject to uncertainty from Doppler noise. The IEC method requires each 1 s sample at each depth cell to be cubed, as shown in Equation (A1), thus cubing the measurement error from Doppler noise. To correct for this, the Doppler noise bias correction method has been developed. To demonstrate the analytical derivation for the Doppler noise bias correction; let us take m as the mean and σ as the standard deviation of a normal variate. The variable q is introduced, which is related to the variance of the distribution as shown in Equation (A2).

$$\sigma^2 = m^2 q \tag{A2}$$

In the case of power measurements derived from ADP data, it is the distribution of the cube of samples, $(X_k)^3$, which have been obtained from the normal distribution $N(m,\sigma)$ that is important. The effect of cubing a measurement with an included sampling error was first formally examined by Haldane in statistical biology [31]. Haldane showed that the mean of this resulting distribution exceeds m^3 by the ratio R, as follows in Equation (A3).

$$R = (1 + 3q) = \overline{(X_k^3)}/m^3 \tag{A3}$$

The values of q and m, as defined above, are readily obtainable from the ADP data. We have called the ratio R the 'Doppler noise bias', and R can be applied as a correction wherever the cube of a noisy velocity signal is sought and the ratio of m and σ is known. For the derivation of the power-weighted, area-averaged velocity, as previously described in Equation (A1), this can now be re-written to include the bias correction factor derived, as shown in Equation (A3). The outcome of this is an unbiased velocity measurement and reduced uncertainty in the propagation of the performance metrics.

$$\hat{U}_{i,j,n} = \left[\frac{1}{A} \times \sum_{k=1}^{s} \frac{U_{i,j,k,n}^3 \times A_k}{R_{i,k,n}}\right]^{\frac{1}{3}} \tag{A4}$$

The datasets were averaged over periods between 2 and 10 min for the Strangford Lough testing. For CNR-INSEAN, due to the limited length of the tanks, the maximum averaging period was approximately 90 s; however, given the controlled nature of the experiments in the laboratory this was not considered to be an issue. The mean velocity for the data set ($\overline{U}_{i,n}$) is calculated from the power weighted values $\hat{U}_{i,j,n}$ over the time period from j = 1 to j = L, the length of the run is in seconds. Lastly, the average for all the velocities recorded in the given current velocity bin is calculated. The velocity bin increments were set to 0.10 ms^{-1} and only flood phase of the tide is considered. These steps are described and equations defined in previous work [13,23].

The turbine's instrumentation as described earlier includes a torque sensor and rotational encoder on the driveshaft behind the rotor. These outputs provide the mechanical power (P, W) of the rotor in advance of drivetrain losses. Power is calculated using Equation (A5).

$$\acute{P}_i = \frac{1}{N_i} \sum_{n=1}^{N_i} \acute{Q}_{i,n} \times \acute{\omega}_{i,n} \tag{A5}$$

The same velocity bin increments apply here also. When sampled over the same time period as the ADP data, the turbine data has more sample points (n), as it is sampled at 16 Hz as opposed to the 1 Hz sample frequency of the ADP. The mechanical power performance can then be calculated using the non-dimensional performance characteristic, C_P as previously described in Equations (1) and (2) inserting Equations (A4) and (A5). The water density was set to 1000 kg·m^{-3} and 1025 kg·m^{-3} for laboratory and field data respectively.

B.2. Propagation of Uncertainty

Understanding of the propagation of instrument uncertainties is crucial to determining the confidence intervals of derived performance characteristics. Similar studies into the propagation of uncertainty have been conducted in this area before [32]. The previous work showed the significance of uncertainties in the torque, thrust and bending moments when propagated to derive the power performance coefficient. The propagation of inflow uncertainties was not specifically considered by Doman et al., due to the close control afforded by towing tank experiments. For experimental set-ups in real tidal flows, this exception can no longer be made. Therefore, the following section concentrates on the propagation of the velocity uncertainty from the ADP data, through the method of bins, used by the IEC technical specification (TS62600:200).

To ascertain the uncertainty of the derived performance indicators of a tidal turbine, the uncertainty of each variable in the derived performance indicator must be pooled. Equations (A6)–(A8) shows the propagation of uncertainty equation associated with each of the performance indicators.

$$\sigma_{\text{TSR}} = \text{TSR} \times \sqrt{\left(\frac{\sigma_\omega}{\acute{\omega}_i}\right)^2 + \left(\frac{\sigma_r}{r}\right)^2 + \left(\frac{\sigma_{\acute{U}_i}}{\acute{U}_i}\right)^2} \tag{A6}$$

$$\sigma_{C_P} = C_P \times \sqrt{\left(\frac{\sigma_Q}{\acute{Q}_i}\right)^2 + \left(\frac{\sigma_\omega}{\acute{\omega}_i}\right)^2 + \left(\frac{\sigma_\rho}{\rho}\right)^2 + \left(\frac{\sigma_A}{A}\right)^2 + \left(\frac{3 \times \sigma_{\acute{U}_i}}{\acute{U}_i}\right)^2} \tag{A7}$$

$$\sigma_{C_T} = C_T \times \sqrt{\left(\frac{\sigma_T}{\bar{T}_i}\right)^2 + \left(\frac{\sigma_\rho}{\rho}\right)^2 + \left(\frac{\sigma_A}{A}\right)^2 + \left(\frac{2 \times \sigma_{\acute{U}_i}}{\acute{U}_i}\right)^2} \quad \text{(A8)}$$

As Equations (A7) and (A8) show, the propagated uncertainty is most sensitive to uncertainties in the inflow velocity. The uncertainty variables are derived from the Root Mean Squared (RMS) of bias and precision uncertainties, as shown in other work [13,32]. This is the case for all uncertainty parameters derived, with the exception of the inflow velocity uncertainty. The exception to the inflow velocity is due to the correction applied for Doppler noise bias in Equation (A6). This correction accounts for the bias uncertainty, leaving only the precision uncertainty, which is calculated in Equation (A9)

$$\sigma_{\acute{U}_i} = \left(\frac{\sigma}{\sqrt{s}}\right) / \sqrt{n} \quad \text{(A9)}$$

References

1. Department for Buisness, Energy & Industrial Stragety. Contracts for Difference Second Allocation Round Results. 2017. Available online: https://www.gov.uk/government/publications/contracts-for-difference-cfd-second-allocation-round-results (accessed on 20 January 2018).
2. Mason-Jones, A.; O'Doherty, D.M.; Morris, C.E.; O'Doherty, T.; Byrne, C.B.; Prickett, P.W.; Grosvenor, R.I.; Owen, I.; Tedds, S.; Poole, R.J. Non-dimensional scaling of tidal stream turbines. *Energy* **2012**, *44*, 820–829. [CrossRef]
3. Clarke, J.A.; Connor, G.; Grant, A.D.; Johnstone, C.M. Design and testing of a contra-rotating tidal current turbine. *Proc. Inst. Mech. Eng. Part A J. Power Energy* **2007**, *221*, 171–179. [CrossRef]
4. Gaurier, B.; Germain, G.; Facq, J.V.; Johnstone, C.M.; Grant, A.D.; Day, A.H.; Nixon, E.; Di Felice, F.; Costanzo, M. Tidal energy "Round Robin" tests comparisons between towing tank and circulating tank results. *Int. J. Mar. Energy* **2015**, *12*, 87–109. [CrossRef]
5. Mycek, P.; Gaurier, B.; Germain, G.; Pinon, G.; Rivoalen, E. Experimental study of the turbulence intensity effects on marine current turbines behaviour. Part I: One single turbine. *Renew Energy* **2014**, *66*, 729–746. [CrossRef]
6. Jeffcoate, P.; Whittaker, T.; Boake, C.; Elsaesser, B. Field tests of multiple 1/10 scale tidal turbines in steady flows. *Renew. Energy* **2016**, *87*, 240–252. [CrossRef]
7. Jeffcoate, P.; Elsaesser, B.; Whittaker, T.; Boake, C. Testing Tidal Turbines—Part 1: Steady Towing Tests vs. Tidal Mooring Tests. In Proceedings of the ASRANet International Conference on Offshore Renewable Energy, Glasgow, UK, September 2014; Volume 68, pp. 55–87. Available online: https://pure.qub.ac.uk/portal/files/11366206/ASRANet_2014_PJeffcoate.pdf (accessed on 20 January 2018).
8. Atcheson, M.; MacKinnon, P.; Elsaesser, B. A large scale model experimental study of a tidal turbine in uniform steady flow. *Ocean Eng.* **2015**, *110*, 51–61. [CrossRef]
9. Jeffcoate, P.; Salvatore, F.; Boake, C.; Elsaesser, B.; Vallerano, V. Effect of Submergence on Tidal Turbine Performance. In Proceedings of the 11th European Wave and Tidal Energy Conference, Nantes, France, 6–11 September 2015; pp. 3–9.
10. Jeffcoate, P.; Starzmann, R.; Elsaesser, B.; Scholl, S.; Bischoff, S. Field measurements of a full scale tidal turbine. *Int. J. Mar. Energy* **2015**, *12*, 3–20. [CrossRef]
11. Starzmann, R.; Jeffcoate, P.; Scholl, S.; Bischof, S.; Elsaesser, B. Field testing a full-scale tidal turbine Part 1: Power Performance Assessment. In Proceedings of the 11th European Wave and Tidal Energy Conference, Nantes, France, 6–11 September 2015; pp. 1–7.
12. Schmitt, P.; Elsaesser, B.; Bischof, S.; Starzmann, R. Field testing a full-scale tidal turbine Part 2: In-line Wake Effects. In Proceedings of the 11th European Wave and Tidal Energy Conference, Nantes, France, 6–11 September 2015; pp. 1–7.
13. Frost, C.; Benson, I.; Elsäßer, B.; Starzmann, R.; Whittaker, T. Mitigating Uncertainty in Tidal Turbine Performance Characteristics from Experimental Testing. In Proceedings of the European Wave & Tidal Energy Conference, Cork, Ireland, 27 August 2017.
14. Forbush, D.; Polagye, B.; Thomson, J.; Kilcher, L.; Donegan, J.; McEntee, J. Performance characterization of a cross-flow hydrokinetic turbine in sheared inflow. *Int. J. Mar. Energy* **2016**, *16*, 150–161. [CrossRef]
15. Blackmore, T.; Myers, L.E.; Bahaj, A.S. Effects of turbulence on tidal turbines: Implications to performance, blade loads, and condition monitoring. *Int. J. Mar. Energy* **2016**, *14*, 1–26. [CrossRef]

16. European Commission Technology readiness levels (TRL). *Horizon 2020 General Annexes*; European Commission: Brussels, Belgium, 2015; p. 4995.
17. Mason-Jones, A.; O'Doherty, D.M.; Morris, C.E.; O'Doherty, T. Influence of a velocity profile & support structure on tidal stream turbine performance. *Renew. Energy* **2013**, *52*, 23–30. [CrossRef]
18. Payne, G.S.; Stallard, T.; Martinez, R. Design and manufacture of a bed supported tidal turbine model for blade and shaft load measurement in turbulent flow and waves. *Renew. Energy* **2017**, *107*, 312–326. [CrossRef]
19. Harrold, M.J. Experimental and Numerical Assessment of a Tidal Turbine Control Strategy, University of Strathclyde, 2016. Available online: http://ethos.bl.uk/OrderDetails.do?uin=uk.bl.ethos.723008 (accessed on 20 January 2018).
20. Allmark, M.; Grosvenor, R.; Prickett, P. An approach to the characterisation of the performance of a tidal stream turbine. *Renew. Energy* **2017**, *111*, 849–860. [CrossRef]
21. Schlipf, D.; Pao, L.Y.; Cheng, P.W. Comparison of Feedforward and Model Predictive Control of Wind Turbines Using LIDAR. In Proceedings of the 2012 IEEE 51st IEEE Conference on Decision and Control (CDC), Maui, HI, USA, 10–13 December 2012; pp. 3050–3055.
22. Johnstone, C.M.; McCombes, T.; Bahaj, A.S.; Myers, L.E.; Holmes, B.; Koefoed, J.P.; Bittencourt, C. EquiMar: Development of Best Practices for the Engineering Performance Appraisal of Wwave and Tidal Energy Converters. In Proceedings of the 9th European Wave and Tidal Energy Conference, Southampton, UK, 5–9 September 2011.
23. IEC/TS 62600-200 Power Performance Assessment of Electricity Producing Tidal Energy Converters Commissioned by: IEC. 2011. Available online: https://webstore.iec.ch/publication/7242 (accessed on 20 January 2018).
24. Clark, T.; Black, K.; Ibrahim, J.; Minns, N.; Fisher, S.; Roc, T.; Hernon, J.; White, R. MRCF-TiME-KS9b Turbulence: Best Practices for Data Processing, Classification and Characterisation of Turbulent Flows. Ocean Array Systems: Edinburgh, UK. 2015. Available online: http://www.oceanarraysystems.com/publications (accessed on 20 January 2018).
25. Shih, H.H.; Payton, C.; Sprenke, J.; Mero, T. Towing Basin Speed Calibration of Acoustic Doppler Current Profiling Instruments. In *Joint Conference on Water Resource Engineering and Water Resources Planning and Management 2000*; American Society of Civil Engineers: Reston, VA, USA, 2000; pp. 1–10.
26. Oberg, K.; Mueller, D.S. Validation of Streamflow Measurements Made with Acoustic Doppler Current Profilers. *J. Hydraul. Eng.* **2007**, *133*, 1421–1432. [CrossRef]
27. Elsaesser, B.; Torrens-Spence, H.; Schmitt, P.; Kregting, L. Comparison of Four Acoustic Doppler Current Profilers in a High Flow Tidal Environment—Queen's University Belfast Research Portal—Research Directory & Institutional Repository for QUB. In Proceedings of the 3rd Asian Wave and Tidal Energy Conference, Singapore, 24–28 October 2016.
28. Pao, L.Y.; Johnson, K.E. Control of Wind Turbines. *IEEE Control Syst.* **2011**, *31*, 44–62. [CrossRef]
29. Bennett, S. The past of pid controllers. *Annu. Rev. Control* **2001**, *25*, 43–53. [CrossRef]
30. Ziegler, J.G.; Nichols, N.B. Optimum Settings for Automatic Controllers. *J. Dyn. Syst. Meas. Control* **1993**, *115*, 220. [CrossRef]
31. Haldane, J.B.S. Moments of the distributions of power and products of normal variates. *Biometrika* **1942**, *32*, 226–242. [CrossRef]
32. Doman, D.A.; Murray, R.E.; Pegg, M.J.; Gracie, K.; Johnstone, C.M.; Nevalainen, T. Tow-tank testing of a 1/20th scale horizontal axis tidal turbine with uncertainty analysis. *Int. J. Mar. Energy* **2015**, *11*, 105–119. [CrossRef]

© 2018 by the authors. Licensee MDPI, Basel, Switzerland. This article is an open access article distributed under the terms and conditions of the Creative Commons Attribution (CC BY) license (http://creativecommons.org/licenses/by/4.0/).

MDPI
St. Alban-Anlage 66
4052 Basel
Switzerland
Tel. +41 61 683 77 34
Fax +41 61 302 89 18
www.mdpi.com

Energies Editorial Office
E-mail: energies@mdpi.com
www.mdpi.com/journal/energies

www.ingramcontent.com/pod-product-compliance
Lightning Source LLC
Chambersburg PA
CBHW061439210926
51850CB00008B/1554